Palaeomagnetism and plate tectonics

IN THIS SERIES

MINERALOGICAL APPLICATIONS OF CRYSTAL FIELD THEORY
by R. G. Burns

NON-MARINE ORGANIC GEOCHEMISTRY
by Frederick M. Swain

Palaeomagnetism and plate tectonics

M. W. McELHINNY

Senior Fellow in Geophysics
Australian National University

CAMBRIDGE

at the University Press, 1973

Published by the Syndics of the Cambridge University Press
Bentley House, 200 Euston Road, London NW1 2DB
American Branch: 32 East 57th Street, New York, N.Y.10022

© Cambridge University Press 1973

Library of Congress Catalogue Card Number: 72–80590

ISBN: 0 521 08707 4

Printed in Great Britain
at the University Printing House, Cambridge
(Brooke Crutchley, University Printer)

Contents

Contents

Preface

'...an essay in geopoetry' H. H. Hess, 1962.

The study of palaeomagnetism has produced a revolution in the earth sciences over the past decade. After many years in the wilderness, the hypothesis of continental drift was able to be put to a physical test quite independently of the classical, but previously unacceptable, evidence put forward in its favour. Realizing that the palaeomagnetic data were sufficiently compelling, Hess produced the hypothesis of sea-floor spreading, a brilliant and imaginative concept which provided the mechanism for continental drift. Again it was palaeomagnetism, applied this time to marine magnetic anomalies, that showed that this hypothesis was essentially correct. Wilson's concept of transform faults followed as a corollary, and with the aid of observations in seismology, these were incorporated into the theory of plate tectonics as propounded by McKenzie, Morgan and Le Pichon. For the first time the earth sciences has had a unifying theory which accounts for the earth's surface features, tectonic processes and the development of the planet.

This book is designed as a basis for an extended lecture course on palaeomagnetism, and is also an up-to-date review of the state of knowledge of the subject. Since the background of students of the earth sciences varies from the purely geological to the purely physical, it is always difficult to balance the mathematical and non-mathematical approach to topics. I have tried to balance both of these aspects and hope that the one does not detract from the other. Chapter 1 is a simple introduction to the subject and its basis. I have noted that pure geologists often have difficulty grasping many of the facets of magnetic theory, so that in chapter 2 I have provided a brief introduction to magnetism which I hope will assist in following the theory of rock magnetism. The various experimental methods used in palaeomagnetism (chapter 3) have, as far as is possible, been related to the theory outlined in chapter 2. This is an attempt to demonstrate that the techniques used by the palaeomagnetist, especially cleaning techniques, are not just the result of guesswork or wizardry, but have a sound theoretical basis. However those concerned more with results than with the theory and techniques may find it acceptable to omit chapters 2 and 3 altogether. The choice of topics is obviously biased towards my own interests, so that the accent of the succeeding chapters is on the relation between palaeomagnetic information and past movements. Chapter 6 is essentially a world-wide

Preface

review of palaeomagnetic information using the data listed in the appendix. The latter is a current summary of all results which conform with the minimum criteria set out at the end of chapter 3.

A number of people have provided me with pre-publication information incorporated in this book, and in this respect I wish to thank Aziz-ur-Rahman, J. C. Briden, A. Brock, E. R. Deutsch, D. P. McKenzie, W. A. Morris, B. Oversby, J. D. A. Piper, C. K. Scharnberger and H. Spall. Discussions with colleagues, particularly J. C. Briden, B. J. J. Embleton, D. I. Gough, F. E. M. Lilley, I. McDougall and R. L. Wilson, have greatly assisted me during the preparation of the manuscript. My thanks are also due to A. R. Crawford who suffered my persistent questioning and cross-examination on a wide variety of aspects of world geology and their possible interpretation on plate tectonics. I am especially grateful to J. C. Briden and D. Davies who both read the manuscript and made helpful suggestions for its improvement.

It is worth pointing out that, whilst palaeomagnetism has been so successful in demonstrating that large horizontal displacements have occurred in the past, it was Hess' essay in geopoetry that essentially paved the way for the theory of plate tectonics as we know it today. He, more than any other, may be regarded as the Father of Plate Tectonics. For palaeomagnetists the deduction of the past distributions of continental crust still remains as the most challenging application of their techniques to the theory of plate tectonics. A. Holmes (1965) summed up the problem in the following way:

'However the primaeval continents may have been originally distributed it seems probable that they have ever since been subjected to the effects sub-crustal currents, with occasional integrations into larger masses, akin to Wegener's *Pangaea*, alternating with separations of which the geography of today is naturally the most familiar example.'

Canberra, Australia   M. W. McElhinny
September 1971

x

1 Geomagnetism and palaeomagnetism

1.1 GEOMAGNETISM

1.1.1 *Historical*

The properties of lodestone (now known to be magnetite) were known to the Chinese in ancient times. The earliest known form of magnetic compass was invented by the Chinese probably as early as the second century B.C., and comprised a lodestone spoon rotating on a smooth board (Needham, 1962). It was not until the twelfth century A.D. that the compass arrived in Europe, where the first reference to it is made by an English monk, Alexander Neckham, in 1190. During the thirteenth century, it was noted that the compass needle pointed towards the pole star. Unlike other stars the pole star was fixed so that it was concluded that the lodestone with which the needle was rubbed must obtain its 'virtue' from this star. In the same century it was suggested that, in some way, the magnetic needle was affected by masses of lodestone on the earth itself. This produced the idea of polar lodestone mountains, which had the merit at least of bringing magnetic directivity down to earth from the heavens for the first time (Smith, 1968).

Roger Bacon in 1216 first questioned the universality of the north–south directivity of the compass needle. A few years later Petrus Peregrinus questioned the idea of polar lodestone deposits, pointing out that these exist in many parts of the world, so why should the polar ones have preference. Petrus Peregrinus reported a remarkable series of experiments with spherical pieces of lodestone in his *Epistola de Magnete* of 1269 (Smith, 1970). He defined the concept of polarity for the first time in Europe, discovered magnetic meridians and showed several ways of determining the positions of the poles of a lodestone sphere, each of which illustrated an important magnetic property. He thus discovered the dipolar nature of the magnet, that the magnetic force is both strongest and vertical at the poles, and became the first to formulate the law that like poles repel and unlike poles attract. The *Epistola* bears a remarkable resemblance to a modern scientific paper. Peregrinus used his experimental data from which to draw conclusions, unlike his contemporaries who sought to reconcile facts with pre-existing speculation. Although written in 1269 and widely circulated during the succeeding centuries, the *Epistola* was not published in printed form under Peregrinus' name until 1558.

Magnetic declination was known to the Chinese from about A.D. 720

onwards (Needham, 1962; Smith & Needham, 1967), but knowledge of this did not travel to Europe with the compass. It was not rediscovered until the latter part of the fifteenth century. By the end of that century, following the voyages of Columbus, came the great age of exploration by sea, and the compass was well established as an aid to navigation. Magnetic inclination (or dip) was discovered by Georg Hartmann in 1544, but this discovery was not publicized, and consequently was quite independently discovered by Robert Norman in 1576. Mercator however, in a letter in 1546, first realized from observations of magnetic declination that the point which the needle seeks could not lie in the heavens, leading him to fix the magnetic pole firmly on the earth. Norman and Borough subsequently consolidated the view that magnetic directivity was associated with the earth and even shifted the focus from the pole to a region closer to the centre of the earth.

In 1600 William Gilbert published the results of his experimental studies in magnetism in what is usually regarded as the first scientific treatise ever written, entitled *De Magnete*. This credit should probably be given to Petrus Peregrinus for his *Epistola de Magnete* and Gilbert must certainly have leaned heavily on this previous work (Smith, 1970). However Gilbert's work undoubtedly strongly influenced the course of magnetic study. He investigated the variation in inclination over the surface of a piece of lodestone cut into the shape of a sphere. His conclusions are summed up in his statement '*magnus magnes ipse est globus terrestris*' (the earth globe itself is a great magnet). Gilbert's work, again confirming that the geomagnetic field is primarily dipolar, thus represented the culmination of many centuries of thought and experimentation on the subject. His conclusions put a stop to the wild speculations that were then current concerning magnetism and the magnetic needle. Apart from the roundness of the earth, magnetism was the first property to be attributed to the body of the earth as a whole. Newton's gravitation came eighty-seven years later with the publication of his *Principia*.

1.1.2 *Main features of the geomagnetic field*

If a magnetic compass needle is weighted so as to swing horizontally, it takes up a definite direction at each place and its deviation from geographical or true north is called the *declination, D*. In geomagnetic studies D is reckoned positive or negative according as the deviation is east or west of true north. In palaeomagnetic studies, D is always measured clockwise (eastwards) from the present geographic north and consequently takes on any angle between 0° and 360°. The direction to which the needle points is

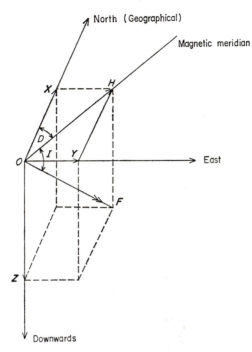

FIGURE I   Main elements of the geomagnetic field. From J. A. Jacobs, *The Earth's core and geomagnetism* (Pergamon, 1963).

called *magnetic north* and the vertical plane through this direction is called the magnetic meridian. A needle perfectly balanced about a horizontal axis (before being magnetized), so placed that it can swing freely in the plane of the magnetic meridian, is called a dip needle. After magnetization it takes up a position inclined to the horizontal by an angle called the *inclination* (or *dip*), I. The inclination is reckoned positive when the north-seeking end of the needle points downwards (as it does in the northern hemisphere) or negative when it points upwards (southern hemisphere).

The main elements of the geomagnetic field are illustrated in fig. 1. The total intensity F, declination D, and inclination I completely define the field at any point. The horizontal and vertical components of F are denoted by H and Z, which is reckoned positive downwards as for I. The horizontal component can be resolved into two components X (northwards) and Y (eastwards). The various components are related by the equations:

$$H = F\cos I, \quad Z = F\sin I, \quad \tan I = Z/H; \qquad (1.1)$$

$$X = H\cos D, \quad Y = H\sin D, \quad \tan D = Y/X; \qquad (1.2)$$

$$F^2 = H^2 + Z^2 = X^2 + Y^2 + Z^2. \qquad (1.3)$$

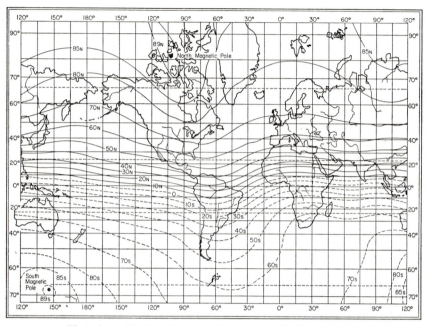

FIGURE 2 Variation of inclination over the surface of the earth for epoch 1945. N and S indicate whether the north-seeking or south-seeking end of a dip needle points downwards, and thus refer to positive or negative values of inclination respectively. From Vestine *et al.* (1947), *Description of the earth's magnetic field and its secular change*, Carnegie Institution of Washington Publication No. 578, figure 147 (A), p. 524.

Variations of the earth's magnetic field over its surface are illustrated by isomagnetic maps, an example of which is shown in fig. 2, giving the variation of inclination over the surface of the earth for the year 1945. The line along which the inclination is zero is called the *magnetic equator*, whilst the *magnetic poles* (or dip poles) are points where the inclination is $\pm 90°$. The *north magnetic pole* is situated where $I = +90°$, and the *south magnetic pole* where $I = -90°$.

The intensity of the earth's field is commonly expressed in gauss using the c.g.s. electromagnetic system of units, whilst in the M.K.S.A. system the tesla (weber m^{-2}) can be used where:

$$1 \text{ gauss} = 10^{-4} \text{ tesla.} \qquad (1.4)$$

For small variations the unit called the gamma (γ) is used where:

$$1 \text{ gamma} = 10^{-5} \text{ gauss} = 10^{-9} \text{ tesla (or 1 nanotesla).} \qquad (1.5)$$

The maximum value of the earth's magnetic field at the surface is currently a little more than 0.7 gauss and occurs in the region of the south magnetic pole.

4

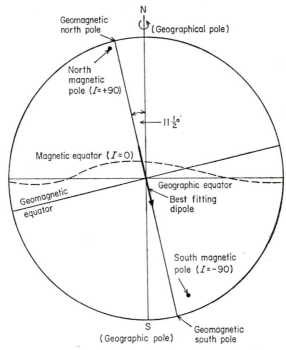

FIGURE 3   To illustrate the distinction between the magnetic,
geomagnetic and geographic poles and equators.

Gilbert's observation that the earth is a great magnet, similar to a
uniformly magnetized sphere, was first put to mathematical analysis by Gauss
in 1839. The earth's magnetic field best approximates to that of a geocentric
dipole inclined at $11\frac{1}{2}°$ to the earth's axis of rotation. The dipole axis, if
extended, intersects the earth's surface at two points situated at $78\frac{1}{2}$N, 70W
(in northwest Greenland) and $78\frac{1}{2}$S, 110E (in Antarctica). These points are
called the *geomagnetic poles* (boreal and austral, or north and south respectively), and must be carefully distinguished from the magnetic poles of the
preceding paragraphs. The circle on the earth's surface coaxial with the
dipole axis and midway between the geomagnetic poles is called the
geomagnetic equator, and is different from the magnetic equator, which is
not in any case a circle. Fig. 3 distinguishes between the magnetic elements
(which are those actually observed at each point) and the geomagnetic
elements (which are related to the best fitting dipole).

In 1634 Gellibrand discovered that the magnetic declination at any
place changed with time. He noted that whilst Borough in 1580 measured
a value of 11.3E for the declination at London, his own measurements in
1634 gave only 4.1E. The difference was far greater than possible experi-

5

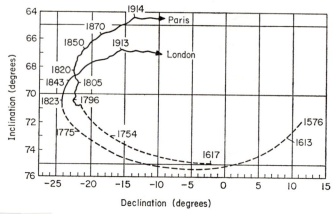

FIGURE 4 Variation of declination and inclination at London and Paris from observatory measurements. After Gaiber-Puertas (1953), *Observ. del. Elso*, Memo. No. 11.

mental error. The change in magnetic field with time is called *secular variation*, and is observed in all the magnetic elements. The secular variation of the direction of the geomagnetic field at London and Paris since about 1580 is shown in fig. 4. At London the declination changed by about 35° in the 240 years between 1580 and 1820, exhibiting a westward drift.

The distribution of the secular variation on the earth's surface can be represented by maps on which lines called *isopors* can be drawn, joining points which show the same annual change in a magnetic element. These isoporic maps show that there are a number of regions on the earth's surface in which isoporic lines form closed loops centred around foci where the secular changes are the most rapid. For example, the total intensity of the earth's field is decreasing most rapidly around isoporic foci situated south-west of the tip of South America and just south of Africa (both at the rate of about 150 gammas per year). Isoporic foci are not permanent, but grow and decay, their lifetime being of the order of 100 years, during which they move on the earth's surface in a somewhat irregular fashion. The movement is not altogether random, but shows a systematic westward drift of 0.2° of longitude per year. Since declination is the most important magnetic element for navigation, records of it were kept by navigators since the early part of the sixteenth century. Records show that the point of zero declination on the equator, now situated in northwest Brazil, was situated in Africa four centuries ago.

The spherical harmonic analysis of the earth's magnetic field (§1.1.3), first undertaken by Gauss in 1839, has been repeated several times since for succeeding epochs. When the best fitting dipole (the main dipole) field is subtracted from that observed over the surface of the earth, the residual

6

TABLE 1   *Geomagnetic dipole moment computed by*
various authors for different epochs

Author	Epoch	Moment (10^{25} gauss cm³)
Erman & Petersen	1829	8.45
Gauss	1835	8.56
Adams	1845	8.49
Adams	1880	8.36
Schmidt	1885	8.35
Fritsche	1885	8.34
Neumeyer & Petersen	1885	8.33
Cain & Hendricks	1900	8.30
Cain & Hendricks	1920	8.20
Dyson & Furner	1922	8.17
Cain & Hendricks	1940	8.10
Jones & Melotte	1942	8.01
Afanasieva	1945	8.01
Vestine and others	1945	8.07
Finch & Leaton	1955	8.06
Cain & Hendricks	1960	8.02
Hurwitz and others	1965	8.00
Leaton and others	1965	8.00

is termed the *non-dipole field*. The secular variation may then be analysed by separately looking at variations in the main dipole and non-dipole fields. These analyses show that the magnetic moment of the main dipole has decreased by about six per cent over the past 130 years (table 1). Since field reversals are now known to have occurred in geological times (see chapter 4), Leaton & Malin (1967) and McDonald & Gunst (1968) have speculated on the demise of the main dipole (about A.D. 3700 to 4000) by extrapolating a linear trend from data such as are given in table 1. Note, however, that there is a tendency towards a reduction in the rate of decrease, so that such an extrapolation to zero must be considered highly speculative. Nagata (1965) has concluded that there is a westward precessional rotation of the main dipole at 0.05° of longitude per year, coupled with a rotation of the dipole towards the geographic axis at 0.02° latitude per year.

The non-dipole field (fig. 5) forms a number of features (foci or anomalies). Bullard *et al.* (1950) have concluded that the non-dipole field was drifting westwards at an *average* rate of about 0.2° of longitude per year, over the first half of this century. Attempts to analyse the behaviour of the main features of the non-dipole field over the past few centuries (Yukutake & Tachimaka, 1968) have shown that some of the features may persist for some time, whilst others change more rapidly. The features appear to form, deform and decay over periods of 100 to 1000 years. Of particular interest is the absence of significant features of the non-dipole field over the Pacific.

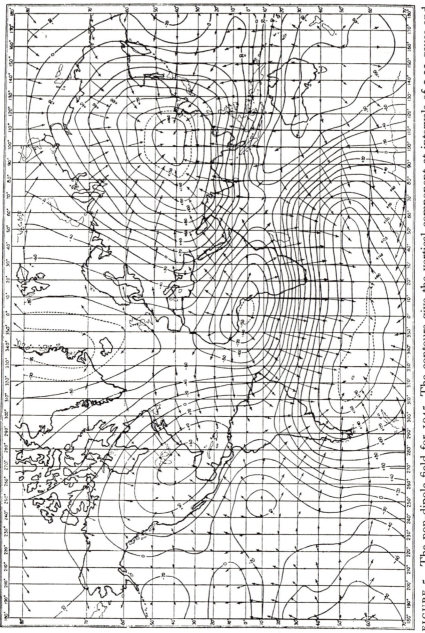

FIGURE 5 The non-dipole field for 1945. The contours give the vertical component at intervals of 0.02 gauss and the arrows give the magnitude and direction of the horizontal component. From Bullard et al. (1950).

8

1.1.3   *Origin of the main field*

The magnetic field observed at the surface of the earth could be produced by sources inside the earth, by sources outside the earth's surface or by electric currents crossing the surface. If it is assumed as a start that there are no electric currents crossing the surface, then the magnetic field at the surface of the earth, assumed to be a sphere of radius a, can be derived from a potential function V, which satisfies Laplace's equations. The potential V can be represented as a series of spherical harmonics in the form:

$$V = a \sum_{n=1}^{\infty} \sum_{m=0}^{n} P_n^m(\theta) \left[\left\{ \left(g_e\right)_n^m \left(\frac{r}{a}\right)^n + \left(g_i\right)_n^m \left(\frac{a}{r}\right)^{n+1} \right\} \cos m\phi \right.$$
$$\left. + \left\{ \left(h_e\right)_n^m \left(\frac{r}{a}\right)^n + \left(h_i\right)_n^m \left(\frac{a}{r}\right)^{n+1} \right\} \sin m\phi \right], \quad (1.6)$$

where θ, ϕ are the co-ordinates of the magnetic colatitude and longitude, r is the distance from the centre of the earth, and $P_n^m(\theta)$ is the spherical harmonic function of Schmidt of degree n and order m. The coefficients g and h are the Gauss coefficients and have subscripts e or i as the sources are external or internal. V itself is not directly observable but two of the three components of the field, Z and either X or Y, allow the coefficients to be determined. Agreement between the coefficients derived through using either X or Y demonstrates the validity of expressing the field as a potential and shows that no electric currents exist crossing the earth's surface. Gauss, in the first spherical harmonic analysis in 1839, showed that not only were there no electric currents but also that the coefficients of external origin g_e and h_e were both zero, from which he concluded that the field was solely of internal origin. In practice the external field is not totally absent, and a small contribution from electric currents in the ionosphere is present usually amounting to perhaps thirty gammas.

If the earth were uniformly magnetized, the average intensity of magnetization of material would be 0.075 emu cm^{-3}. At depths below 25 km or so the temperature would be above the Curie point of any ferromagnetic material, so that to account for the earth's magnetism through permanent magnetization requires that the earth's crust have an average magnetization of about 5 emu cm^{-3}. Strongly magnetized basalts are typically 10^{-2} emu cm^{-3} whilst sediments and granites rarely exceed 10^{-4} emu cm^{-3}. Chatterjee (1956) has supposed that, due to the effects of high pressure and temperature, the outer 20 km-thick shell would behave as a material of high permeability, which becomes saturated in low fields of the order of 0.1 gauss or even less. The magnetizing field on Chatterjee's model is provided by the ring current which circulates in the upper atmosphere during magnetic storms, as a result of the interaction of solar particles with the earth's magnetic field. Given an initial field of only one per cent of the present field, Chatterjee suggests that this would be sufficient to build up the magnetization of the crust to saturation. Apart from the somewhat optimistic values chosen for the abundance of magnetite, temperatures, depths, and so on, there seems to be no obvious reason why the magnetic field produced in this way should have any relationship with the earth's axis of rotation. Furthermore, this and all theories relying on permanent magnetization fail to account for the observation that the earth's magnetic field has reversed itself frequently in the past (see chapter 4). It is necessary therefore to look for some means whereby the earth can generate and maintain its own magnetic field.

Studies of the passage of seismic waves through the earth's interior show that there is a region below 2900 km depth through which the propagation of transverse waves (*S* waves) has never been detected. The spherical boundary at this depth encloses a region of radius 3500 km which must be largely liquid and is termed the core. Further seismic evidence indicates that the core is made up of two regions known as the inner and outer core, the latter being liquid and the former probably solid. Density and geochemical considerations suggest the core is composed largely of iron and nickel, which would be good electrical conductors, and this leads to the possibility that motions in the earth's liquid core are responsible for the magnetic field.

The dynamo theory of the earth's magnetic field originates from a suggestion of Larmor (1919) that the magnetic field of the sun might be maintained by a mechanism analogous to that of a self exciting dynamo. Elsasser (1946) and Bullard (1949) followed up this suggestion proposing that the electrically conducting core of the earth acts like a self exciting dynamo, and produces electric currents necessary to maintain the geomagnetic field. The action of such a dynamo is simply illustrated by the disk dynamo of fig. 6. If a conducting disk is rotated in a small axial magnetic field, a radial electromotive force is generated between the axis and the edge of the disk. A coil in the external circuit is placed coaxial with the disk to provide positive feedback, so that the magnetic field it produces reinforces the initial axial field. This causes a larger current to flow because of the increased e.m.f. and the axial field is increased further, being ultimately limited by Lenz's law and the electrical resistance of the circuit. The main point, however, is that starting from a very small field, perhaps a stray one, it is possible to generate a much greater field.

In the earth's interior, the mechanisms that correspond to the rotating disk and coil are provided by convection currents in the fluid core, although

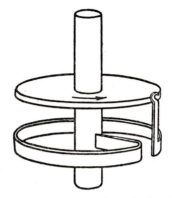

FIGURE 6   Disk dynamo. From Bullard (1968).

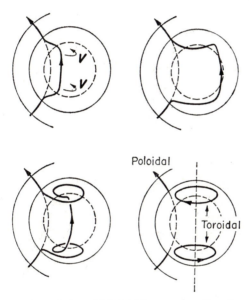

FIGURE 7   Generation of a toroidal field in a non-uniformly rotating sphere.
From Elsasser (1955), *Amer. J. Phys.* **23**, 590–609.

it should be noted that, unlike the disk, the electrical conductivity in the
core is probably homogeneous. However, if convection alone were active,
the fluid particles would describe closed loops confined to planes and no
dynamo would result. When the system rotates, the paths of the particles
are twisted into three-dimensional shapes by the action of the Coriolis
force. If a magnetic field existed originally in meridional planes, then the
lines of force will be twisted as shown in fig. 7 and a *toroidal* field will have
been produced from an original *poloidal* field (Elsasser, 1955). Toroidal

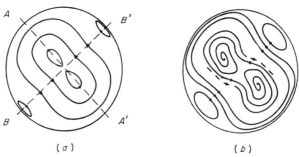

FIGURE 8   Streamlines in the equatorial plane. (*a*) The Bullard–Gellman dynamo with *A–A'* and *B–B'* marking the planes of symmetry in the flow pattern. (*b*) The Lilley dynamo with the planes of symmetry destroyed. From Lilley (1970*b*), *Nature*, **227**, 1336–7.

fields are characterized by the absence of radial components, that is they are confined within the rotating core. The poloidal fields are the only ones observed outside. Purely toroidal fluid motions do not interact with a toroidal field and a purely poloidal fluid motion merely rearranges the circular lines of force without generating a poloidal field. Considerations of the different interactions possible between different types of fluid motions and a given field show that in order to produce the poloidal field outside the earth a much more powerful toroidal field must exist in the earth's core.

Theoretical analysis of homogeneous dynamos involves selecting a possible flow pattern and testing numerically to see whether this can act as a dynamo. This was first undertaken by Bullard & Gellman (1954) for one particular flow pattern, but the storage capacity of computers at that time severely limited the application of their method. Gibson & Roberts (1969) have subsequently shown that the numerical process upon which the Bullard–Gellman dynamo depends does not converge. This is in all probability due to the flow pattern having too great a symmetry for it to act as a dynamo (fig. 8*a*). Recently Lilley (1970*a*) has modified the Bullard–Gellman dynamo and, by destroying the planes of symmetry, produced a flow pattern (fig. 8*b*) which he has shown may act as a dynamo. The streamlines of Lilley's flow pattern form a distinct whorling pattern in three dimensions and this may be a physical characteristic necessary for dynamo action. The main magnetic fields of the dynamo are all toroidal so that the dominant components would all be completely contained within the core. Lilley shows that variation in the subsidiary poloidal components of the field may then produce secular variation and even dipole reversals, but without major change in the series of interactions between the toroidal components that form the basic dynamo.

Lilley (1970*b*) proposes that the present deviation of the geomagnetic dipole axis from the geographic axis (the expected axis of rotation) is in fact an expression of asymmetric motions in the earth's core. It follows that a movement of the geomagnetic poles to coincide with the geographic poles may be an indication of the flow in the core becoming symmetrical. If this should happen dynamo action will be lost and the dipole field will decay, perhaps to grow again in the opposite direction (that is, reverse), when the flow pattern once again becomes asymmetric. From secular variation studies Nagata (1965) suggests a decrease in strength of the dipole field of 0.05 per cent per year, accompanied by a rotation of the dipole towards the geographic axis of 0.02° of latitude per year (§1.1.2). These figures are reasonable if a 100 per cent decrease were to accompany a $11\frac{1}{2}°$ rotation of the present dipole axis.

1.1.4 *Models of the geomagnetic field*

The dynamo theory of the geomagnetic field is unaffected by the polarity of the field. Thus if a dynamo can maintain a magnetic field in one direction, it can do so equally well in the opposite direction. The discovery that the earth's magnetic field has reversed itself many times in the past (see chapter 4) has led to the investigation of several simple models that bear some remote resemblance to a homogeneous fluid dynamo. Bullard (1955) has shown that reversals in current flow do not occur in a single-disk dynamo as illustrated in fig. 6. The double-disk dynamo of Rikitake (1958) has proved to be a very useful analogue of the behaviour of the earth's magnetic field (fig. 9). Computations from the differential equations which govern the electromagnetic induction in this model have been made by Rikitake (1958), Allan (1958, 1962) and Mathews & Gardner (1963). Solutions show oscillations about the mean field and complete reversals, not unlike the apparent behaviour of the earth's field. Typical solutions obtained by Mathews & Gardner (1963) are shown in fig. 10.

Lowes & Wilkinson (1963, 1968) have constructed a laboratory model of geomagnetic dynamo action using cylinders rotating inside a casting, both of which are made of high permeability ferromagnetic material (fig. 11). Electrical connection of the cylinders to the casting was accomplished by filling the spaces with mercury. The dynamo is similar in principle to the double-disk dynamo. An initial axial field in one cylinder causes it to act as a homopolar disk, generating an e.m.f. between its axis and periphery. The resulting circulating current produces a toroidal field which links with the other cylinder, from which a further current flow is initiated. Positive feedback then occurs when this current flow produces a second toroidal

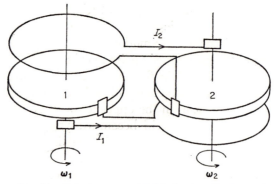

FIGURE 9   Double-disk dynamo of Rikitake (1958). From Cox (1968),
J. Geophys. Res. **73**, 3248.

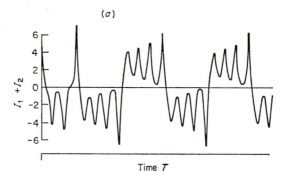

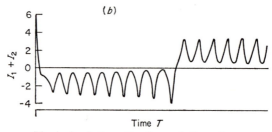

FIGURE 10   Typical solutions for the variation of current with time in a
double-disk dynamo. After Mathews & Gardner (1963), from Cox (1968),
J. Geophys. Res. **73**, 3248.

field linking with the first cylinder, provided that the rotations are in the
correct sense. A field of either polarity may be self-maintained using the
correct combination of cylinder rotations.

The foci of the non-dipole field of fig. 5 are suggestive of an array of
deep-seated, radially oriented dipoles. Such a model of the geomagnetic
field has been examined by Lowes (1955) and Alldredge & Hurwitz (1964).

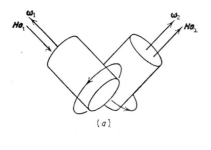

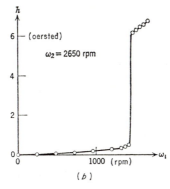

FIGURE 11 Laboratory model of geomagnetic dynamo action due to Lowes &
Wilkinson (1963), *Nature*, **198**, 1158–60. (*a*) Schematic diagram of the rotating
cylinders. (*b*) Observed external field as the speed of one of the cylinders is increased, the other rotating at constant speed.

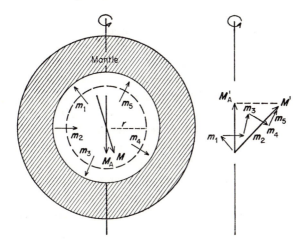

FIGURE 12 Eccentric dipole model of the geomagnetic field.
From Cox (1968), *J. Geophys. Res.* **73**, 3249.

TABLE 2   *Parameters of the three dipole model of the geomagnetic field due to Bochev (1969). The virtual poles are the points where the axes of the dipoles intersect the earth's surface*

	Dipole 1	Dipole 2	Dipole 3
Longitude (°)	239	352	93
Colatitude (°)	119	125	78
Distance from geocentre (km)	485	2810	1995
Moment (10^{25} gauss cm^3)	7.05	1.05	1.94
Virtual north pole	73N, 302E	79S, 206E	76N, 101E
Virtual south pole	76S, 150E	28N, 359E	68S, 88E

The dipole and non-dipole fields are represented by a central dipole M with an array of n smaller eccentric dipoles m_i at radius r as shown in fig. 12. The dipole term found from spherical harmonic analysis of this system corresponds to a geocentric dipole of moment $M + M'$, where

$$M' = \sum_{i=1}^{n} m_i. \tag{1.7}$$

Lowes (1955) represented the non-dipole field by eight to ten dipoles of strength 0.2 to 0.5×10^{25} gauss cm^3 situated about 1000 km below the core–mantle boundary. Alldredge & Hurwitz (1964) represented the field by eight dipoles situated just inside the liquid core, above the boundary of the inner core at about 0.25 earth radii. Five of these dipoles had moments between 2 and 5×10^{25} gauss cm^3 whilst the moments of the other three were less than 1×10^{25} gauss cm^3. On Alldredge and Hurwitz' model, however, the central dipole is about twice as great as that obtained from spherical harmonic analysis.

Bochev (1969) has constructed a model of only three dipoles to represent the geomagnetic field, arguing that such a representation of the field is more suggestive of its physical sources. He finds that the field and its secular variations can be accounted for quite satisfactorily in this way. The positions of the three dipoles which give the best fit to the 1960 field are listed in table 2 together with the points where the axes of these dipoles intersect the earth's surface. Dipoles 1 and 3 are thus approximately parallel to the axis of rotation and to one another, and are therefore more or less equivalent to the axial dipole and multipoles respectively obtained from spherical harmonic analysis. Dipole 2 is aligned obliquely to the earth's axis and approximately represents the equatorial dipole and multipoles.

1.2 PALAEOMAGNETISM

1.2.1 *Magnetism in rocks*

The study of the history of the earth's magnetic field prior to a few centuries ago relies on the record of the field preserved as fossil magnetization in rocks. Although most rock forming minerals are non-magnetic, all rocks exhibit some magnetic properties due to the presence of various iron oxides as accessory minerals making up only a few per cent of the rock. The magnetization of the accessory minerals is termed the fossil magnetism, which, if acquired at the time the rock was formed, may act as a fossil compass and be used to determine both the direction and intensity of the geomagnetic field in the past. The study of fossil magnetism in rocks is termed palaeomagnetism, and is a means of investigating the history of the geomagnetic field over the geological time-scale. The study of pottery and baked hearths from archaeological sites has been successful in tracing secular variation in historic times. This type of investigation is usually distinguished from palaeomagnetism and is generally termed *archaeomagnetism* (see §1.2.2).

The fossil magnetism of rocks measured initially (after preparation into suitably sized specimens) is termed the *natural remanent magnetization* or simply NRM. The mechanism by which the NRM was acquired depends upon the mode of formation and subsequent history of rocks as well as the characteristics of the magnetic minerals. Magnetization acquired by cooling from high temperatures through the Curie point(s) of the magnetic mineral(s) is called *thermoremanent magnetization* (TRM). If the magnetization is acquired by chemical action during the formation of iron oxides at low temperatures, then it is termed *chemical remanent magnetization* (CRM). The alignment of detrital magnetic particles that might occur in a sediment gives rise to *detrital remanent magnetization* (DRM).

The component of NRM acquired when the rock was formed is termed the *primary magnetization* and this may represent all, part, or none of the total NRM. Subsequent to formation the primary magnetization may decay either partly or wholly and further components may be added by a number of processes. These subsequent magnetizations are called *secondary magnetization*. A major task in all palaeomagnetic investigations is to recognize and either correct for or preferably eliminate secondary components, whilst selectively preserving sufficient primary component for measurement. A special class of secondary magnetization is that acquired between collection in the field and measurement in the laboratory, this is termed *temporary magnetization*. A full discussion of all these aspects of magnetism in rocks is given in chapter 2.

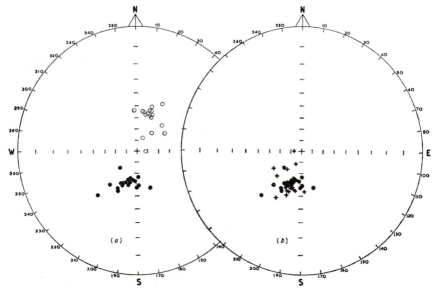

FIGURE 13   Stereographic (equal angle) projection of the directions of magnetization observed in 36 Lower Oligocene lava flows from the Liverpool Volcano, New South Wales (Wellman *et al.*, 1969). (*a*) Solid circles on the lower hemisphere, open circles on the upper hemisphere. (*b*) Directions all plotted on the lower hemisphere, so that the open circles in (*a*) become south-seeking directions represented by the crosses in the opposite quadrant.

Directions of magnetization measured in rocks are specified in the usual way by the declination (*D*) eastwards from true north, and the inclination (*I*), reckoned positive downwards. These directions are represented by vectors at the centre of a sphere, and are plotted graphically as points where the vector cuts the surface of the sphere. Either an equal area (Schmidt) projection or equal angle (Wulf) projection is used. The choice is normally a matter of personal preference. Whereas the stereographic projection becomes distorted at the edges, it has the advantage that a circle on a sphere plots as a circle on the projection and this is sometimes convenient in palaeomagnetic work. Where the clustering of points at different parts of the projection needs to be compared, then the equal area projection is more useful.

There are two ways of presenting directions of magnetization on these projections as illustrated in fig. 13. This arises because of the presence of reversals of magnetization. The more common method is illustrated in fig. 13*a*. Directions are distinguished as they fall on the lower (solid symbols) or upper (open symbols) hemisphere of the projection. The vectors of one set are directed upwards in the northeast quadrant, whilst those of the other set are directed downwards in the southwest quadrant. The two groups are directed approximately 180° apart due to the presence of reversals. An alternative presentation of the same data is shown in fig. 13*b*, where only one half of the projection (either the upper or lower hemisphere) is used. Vectors are then distinguished as they are 'north-seeking' or 'south-seeking' on that particular hemisphere. In fig. 13*b* the lower hemisphere has been used so that the upwards directed vectors in the northeast quadrant of fig. 13*a* become 'south-seeking' vectors downwards on the lower hemisphere in the southwest quadrant. In fig. 13*b*, both 'normal' and 'reversed' directions plot in the same region and it is relatively a simple matter to judge whether or not they are nearly opposed directions. There are difficulties, however, with this system when the points plot near the edge of the stereonet and it is thus not very often used.

1.2.2   *Archaeomagnetism*

Pottery, and more usefully, bricks from pottery kilns and ancient fireplaces, whose last dates of firing can be estimated from carbon-14 contents of ashes, have a thermoremanent magnetization dating from their last cooling. Samples used in these studies often have awkward shapes, but they can be measured by the usual techniques of palaeomagnetism. Pioneer work in this field was undertaken by Folgheraiter (1899) and Thellier in France. The techniques commonly in use are those developed by Thellier and have been reviewed by Thellier (1966). Aitken and his co-workers (Aitken & Weaver, 1962; Aitken & Hawley, 1967) have made systematic studies of oriented samples collected from archaeological kilns, ovens, and hearths, mainly from the southern part of Britain. The variation in declination and inclination from archaeomagnetic studies (Aitken, 1970) have been combined to give an extension of Bauer's (1899) representation of the secular variation at London from direct observations (fig. 14). From the strong easterly declination *circa* A.D. 1000, there is a movement to the shallow inclination of *circa* A.D. 1300. This is followed by a return to the steeper inclinations to join the Bauer curve at the end of the sixteenth century. There are only a few measurements for the pre-Roman era in Britain covering the period from about A.D. 100 to 300.

Studies of the intensity of the field over archaeological times can be undertaken with a much wider variety of materials, such as pottery, because the precise orientation of the samples need not be known. The method used is that developed by Thellier & Thellier (1959). Whereas the secular variation of declination and inclination always refers to a particular area, the measurements of ancient geomagnetic intensity are essentially a function

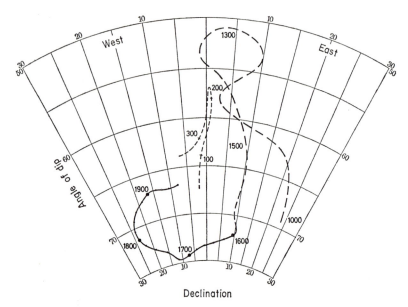

FIGURE 14 Variation of declination and inclination at London from archaeo-magnetic investigations in Britain, from Aitken (1970). The dashed curve deduced from the archaeomagnetic data joins the observed values since 1580, shown by the solid curve due to Bauer (1899).

of latitude at each place (§1.2.4). Measurements from all over the world can be normalized to an equivalent reduced dipole moment of the earth (Smith, 1967). Rapid variations of the non-dipole field give an additional noise signal at each place, so that even the most accurate measurements at one locality may not give the true rate of change of the geomagnetic dipole moment. To remove these variations the dipole moments must be averaged not only from different parts of the world but also in class intervals of a few hundred years. Cox (1968) has analysed the data covering the past 8000 years by averaging over class intervals of 500 years. The results of this analysis are shown in fig. 15. During the past 4000 years the dipole intensity appears to have undergone a well-defined half-cycle with peak value of 12×10^{25} gauss cm^3, which is an increase of fifty per cent over the present value of 8×10^{25} gauss cm^3. Prior to 4000 years ago, there is the suggestion of part of a cycle corresponding to a period of about 8000 years, although the number of samples is too small and the sampling area too restricted to define the period accurately. The time spanned by the observations is too short to determine whether fluctuations in dipole moment are periodic, but if they are, then the data at present indicate a period of the order of 10^4 years.

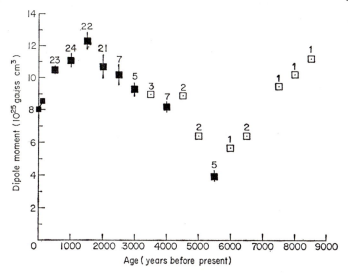

FIGURE 15 Variation in the dipole moment of the geomagnetic field from archaeomagnetic intensity studies. Changes during the past 130 years deduced from spherical harmonic analyses are shown by the short bar. The archaeomagnetic results are averages over the indicated 500 year class interval. The number of values in each class interval is given above each point and the standard error shown by a vertical line. Unshaded squares refer to intervals containing too few data to give a reliable value. From Cox (1968), *J. Geophys. Res.* **73**, 3250.

1.2.3 *Early work in palaeomagnetism*

The fact that some rocks possessed extremely strong remanent magnetization was noted as early as the late eighteenth century from their effect on the compass needle. Von Humboldt in 1797 attributed these effects to lightning strikes. Further investigations during the nineteenth century of these intense magnetizations occasionally found in rock exposures were also generally explained in this way. These were the first palaeomagnetic phenomena to attract attention. The first studies of the direction of NRM of rocks were made by Delesse in 1849 and Melloni in 1853. They both found that certain recent lavas were magnetized parallel to the earth's magnetic field. The work of Folgheraiter (1899) both extended and confirmed these earlier investigations. Later Chevallier (1925), from studies of the historical flows of Mt Etna, was able to trace the secular variation of the field during the past 2000 years.

David (1904) and Brunhes (1906) first investigated the material baked by lava flows, comparing the direction of NRM of the flows with those of the underlying baked clay. They reported the first discovery of NRM directions roughly opposed to that of the present field. Confirmation that the baked clays were also reversely magnetized led to the first speculation that the earth's magnetic field had reversed itself in the past. Mercanton (1926) then argued that if the earth's magnetic field had reversed in the past, reversals should be found in rocks from all parts of the world. In studies of rocks of various ages from Spitsbergen, Greenland, Iceland, the Faroes, Mull, Jan Mayen Land and Australia, he found that some were magnetized in the same sense as the present field and some in the opposite sense. Concurrently Matuyama (1929) observed similar effects in Quaternary lavas from Japan and Manchuria, but noted that the reversed lavas were always older than those directed in the same sense as the present field (normal lavas). He concluded that during the early Quaternary the earth's magnetic field was directed in a sense opposite to that of the present, and gradually changed to its present sense later in the Quaternary.

By 1930 a number of important aspects of palaeomagnetism in rocks had been established, culminating in the suggestion by Mercanton (1926) that, because of the approximate correlation of the geomagnetic and rotational axes at the present, it might be possible to test the hypotheses of polar-wandering and continental drift. This inspiration was not put into practice until the mid 1950s and the first review of palaeomagnetic data that takes this renewed activity into account is given by Cox & Doell (1960). Since then a major work on the subject has been published by Irving (1964). Improved techniques and the undertaking of extensive palaeomagnetic surveys by workers in many parts of the world has increased the amount of palaeomagnetic information seven-fold over the past ten years. The number of independent investigations listed by Cox & Doell (1960) for the period up to the end of 1959 was about 200, and by the end of 1963 this had increased to about 550 as listed by Irving (1964). By the end of 1970 this figure had risen to about 1500, but what is more important the standard demanded of palaeomagnetic investigations has increased dramatically over the past seven years. It is now possible to review critically many of the older measurements, both in the light of new ones and through the results of the use of improved techniques. In the final chapters of this book such a review of the palaeomagnetic data is presented in so far as they are relevant to the problem of polar-wandering and continental drift.

1.2.4 *Axial geocentric dipole hypothesis*

On the geological time-scale the study of the geomagnetic field requires some model for use in analysing palaeomagnetic results, so that measurements from different parts of the world may be compared. The model should be one that expresses the long-term average behaviour of the field

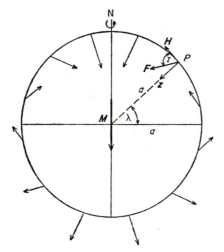

FIGURE 16   Field of an axial geocentric dipole.

rather than its more detailed short-term behaviour. The model used is termed the *axial geocentric dipole field* and its use in palaeomagnetism is essentially an application of the principle of uniformitarianism. It is known from palaeomagnetic measurements (see chapter 6) that for the past few million years the earth's magnetic field, *when averaged over periods of several thousands of years*, has conformed to this model, so that it is then used as a working hypothesis through geological time. The model is a simple one and corresponds to the field of a geocentric dipole directed along the rotational axis as shown in fig. 16. For such a model the geomagnetic and geographic axes coincide as do the geomagnetic and geographic equators. For any point on the earth's surface, the geomagnetic latitude λ equals the geographic latitude.

If M is the magnetic moment of the dipole and a the radius of the earth, the horizontal (H) and vertical (Z) components of the field at latitude λ can be simply derived from the geometry of fig. 16 as

$$H = \frac{M\cos\lambda}{a^3}, \quad Z = \frac{2M\sin\lambda}{a^3}, \qquad (1.8)$$

whilst the total field F and its inclination I are given by:

$$F = (H^2 + Z^2)^{\frac{1}{2}} = \frac{M}{a^3}(1 + 3\sin^2\lambda)^{\frac{1}{2}}, \qquad (1.9)$$

$$\tan I = Z/H = 2\tan\lambda. \qquad (1.10)$$

By definition $D = 0° \qquad (1.11)$

23

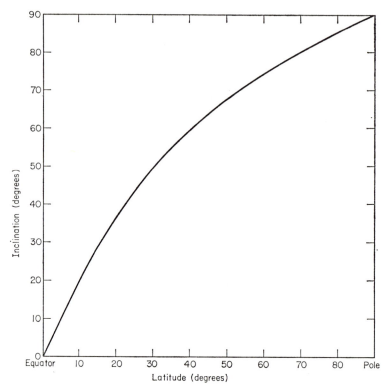

FIGURE 17 Variation of inclination with latitude for an axial
geocentric dipole from (1.10).

and the colatitude p is given by

$$\tan I = 2\cot p \quad (0° \leqslant p \leqslant 180°). \tag{1.12}$$

The relationship given in (1.10) is an important one in palaeomagnetic investigations. It indicates that the axial geocentric dipole model, when applied to results from different geological periods, enables the palaeomagnetic latitude to be simply derived from the mean inclination. This relationship is illustrated in fig. 17.

In order to compare palaeomagnetic results from widely separated localities, it is necessary to calculate some parameter, which, on the basis of the axial geocentric dipole model, should have the same value at each observing locality. The parameter used is the *palaeomagnetic pole* and represents the position where the dipole axis cuts the surface of the earth. In general the dipole axis may be inclined to the present axis of rotation, so that the position of the palaeomagnetic pole is referred to the present latitude–longitude grid. Thus, if the palaeomagnetic mean directions

24

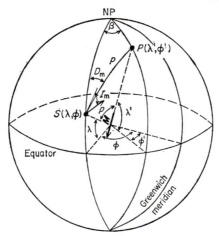

FIGURE 18   Calculation of palaeomagnetic pole from average directions of magnetization (D_m, I_m). The sampling site S has latitude and longitude (λ, ϕ) and the pole P has co-ordinates (λ', ϕ').

D_m, I_m are known at some sampling locality S, whose latitude and longitude are respectively (λ, ϕ) the position of the palaeomagnetic pole $P(\lambda', \phi')$ can be calculated from the following equations and by reference to fig. 18:

$$\sin\lambda' = \sin\lambda\cos p + \cos\lambda\sin p\cos D \quad (-90° \leqslant \lambda' \leqslant +90°), \quad (1.13)$$

$$\phi' = \phi + \beta \quad \text{when} \quad \cos p \geqslant \sin\lambda\sin\lambda',$$

or $\qquad \phi' = \phi + 180 - \beta \quad \text{when} \quad \cos p < \sin\lambda\sin\lambda', \qquad (1.14)$

where $\quad \sin\beta = \sin p\sin D/\cos\lambda' \quad (-90° \leqslant \beta \leqslant +90°). \qquad (1.15)$

Latitudes are positive for the northern hemisphere and negative for the southern hemisphere, whilst longitudes are measured eastwards from the Greenwich meridian and lie between 0 and 360°, otherwise the symbols are as defined in fig. 16. The colatitude p is defined in (1.12).

The term palaeomagnetic pole carries with it the implication that the data from which it has been derived are truly representative of the axial dipole model. That is the data are sufficient so that it can be presumed that the field has been averaged over several thousands of years and that the secular variation has been averaged out. Palaeomagnetic sampling should be designed so as to produce a series of spot measurements of the field in the past. It is only when a suitable number of these measurements are averaged that a palaeomagnetic pole can be calculated from the *average* field values (D_m, I_m). For example a single lava flow can represent only one point in time and provides virtually an instantaneous record of the geomagnetic field at the time the lava cooled. Many lava flows have to be

TABLE 3 *Summary of poles used in geomagnetism and palaeomagnetism*

North (south) magnetic pole	Point on the earth's surface where the magnetic inclination is *observed* to be $+90°$ $(-90°)$. The poles are not exactly opposite one another and for epoch 1965 lie at $75\frac{1}{2}$N, 101W and 66s, 140E.
Geomagnetic north (south) pole	Point where the axis of the *calculated* best fitting dipole cuts the surface of the earth in the northern (southern) hemisphere. The poles lie opposite one another and for epoch 1965 are calculated to lie at $78\frac{1}{2}$N, 70w and $78\frac{1}{2}$s, 110E.
Virtual geomagnetic pole (VGP)	The position of the equivalent geomagnetic pole calculated from a spot reading of the palaeomagnetic field. It represents only an instant in time, just as the present geomagnetic poles are instantaneous.
Palaeomagnetic pole	The average palaeomagnetic field over periods sufficiently long so as to give an estimate of the geographic pole. Averages over times of 10^4 to 10^5 years are estimated to be sufficient. The pole may be calculated from the average palaeomagnetic field or from the average of the corresponding VGPs.

sampled and their directions averaged before a palaeomagnetic pole can be determined. However, it is possible of course to calculate a 'pole' from (1.13), (1.14) and (1.15) using only a single spot measurement of the field (D, I) from a single lava flow. The term *virtual geomagnetic pole* (VGP) is then used to distinguish this type of pole from a palaeomagnetic pole, and can be regarded as the palaeomagnetic analogue relating to the geomagnetic poles of the present field, which are a function of the inclined dipole (or geomagnetic) axis. Essentially it takes a number of virtual geomagnetic poles to determine a single palaeomagnetic pole. A summary of the geomagnetic and palaeomagnetic terminology relating to poles is presented in table 3.

Tests for the validity of the axial geocentric dipole model over geological time can be made, at least theoretically, in a number of ways. The simplest test is that the field when viewed from regions of continental extent, should be consistent with that of geocentric dipole. This is true if the palaeomagnetic poles obtained from different rock units belonging to the same geological epoch are in close agreement, at least as good as that observed over the past few million years. A study of the palaeointensity of the palaeomagnetic field as a function of palaeolatitude should conform with (1.9), whilst models for palaeosecular variation based primarily on the dipole hypothesis should conform with palaeomagnetic data (§1.1.5). However, all these tests do not necessarily indicate that the geocentric dipole was *axial*. Testing for an axial geocentric dipole appeals to palaeoclimatic

26

evidence. The earth's climate is controlled by the rotational axis and has an equator to pole distribution. It is warmer at the equator than at the poles. The palaeolatitude spectra of various palaeoclimatic indicators should all be latitude dependent to be consistent with the hypothesis of an axial geocentric dipole field. Results from this type of investigation are discussed more fully in chapter 6.

It is, of course, possible to explain variations in the palaeomagnetic field in the past by proposing that the field was irregular and similar in form, say, to the present non-dipole field. This has been referred to as the *non-dipole hypothesis* and it can be argued (Irving, 1964) that this is fundamentally a non-scientific hypothesis. It is a feature of all scientific investigations, indeed a basic part of the scientific method, that a hypothesis is one which is amenable to or potentially amenable to scientific test. The non-dipole hypothesis is one which is irrefutable, all observations can be made to conform to it by convenient *ad hoc* adjustments to suit the situation. It is really no hypothesis at all, it does not predict observable geophysical effects and cannot be checked by independent observations. The axial geocentric dipole hypothesis, however, has the merit of being a refutable hypothesis, for which the evidence to date suggests that for the Phanerozoic the model stands up to the observations remarkably well. For the Precambrian there are now some investigations that suggest the geocentric dipole field is valid, but as yet no attempt has been made to apply palaeoclimatological information to these palaeomagnetic results in a systematic way.

1.2.5 *Palaeosecular variation*

The axial geocentric dipole model takes no account of secular variation, although its effect must be averaged out before palaeomagnetic measurements are said to conform with the model. The secular variation in the past is thus expressed by the scatter in directions obtained in palaeomagnetic studies, usually expressed in terms of the angular dispersion of directions (§3.2.1). The information will not be in the form of a continuous record of the field, but will merely be a statistical record of the short-term variations. Palaeosecular variation may therefore be simulated by an extension of the axial geocentric dipole model and can be considered to be due to the following factors:

(*a*) changes in the strength and direction of the non-dipole field,

(*b*) changes in the strength of the central dipole (dipole oscillations),

(*c*) changes in the orientation of the central dipole, such that on average the dipole axis coincides with the axis of rotation (dipole wobble).

All these factors will give rise to values of angular dispersion which show

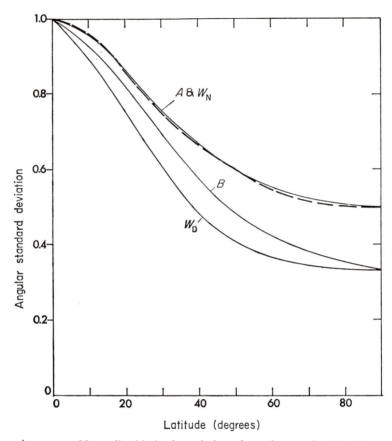

FIGURE 19   Normalized latitude variation of angular standard deviation given by model A, model B, and the W_D term of model D. The W_N term of model D follows the curve for model A. The best fit of world-wide palaeomagnetic data to model D is given by the dashed curve after Brock (1971). This shows that the W_N term predominates and that model A describes the data almost precisely. On Brock's (1971) analysis the average equatorial value of the angular standard deviation is 18.9°.

latitude variation. Using the nomenclature of Irving (1964), there are four models that have been proposed.

Model A, due to Irving & Ward (1964), considers an axial geocentric dipole field of fixed dipole moment. Secular variation is due to a component of fixed magnitude but random direction which perturbs the dipole field. The model predicts a latitude variation in which the angular dispersion at the poles is one half that at the equator (fig. 19). In this model there is no dipole wobble. Model B is due to Creer *et al.* (1959) and postulates a wobble of the main dipole which follows the Fisher (1953) distribution

(§3.2.1). No appeal is made to non-dipole components, so that the model predicts a latitude variation in which the angular dispersion at the poles is one-third its equatorial value (fig. 19). Model C is due to Cox (1962) in which dipole wobble is combined with non-dipole components, but this has been superseded by model D proposed by Cox (1970). In this model the latitude variation of the angular dispersion due to non-dipole components and dipole oscillations (W_N) is the same as for model A, whilst the variation due to dipole wobble (W_D) is as shown in fig. 19, predicting polar values of dispersion one-third of the equatorial value. The latitude variation of model D will lie somewhere between that of W_N and W_D depending on the relative contributions of each.

Brock (1971) has analysed selected palaeomagnetic data from igneous rocks in terms of model D, from which he has deduced the relative contributions of dipole wobble and non-dipole components (including dipole oscillations) to palaeosecular variation. His analysis suggests that non-dipole components together with dipole oscillations predominate over dipole wobble. Dipole wobble is small for separate analyses of Tertiary and pre-Tertiary data, but secular variation is significantly smaller in pre-Tertiary times than in the Tertiary.

1.2.6 *Palaeointensity*

Attempts have been made to extend the variation in the strength of the geomagnetic field in historic and archaeological times to cover the whole of geological time. The problems of determining the palaeointensity of the geomagnetic field are much more complex than those associated with palaeodirectional measurements. These problems have been discussed by Smith (1967), who has concluded that the geomagnetic dipole is not constant within any given polarity, but fluctuates in strength, possibly with a period of 10^4 years as suggested by the archaeomagnetic data (§1.2.2). The standard deviation of measurements for a given epoch is about fifty per cent, comparable with the fluctuations observed in the archaeomagnetic data, so that many measurements are necessary to define the mean dipole moment.

A number of measurements have been reported for the past 500 My by Smith (1967), Briden (1966) and Carmichael (1967) and these are summarized in fig. 20. Smith (1967) reduced all measurements to a *virtual dipole moment* (VDM) found by substituting the deduced palaeofield strength and palaeolatitude in (1.9). The scatter in the data is high as might be expected from data having large standard deviations and too few measurements. However the measurements seem to indicate a clear decrease in the

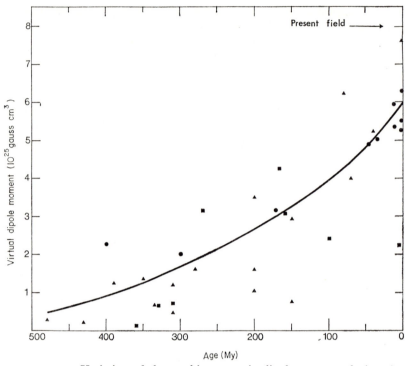

FIGURE 20　Variation of the earth's magnetic dipole moment during the past 500 My, drawn from the data of Smith (1967) (circles), Briden (1966) (squares), and Carmichael (1967) (triangles).

calculated geomagnetic dipole moment during the early Palaeozoic. A decrease to only ten per cent of the present value is indicated around 500 My ago. It might therefore be supposed that a gradual decay of the natural remanence could explain these observations. However this decrease does not extend back into the Precambrian. Data produced by Carmichael (1967) and Schwarz & Symons (1969) for a range of dated Precambrian rocks show that the virtual dipole moment, although subject to considerable short term fluctuations, was on average very similar to the present value. The minimum observed during the early Palaeozoic thus appears to be a real effect.

Of particular interest are palaeointensity measurements made on the oldest Precambrian rocks. McElhinny & Evans (1968) deduced a value of 1.5 times the present from the Modipe gabbro of Botswana dated at around 2650 My, whilst Bergh (1970) reports a value about two-thirds the present from the Stillwater Complex of Montana, now dated at 2450 My (Fenton & Faure, 1969). Data from rocks of age between 2400 and 2500 My by

Carmichael (1967), Kobayashi (1968) and Schwarz & Symons (1969) also suggest values similar to the present. These measurements all indicate that the earth's core must have been of sufficient size at least 2700 My ago to maintain a geomagnetic field equal to its present-day value. Elsasser (1963) proposed that the core was formed early in the earth's history and would have been essentially complete 3000 My ago. On the other hand, Runcorn (1964) has suggested that the core has been growing to its present size throughout geological time and that 3000 My ago was of negligible size. The palaeointensity measurements from these very old rocks thus support the view that the formation of the core was essentially completed at an early stage in the earth's history.

2 Rock magnetism

2.1 BASIC PRINCIPLES OF MAGNETISM

2.1.1 *Remanent and induced magnetism*

The study of magnetism originated from observation of the behaviour of natural permanent magnets, the earliest known of which were used as magnetic compass needles. A permanent magnet is usually described by its 'north' and 'south' magnetic 'poles', imagined to reside at the opposite ends of the magnet. The concept of magnetic poles has been of considerable use in analysing the behaviour of magnets, but since the discovery by Oersted in 1820 that an electric current flowing in a wire deflected a compass needle placed near it, it has been found that all magnetic effects are more appropriately described in terms of electric currents. The magnetic poles at the end of a magnet cannot be isolated and are just convenient fictions for the purpose of simple analysis. An electron in orbit around a nucleus is essentially the same as a current flowing in a loop, and has associated with it a *magnetic dipole moment*, which is given by the product of the current in the loop and the area of the loop. The magnetic effects of materials can all be described in terms of these elementary current loops.

Magnetic dipole moment is generally quoted in 'emu' in the c.g.s. electromagnetic system of units, which will be used throughout this book. The *intensity of magnetization* (or more strictly *magnetization*) of a material is its magnetic moment per unit volume, often given in units of emu cm^{-3}.†
The magnetization of any material is generally made up of two components, the *remanent magnetization* (or simply *remanence*), which is that remaining after removal of an applied field, and the *induced magnetization*, which is that induced by an applied field but which disappears after removal of the field. When dealing with rocks, the total magnetization J is made up of the vector sum of the remanence J_n, and the magnetization induced in the earth's magnetic field J_i, where

$$J = J_i + J_n. \qquad (2.1)$$

† The situation becomes complicated if S.I. (or rationalized M.K.S.A.) units are used, because there are two alternative proposals for the definition of the magnetic moment of a current loop. In the Kennelly proposal, used for example by Brailsford (1966), magnetic moment has units of weber m, whilst magnetization is in weber m^{-2}. On the other hand the Sommerfeld proposal, used for example in a text by Kip (1962), results in units of amp m^2 for magnetic moment and amp m^{-1} for magnetization.
For a text on magnetism using c.g.s. units, readers are referred to Morrish (1965).

The induced magnetization J_i is proportional to the applied H, that is

$$J_i = \chi H, \qquad (2.2)$$

where χ is a constant of proportionality called the *magnetic susceptibility*. If H is measured in gauss and χ is a dimensionless constant, then J_i should also be measured in gauss. Thus intensity of magnetization is sometimes quoted in gauss using the c.g.s. system and this is equivalent to emu cm^{-3}, so that the unit of magnetic moment becomes gauss cm^3. Some idea of the magnitude of these units can be gauged from a simple example. A sphere of magnetic material of radius 1 cm with intensity of magnetization of 1 gauss has a magnetic moment of roughly 4 gauss cm^3 and produces a magnetic field of strength varying between 4 and 8 gauss near its surface and between 0.5 and 1 gauss at a distance of 1 cm from its surface. The latter is comparable with the magnitude of the geomagnetic field at the earth's surface. In isotropic materials the direction of the induced magnetization J_i lies along the direction of the applied field H. Some banded sediments, layered intrusions and foliated metamorphic rocks are magnetically anisotropic and have greater susceptibility in the plane of layering. These are special cases and most rocks used in palaeomagnetism such as basalts, dolerites and redbeds are magnetically isotropic (see also §2.3.7).

The *Koenigsberger ratio* (Q) has been defined as the ratio of the remanent to induced magnetization (Koenigsberger, 1938) and is given by

$$Q_n = J_n/\chi H \qquad (2.3)$$

or

$$Q_t = J_t/\chi H \qquad (2.4)$$

where Q_n is the ratio of the intensity of NRM (J_n) to that induced by the earth's field at the sampling site. Q_t on the other hand is the ratio of the thermoremanence (J_t) acquired in the field H to the magnetization induced by the same field at room temperature.

2.1.2 *Diamagnetism and paramagnetism*

An electron in its path around the nucleus creates a dipole moment referred to as the *orbital magnetic moment*. Also the electron spins about its own axis, creating an additional magnetic moment known as the *spin magnetic moment*. The magnetic dipole moment of an atom as a whole will be the resultant of all the orbital and spin magnetic moments of its electrons, and in some atoms this resultant is zero. In other atoms, such as in the case of the simple hydrogen atom, the formation of a molecule occurs when two atoms containing electrons of opposite spins are brought together. In this case, although individual hydrogen atoms have a resultant magnetic

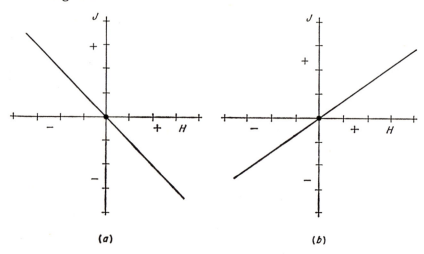

FIGURE 21   Variation of magnetization with applied field for
(a) diamagnetic and (b) paramagnetic material.

moment, the forces (known as exchange forces) which tie the atoms together as a hydrogen molecule, also give the molecule a zero magnetic moment.

When these atoms or molecules with zero magnetic moment are placed in a magnetic field, the individual atomic current loops are acted on by a couple and the orbits of the electrons execute a precessional motion about the applied field; that is the plane of the orbit itself describes an orbit. The precession of the orbit can be imagined as equivalent to the electron (in addition to its other motions) rotating in an orbit in the *opposite* direction. Thus although the individual current loops are aligned by the applied field and the net atomic or molecular magnetic moment is zero, the orbital precessions cause a magnetic moment to be induced in each atom in the *opposite* direction to that of the applied field. This phenomenon is known as *diamagnetism*. It is a fundamental property of all matter and independent of temperature. Because the induced magnetization is in the opposite direction to that of the applied field, the susceptibility is negative (fig. 21 *a*) and typically of the order of 10^{-6}.

If an atom has a resultant magnetic moment, however, the application of a magnetic field tends to align these dipole moments along the direction of the field. Although the diamagnetic effect still occurs, it is swamped by the alignment of the atomic dipole moments. Substances which exhibit this effect are called *paramagnetics*, and the induced magnetization is in the same direction as the applied field, giving a positive susceptibility (fig. 21 *b*) which lies typically between 10^{-4} and 10^{-6}. In metallic substances a further situation arises because the individual atoms and their inner orbital electrons are closer together in the solid state than the virtual radii of the valence electrons. The outer valence electrons are thus no longer associated with individual atoms and they wander freely through the metal. In an atom devoid of its valence electrons, the net atomic dipole moment is zero. The application of a magnetic field causes the 'free' electrons, equal numbers of which have opposite spins, to have their spins aligned parallel to the magnetic field. The substance thus acquires a dipole moment and paramagnetism results.

2.1.3   *Ferro, antiferro and ferrimagnetism*

Diamagnetic and paramagnetic substances only exhibit weak magnetic effects, for the dipole moments involved are relatively small. However, substances like iron, cobalt and nickel exhibit strong magnetic effects known as *ferromagnetism*. These ferromagnetic substances are distinguished by the fact that the individual atoms, and their inner orbital electrons, are very much closer together than the virtual radii of the valence electron orbits when compared to the metallic paramagnetics. Also, there are more valence electrons available to move freely through the metal, so that they become crowded together and react strongly with one another. The exchange forces between these electrons are such as to cause their spins to become aligned *even in the absence of an applied magnetic field*. Ferromagnetic substances thus exhibit *spontaneous magnetization*, and may have a permanent magnetic dipole moment in the absence of an applied field. As the temperature is increased thermal agitation may destroy the alignment process. This occurs at a critical temperature for each substance called the *Curie temperature* or *Curie point*. The spontaneous magnetization reduces to zero at the Curie point (fig. 22) and above this temperature the substance behaves like an ordinary paramagnetic.

Some substances are characterized by a sub-division into two sub-lattices (usually designated A and B). The atomic moments of A and B are each aligned, but antiparallel to one another. The ferromagnetic effects cancel one another out when the moments of the two sublattices are equal (fig. 23) and there is no net magnetic moment. This phenomenon is known as *antiferromagnetism*. Such substances do not have a Curie temperature, because there is no net ferromagnetism. In this case the ordering of the atomic moments is destroyed at a critical temperature called the *Néel temperature*, above which the substances behave like ordinary paramagnetics. If the atomic moments of the A and B sub-lattices are unequal, then there is a net spontaneous magnetization and a weak ferromagnetism results, known as *ferrimagnetism*. Alternatively the equal atomic moments in the two sublattices may not be exactly antiparallel and a small spon-

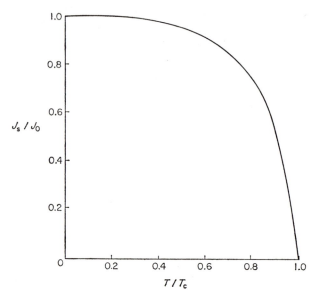

FIGURE 22 Variation of spontaneous magnetization of a typical ferromagnetic with temperature. The spontaneous magnetization J_s is normalized to its value J_0 at absolute zero, and the temperature T is given as a fraction of the Curie temperature T_c.

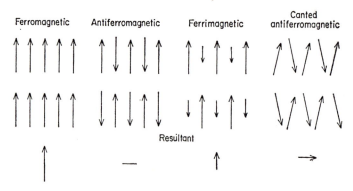

FIGURE 23 To illustrate the spontaneous magnetization vectors in crystals. The resultant spontaneous magnetization is given at the bottom.

taneous magnetization results (fig. 23). This substance is called a *canted antiferromagnetic*. Both the ferrimagnetic and canted antiferromagnetic substances behave as ordinary ferromagnetics; they have a Curie temperature and all the properties of ferromagnetics. The important minerals in rock magnetism are of these two types, to which the basic theories of ferromagnetism may be applied.

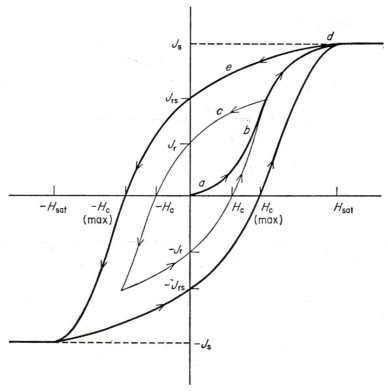

FIGURE 24   Initial magnetization curve and hysteresis loops
(*J–H* loops) for a ferromagnetic.

2.1.4   *Hysteresis*

When a ferromagnetic substance, initially in a demagnetized state, is placed
in an applied field H, the specimen follows the magnetization curve from
the origin as in fig. 24. As H is increased from zero, the intensity of
magnetization J rises linearly as shown by the portion a of the curve. If H
is reduced to zero at this point, the process is reversible and J also falls to
zero. The slope of the *J–H* curve here gives the *initial susceptibility* of the
ferromagnetic. As H is further increased so the slope of the curve increases
(in the region b); if H is now reduced to zero, J does not fall to zero, but
follows the path c, and an *isothermal remanent magnetization* (IRM), given
by J_r, results. On the magnetizing curve further increases in H beyond
point d would produce no further increase in J, and a *saturation magnetization*, J_s, is reached at the saturating field H_{sat}. On reducing the field to zero
(along portion e) the *saturation* IRM, J_{rs} occurs. On applying a field in the
opposite direction, the IRM is overcome and J is reduced to zero in a field

37

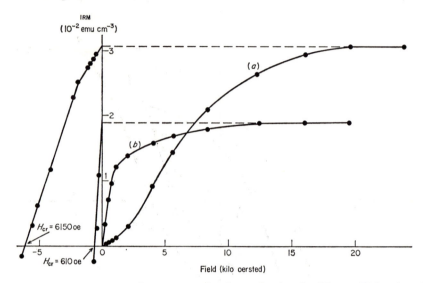

FIGURE 25   IRM curves for two samples from the Antrim Plateau Volcanics of northern Australia. (*a*) Haematite rich sample, (*b*) magnetite rich sample. Data of Luck (1971).

H_c, called the *coercivity* or *coercive force*. Further increases in H in the negative direction causes saturation to occur again and repeated cycling of the field causes the intensity of magnetization to follow a hysteresis loop as in fig. 24. The largest hysteresis loop occurs when the field is cycled with saturation being reached, and in this case H_c has its maximum value, called the maximum coercivity. If the field is cycled without saturation being reached, then a smaller hysteresis loop results as shown.

It is usually more convenient when dealing with rocks to study the IRM (J_r) rather than J, in which case the typical IRM curves of fig. 25 are produced. If successively increasing fields are applied in the opposite direction to J_{rs}, until the saturation IRM is reduced to zero, then the 'back' field required to reduce this to zero is called the *coercivity of maximum remanence* (H_{cr}). When this process is applied to the NRM of rocks, the 'back' field required to reduce the NRM to zero is called the *coercivity* of NRM (H_{cr}').

2.2   MAGNETIC MINERALS IN ROCKS

2.2.1   *Mineralogy*

The minerals that are largely responsible for the magnetic properties of rocks are within the ternary system FeO–TiO$_2$–Fe$_2$O$_3$) fig. 26). In this system it is generally sufficient to distinguish between the two types of

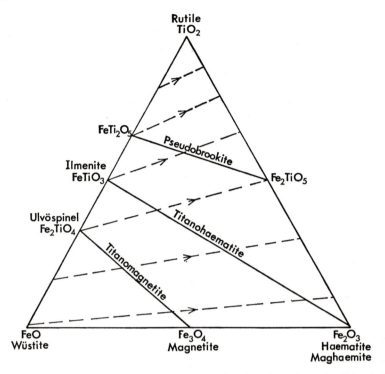

FIGURE 26   FeO-TiO$_2$-Fe$_2$O$_3$ ternary system, showing the principal solid solution series (solid lines). The dashed lines are some lines of constant Fe:Ti ratio along which oxidation may proceed in the direction of the arrows.

magnetic mineral. There are the strongly magnetic cubic oxides *magnetite* (Fe$_3$O$_4$), *maghaemite* (γ-Fe$_2$O$_3$), and the solid solutions of magnetite with *ulvöspinel* (Fe$_2$TiO$_4$), which are known as titanomagnetites. The more weakly magnetic rhombohedral minerals are based on *haematite* (α-Fe$_2$O$_3$) and its solid solutions with *ilmenite* (FeTiO$_3$), which are known as titano-haematites. The members of the orthorhombic pseudobrookite series are all paramagnetic above liquid oxygen temperatures and need not be considered further. There is only complete solid solution at high temperatures, in which case oxidation proceeds along the dashed lines in fig. 26 from left to right. Examples of the progressive oxidation of minerals in the two main solid solution series are given in plate 1 (facing pp. 64 and 65) from Wilson & Haggerty (1966) and are discussed below.

Pyrrhotite (FeS$_{1+x}$, where $0 < x \leqslant 0.14$) is the only iron sulphide that is ferrimagnetic. The Néel transition to the paramagnetic state occurs at about 300 °C, whilst below this temperature it is antiferromagnetic when $0 < x \leqslant 0.09$ and ferrimagnetic when $0.09 < x \leqslant 0.14$. Two important

39

naturally occurring oxyhydroxides of iron, which dehydrate to oxides at 100 to 300 °C are *goethite* (α-FeOOH) and *lepidocrocite* (γ-FeOOH). The chemically precipitated ferric hydroxide in a sediment may suffer dehydration during compaction or later and give rise to the corresponding form of Fe_2O_3 (haematite from goethite, maghaemite from lepidocrocite). The mineralogy of the rock magnetic minerals has been reviewed by Nicholls (1955).

2.2.2 *Magnetite–ulvöspinel series*

Magnetite is cubic and a member of the spinel group (inverse type) having a fairly high saturation magnetization of 480 emu cm^{-3}. As in all spinels, the cations are located in two lattices A and B, the A sites in four-fold co-ordination with oxygen ions and the B sites in six-fold co-ordination. There are two B cations for each A cation so that the two interacting sub-lattices are unequal giving rise to the observed ferrimagnetism. In normal spinels the divalent metal ion occupies the A sites, whilst the two trivalent metal ions occupy the B sites. In inverse spinels the divalent ion and one trivalent ion exchange places (fig. 27).

The Curie point of magnetite is 578 °C and it has cell dimensions $a = 8.39$ Å. In the magnetite–ulvöspinel series there is complete solid solution at temperatures in excess of 600 °C (fig. 28). At lower temperatures, the solid solution is much more restricted and there is a tendency for the two phases to exsolve. The ionic replacement in this solid solution series takes the form

$$2Fe^{3+} \rightleftarrows Fe^{2+} + Ti^{4+}. \qquad (2.5)$$

Thus the end member of the series, ulvöspinel, is antiferromagnetic (fig. 27). As the proportion of ulvöspinel increases the cell dimensions increase and the Curie point decreases (fig. 29). Ulvöspinel is paramagnetic at room temperature and antiferromagnetic at low temperatures, the Néel temperature being 120 °K (-153 °C). The cell dimensions are $a = 8.53$ Å. In practice the composition of naturally occurring spinels tends to be displaced towards the ilmenite–haematite series in the direction of increased oxidation (fig. 26, plate 1). The oxidation takes place at temperatures of 600–1000 °C during initial cooling, and is not due to weathering. Oxidation of the titanomagnetite series progresses along the dashed lines in fig. 26 towards the titanohaematite series. On cooling the solid solution is restricted to compositions near the end members (see figs. 28 and 30). Thus the oxidation during cooling tends to proceed first through the production of ilmenite lamellae (plate 1B) and then progresses towards the pseudo-brookite series (plates 1C–F).

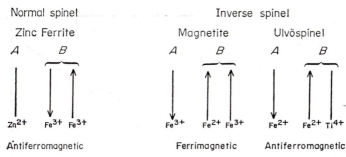

FIGURE 27   Magnetization vectors in the two lattices A and B in
                      normal and inverse spinels.

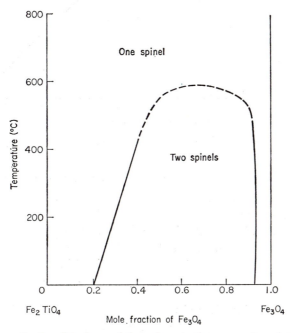

FIGURE 28   Possible form of the solvus in the magnetite–ulvöspinel
                      series. After Basta (1960) from Nagata (1961).

Maghaemite (γ-Fe_2O_3) is also a cubic mineral with cell dimension
$a = 8.35$ Å. It has an inverse spinel structure similar to magnetite but has
a defective lattice; one ninth of the Fe positions in the magnetite lattice are
vacant. It is metastable and reverts to α-Fe_2O_3 (rhombohedral) irreversibly
on being heated to temperatures in the range 300 to 700 °C. The Curie
point, obtained by extrapolation, is 675 °C as for α-Fe_2O_3, and the saturation
magnetization is about 450 emu cm^{-3}. Titanomaghaemite can be produced
from titanomagnetite during low-temperature oxidation. Maghaematization

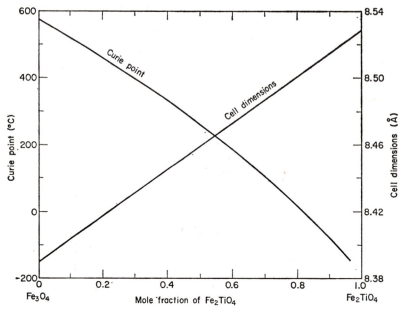

FIGURE 29   Variation of Curie temperature and unit cell size with composition in the magnetite–ulvöspinel series.

can take place either during the late stages of the initial cooling (deuteric alteration) or perhaps during subsequent weathering. The oxidation lines are those close to the magnetite–maghaemite series at the base of the ternary diagram in fig. 26. Note that this process cannot take place at high temperatures because maghaemite can only exist at low temperatures. Two stages of such low temperature oxidation are illustrated in plates 1 K and L.

2.2.3   *Haematite–ilmenite series*

Haematite is rhombohedral with cell dimension $a_{rh} = 5.427$ Å. The lattice is made up of layers of oxygen ions and layers of cations in six-fold co-ordination parallel to the triad axis. The oppositely magnetized Fe^{3+} ions in the two sub-lattices are canted at a small angle and give rise to a small saturation magnetization of 2.2 emu cm^{-3}. The Curie point is 680 °C, whilst at temperatures *below* about -20 °C (the Morin transition) this intrinsic weak ferromagnetism disappears.

In addition to this spin-canted ferromagnetism, there is also an additional component which arises from the interaction between the antiferromagnetism and lattice defects or impurities. This defect ferromagnetism is observable below -20 °C and between 680 °C and the Néel point of

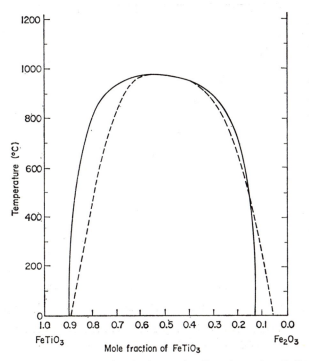

FIGURE 30   Solvus curve for the haematite–ilmenite series. Solid line after
Uyeda (1958), dotted line after Carmichael (1961). From Nagata (1961).

725 °C. Because defect ferromagnetism is sensitive to structure, it is altered
by stress or heating and could provide spurious palaeomagnetic information. Dunlop (1970) has shown, however, that the defect remanence of
fine-grained haematite, such as occurs in red sediments, is magnetically
softer than the spin-canted remanence and can be erased by partial demagnetization (see §3.4.2). In red sediments it is usually observed that
haematite occurs in two forms, either as very fine-grained ($\lesssim 1 \mu$) red
pigment or cement or else as large specularite grains (usually $\geqslant 10 \mu$),
sometimes with small inclusions of magnetite. The relative importance of
these two forms of haematite is discussed more fully in §3.4.3.

In the haematite–ilmenite series there is complete solid solution above
temperatures of 1050 °C (fig. 30). At lower temperatures the solid solution
is restricted and the intermediate compositions are represented by intergrowths of the end members. Ionic replacement in the solid solution series
is in the same form as that in the titanomagnetite series (2.5). As the proportion of ilmenite increases, the cell dimensions increase and the Curie
point decreases (fig. 31). Ilmenite is paramagnetic at liquid nitrogen

43

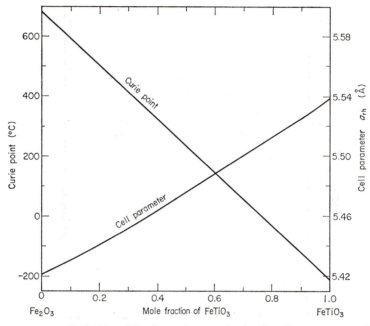

FIGURE 31 Variation of Curie temperature and unit cell parameter (a_{rh}) with composition of the haematite–ilmenite series.

temperatures and antiferromagnetic at very low temperatures with cell dimension $a_{rh} = 5.523$ Å. In the composition range 0 to 50 per cent ilmenite, the solutions behave as a canted antiferromagnetic, whilst between about 45 and 95 per cent ilmenite they become ferrimagnetic, reverting to being antiferromagnetic near 100 per cent ilmenite. Between compositions of 45 to 60 per cent ilmenite, synthetic specimens show self-reversing properties (see §4.1.1). Since pure haematite is already in its highest oxidation state, progressive high temperature oxidation in the temperature range 600–1000 °C is observed in the other end member of the series-ilmenite. The progressive high temperature oxidation in discrete ilmenite grains as illustrated in plates 1 G–J, and proceeds towards the rutile–haematite side of the ternary diagram (fig. 26).

2.3 PHYSICAL THEORY OF ROCK MAGNETISM

2.3.1 *Magnetic domains*

When the magnetization of a body gives rise to an external field (that is it exhibits a remanence), it has a certain *magnetostatic energy* or *energy of self-demagnetization*. This arises from the shape of the body, because it is more

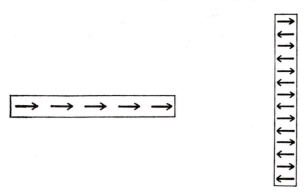

FIGURE 32   To illustrate the ease with which a rod can be magnetized
along its length compared with at right angles to its length.

easily magnetized in some directions than in others. The internal field
tends to oppose the magnetization. Two extreme examples illustrate this
point in fig. 32. A long thin rod prefers to be magnetized along its length
rather than across it. When magnetized along its length elementary magnets
can be imagined to be lined up inside, so that the north pole of one lies
next to the south pole of its neighbour, and the magnetization is aided by
their mutual attraction. In this case there is no internal field opposing the
magnetization along the axis, and there is no magnetostatic energy. To
magnetize the rod at right angles to the axis requires that the elementary
magnets be lined up so that the north poles of neighbours are adjacent as
also are their south poles. The elementary magnets then tend to rearrange
themselves so that each alternate one points in the opposite direction
(fig. 32). This is the same as saying there is a large internal field tending to
oppose the magnetization.

The demagnetizing field is proportional to a factor related to the shape
of the body called the *demagnetizing factor*, N. Demagnetizing factors have
been tabulated by Stoner (1945) and Osborn (1945) for ellipsoids and have
values between o and 4π. The long thin rod has $N = 0$ along its axis and
$N = 2\pi$ across it, whilst across a large flat plate or sheet $N = 4\pi$. The rod
can be considered as a greatly elongated ellipsoid and the plate a greatly
flattened one. In the general case, an ellipsoid uniformly magnetized with
intensity J, has a magnetostatic energy per unit volume given by:

$$E_m = \tfrac{1}{2}NJ^2, \tag{2.6}$$

where N is the demagnetizing factor in the direction of magnetization.

Suppose a ferromagnetic grain is magnetized to saturation as in fig. 33 a.
The internal field creates a large magnetostatic energy, but if the grains
were subdivided into two regions oppositely magnetized as in fig. 33 b, the

45

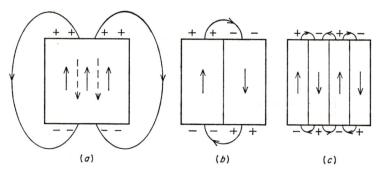

FIGURE 33   Subdivision of a ferromagnetic grain into domains. Domain magnetizations are indicated by the straight solid arrows. The dashed arrows indicate the
demagnetizing field. After D. S. Parasnis (1961), *Magnetism*, Hutchinson.

internal field and consequently the magnetostatic energy is decreased.
However a boundary or wall must be formed between the two oppositely
magnetized regions and magnetic energy (the wall energy) is stored in this
wall. The system assumes the state of lowest total energy. The process of
subdivision will therefore continue as in fig. 33 c until the energy required for
the formation of an additional boundary is greater than the consequent reduction in magnetostatic energy. The subdivided regions are called *magnetic
domains* and the boundaries between them are referred to as the *domain walls*.

The change in the direction of magnetization between one domain and
the next does not occur abruptly across a single atomic plane. The domain
walls are generally between 100 and 1000 Å thick with energy E_w proportional to the area of the wall. The wall energy per unit area, $w \approx 0.5$ erg
cm^{-2}, is a well-determined quantity in domain theory. For small grains
no domain walls can occur, and these are termed *single domain grains*, but
at some critical size the grain will subdivide into two or more domains and
form *multidomain grains*. The magnetic behaviour of single domain and
multidomain grains is quite different so that it is of some importance to
determine which of the configurations are relevant to the magnetic minerals
in rocks. It is simplest at first to consider spherical grains for which possible
domain structures are illustrated in fig. 34.

In addition to the magnetostatic energy of a grain which arises from its
shape anisotropy (that is it is easier to magnetize along the long axis than in
other directions), there is *magnetocrystalline anisotropy energy* which arises
because it is easier for the domain magnetizations to lie along certain
crystallographic axes than along others. For example in a magnetite crystal
the easy direction of magnetization is along the [111] axis, whilst the difficult
or hard direction is along the [100] axis. The magnetocrystalline anisotropy energy is the difference in magnetization energy between the hard and

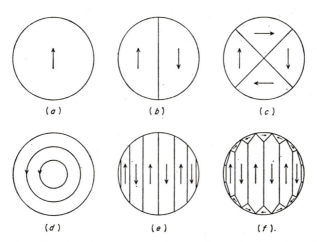

FIGURE 34   Possible domain structures of magnetite grains in rocks. (*a*) is a single domain grain, (*b*) is a special case of (*e*), and (*c*) a special case of (*f*). Anisotropy is sufficiently strong to preclude (*d*). From Stacey (1963).

easy directions. Obviously this contribution to the total energy will be a minimum when the various domain magnetizations all lie along easy directions. The energy of the single domain configuration of fig. 34*a* is magnetostatic energy only, but if the crystal has a high uniaxial anisotropy, configuration *b* is stable and the magnetostatic energy decreases to about half. The circular spin configuration of fig. 34*d* can only arise if the magnetocrystalline anisotropy is not too high, but is sufficiently strong in rock magnetic minerals that this configuration is precluded.

The magnetization of ferromagnetics is usually accompanied by their mechanical deformation, giving rise to *magnetostriction*, which is due to strain arising from magnetic interaction along the atoms forming the crystal lattice. Thus the presence of some impurity in the crystal lattice, causing dislocations, produces internal stress which then acts as a barrier to changes in magnetization. The lattice distortions of adjacent atoms whose magnetizations are oppositely directed are identical. Magnetostrictive strain is thus avoided by the domain structures of figs. 34*b* and *e*. Strain occurs when closure domains are formed as in fig. 34*f*.

Following the method of Kittel (1949) the magnetostatic energy of a spherical grain of diameter d and saturation magnetization J_s in the single domain (E_1) and two domain states (E_2) as in figs. 34*a* and *b* are:

$$E_1 = \tfrac{1}{2}NJ_s^2(\tfrac{1}{6}\pi d^3) \approx 2E_2, \qquad (2.7)$$

where N is the demagnetizing factor ($\tfrac{4}{3}\pi$ for a sphere). The critical diameter d_0 for the transition between the two states arises when the decrease in

47

magnetostatic energy exactly balances the wall energy that is,

$$E_1 = E_2 + E_w, \tag{2.8}$$

where

$$E_w = \tfrac{1}{4}\pi d^2 w, \tag{2.9}$$

and hence

$$d_0 = \frac{9}{2\pi} \cdot \frac{w}{J_s^2}. \tag{2.10}$$

Putting $w = 0.5$ erg cm^{-2}, then for magnetite $J_s = 480$ emu cm^{-3} and hence $d_0 = 0.03\,\mu$, whilst for haematite $J_s = 2.2$ emu cm^{-3} and hence $d_0 = 0.15$ cm. Grains of haematite larger than this are rarely encountered in rocks used for palaeomagnetic investigations so that haematite grains always have single domain structure. However magnetite grains as small as $0.03\,\mu$ are of little interest in palaeomagnetic work for they are super-paramagnetic (see §2.3.3). It was for this reason both Néel (1955) and Stacey (1963) considered that to describe the magnetic properties of rocks containing grains of magnetite or titanomagnetite, it is necessary to use multidomain theory.

The calculation outlined above is for a spherical grain, which is the least favourable case for single domain size. The critical size may be increased considerably if the particles are elongated. Morrish & Yu (1955) have shown that magnetite particles of axial ratio 10:1 or more and lengths of several microns may still be single domained. Suppose a grain of length d is a prolate spheroid of axial ratio 10:1 (that is width 0.1d, and volume $\approx 0.01d^3$). Then if the calculation of (2.7) to (2.9) above is repeated, using the area of the domain wall as 0.1d^2, and $N = 0.25$, the critical value of d is given by

$$d_0 = 160 \frac{w}{J_s^2}, \tag{2.11}$$

which gives for magnetite $d_0 \approx 3\,\mu$. This simple calculation gives much the same value as that obtained more rigorously by Morrish & Yu (1955) and shows that the critical size for elongated particles is two orders of magnitude greater than for spheres. There is now increasing evidence showing that single domain magnetite not only occurs very commonly, but may be of prime importance in palaeomagnetism (Evans & McElhinny, 1969; Dunlop & West, 1969; Murthy *et al.*, 1971; Evans & Wayman, 1970).

Stacey (1963) has shown that for multidomain magnetite grains, the domain structure of fig. 34*f* is favoured since the total anisotropy and magnetostrictive energies of *f* is considerably less than the magnetostatic energy of *e*. This means that in the transition region between true single domain and multidomain grains, the domain structure of fig. 34*c* is also favoured. In multidomain grains the potential barriers opposing changes in remanence arise from crystal imperfections, which cause local stresses and variations in spontaneous magnetization. The energy of a domain wall is thus a minimum at a number of discrete positions. Changes in magnetization take place when the domain wall moves in so-called Barkhausen jumps from one potential minimum to the next. In a large multidomain grain, the domain walls can generally find suitable positions so that the magnetostatic energy is zero. The grain can thus be demagnetized so as to have zero magnetic moment. This may be contrasted with single domain grains whose moments may be reversed, but cannot be destroyed (except of course by heating above the Curie point). In very small multidomain grains, such as those with the four domain structure of fig. 34c, the Barkhausen discreteness of the positions of the domain walls prevents them occupying the precise positions necessary to give the grain zero magnetic moment. The moments of these grains may thus be reversed, but they cannot be demagnetized. Such grains therefore behave very similarly to single domain grains, and Stacey (1963) has therefore termed them *pseudo single domain* (p.s.d.) grains. He estimates that the critical size for the four domain structure of these grains is about 17 μ. Their magnetic behaviour may be described in terms of the theory of single domain grains.

Subsequently, Stacey (1967) has concluded that pure multidomain thermoremanence in rocks is probably rare, and that virtually all rocks of interest in palaeomagnetism contain sufficient fraction of magnetic material in grains less than about 20 μ (in the case of magnetite) for these grains to dominate their remanence. It now seems clear that both single domain and pseudo single domain grains are the only ones which need to be considered when dealing with magnetite grains. Single domain theory therefore provides an adequate explanation of the magnetic properties of both magnetite and haematite bearing rocks.

2.3.2 *Theory for single domain grains*

The theory of the magnetization of an assemblage of single domain particles is due to Néel (1949, 1955). Although it appears to have wide applicability, it is based upon the assumption that there are no grain interactions and this leads to various shortcomings in certain aspects of the theory. Consideration of grain interactions has led to the Preisach–Néel theory, which has been developed by Dunlop & West (1969). A discussion of the details of this theory is beyond the scope of this book, but the essential features of the behaviour of magnetic grains in rocks over the geological time-scale can be adequately described in terms of Néel's simple theory.

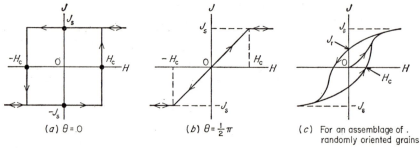

(a) $\theta = 0$ (b) $\theta = \frac{1}{2}\pi$ (c) For an assemblage of . randomly oriented grains

FIGURE 35 Hysteresis loops of single domain particles. θ is the angle between the applied field and the particle axis. After Néel (1949, 1955). From Irving, *Paleomagnetism and its application to geological and geophysical problems.* © John Wiley & Sons, Inc. 1964.

Imagine a set of identical grains with uniaxial symmetry; that is their magnetic moments may be oriented either one way or in the opposite direction, but not in between. The axes of the grains are randomly oriented so that a specimen may be demagnetized with zero magnetic moment if the magnetizations are directed so as to cancel one another. On application of a magnetic field in any direction, the specimen acquires a magnetic moment because the net magnetizations will be directed in that direction. Thus although individual magnetic grains are themselves magnetically anisotropic, a random assemblage of grains making up a specimen is magnetically isotropic.

The magnetic properties of a grain depend upon its orientation with respect to the applied field. When the field is parallel to the axis of the particle a rectangular hysteresis loop results as in fig. 35a. The height is twice the saturation magnetization J_s, and the width is twice the coercivity H_c. At $H = +H_c$ and $H = -H_c$ there are discontinuities in the magnetization. At the other extreme, when the axis of the grain is perpendicular to the applied field, there is no hysteresis (fig. 35b). For $H > H_c$ and $H < -H_c$, the magnetization is J_s and $-J_s$ respectively. As H changes from $-H_c$ to $+H_c$, J_s varies linearly from $-J_s$ to $+J_s$. In a randomly oriented assemblage of grains, the average limiting hysteresis cycle is as shown in fig. 35c. There is a remanence J_r equal to $\frac{1}{2}J_s$ and a coercivity of approximately $\frac{1}{2}H_c$.

In a single domain grain the internal magnetization energy depends only on the orientation of the magnetic moment with respect to certain axes in the grain. For magnetically uniaxial grains, the energy E is given by

$$E = Kv\sin^2\theta, \qquad (2.12)$$

where v is the volume of the grain and θ the angle between the magnetic moment and the axis. K is the anisotropy constant and can arise from three

factors that contribute to the magnetic anisotropy of a single domain grain –
magnetocrystalline anisotropy, shape anisotropy and anisotropy arising
from mechanical stress. The coercivity H_c is simply related to the anisotropy constant by the relation

$$H_c = \frac{2K}{J_s}. \tag{2.13}$$

For the three cases the anisotropy constant and coercivity are:

(*a*) *Magnetocrystalline*

$$K = K_1, \quad H_c = \frac{2K_1}{J_s}, \tag{2.14}$$

where K_1 is the first magnetocrystalline anisotropy constant.

(*b*) *Shape*

$$K = \tfrac{1}{2}(N_b - N_a) J_s^2, \quad H_c = (N_b - N_a) J_s, \tag{2.15}$$

where N_b and N_a are the demagnetizing factors along the equatorial and
polar axes of the prolate spheroid.

(*c*) *Stress*

$$K = \tfrac{3}{2}\lambda\sigma, \quad H_c = \frac{3\lambda\sigma}{J_s}, \tag{2.16}$$

where λ is the average magnetostriction coefficient and σ is the internal
stress amplitude.

Note that the widely different values of J_s for haematite (2.2 emu cm^{-3})
and magnetite (480 emu cm^{-3}) mean that different forms of anisotropy are
important in the two cases. For haematite, it is obvious that shape anisotropy is of no significance compared with either magnetocrystalline or
stress induced anisotropy. Coercivities are at least 1000 oersted. For magnetite $K_1 = 1.1 \times 10^5$ erg cm^{-3}, so that magnetocrystalline anisotropy cannot give rise to coercivities in excess of 450 oersted. Mechanical stress may
increase this value, but inordinately high stresses ($\sigma > 3 \times 10^9$ dyne cm^{-2})
must be invoked to explain coercivities of more than 1000 oersted, given
that $\lambda \approx 55 \times 10^{-5}$. Shape anisotropy, however, can give rise to coercivities
considerably greater than this. The theoretical maximum (for infinitely
long needles) is given by $2\pi J_s \approx 3000$ oersted. The coercivity of pseudo
single domain grains is, however, related to the theory of multidomain
grains rather than single domain grains, since it involves the movement of
domain walls. Evans & McElhinny (1969) have shown that multidomain
grains of magnetite cannot have coercivities in excess of 900 oersted. Thus
the observation of coercivities in magnetite in excess of 1000 oersted must
arise from the shape anisotropy of single domains (Evans *et al.*, 1968).

Some authors have suggested that single domain-like regions can occur as a result of the subdivision of larger grains by means of stress centres (Verhoogen, 1959), microscopic grain boundaries (Ozima & Ozima, 1965) or intergrowths (Strangway *et al.*, 1968*a*). Objections to this type of proposal have been made by Stacey (1963) and Dickson *et al.* (1966). Both optical (Evans *et al.*, 1968; Evans & McElhinny, 1969) and electron microscopy (Evans & Wayman, 1970) confirm, however, that single domain particles occur naturally within an appropriate size–shape range (see §2.3.3, fig. 36).

The magnetic susceptibility χ_s (initial susceptibility) of a random assemblage of single domain grains (fig. 35*c*) is given by

$$\chi_s = \frac{J_s^2}{3K}. \tag{2.17}$$

For magnetite χ_s therefore lies between 0.1 (infinite needles) and about 1.0, but for haematite the very much lower value of J_s makes χ_s a factor of 10^4 smaller. For multidomain grains the susceptibility χ_m is controlled by the demagnetizing factor N and is not a property of the magnetic material (Stacey, 1963)

$$\chi_m = 1/N. \tag{2.18}$$

2.3.3 *Magnetic viscosity*

The magnetic moment of a uniaxial single domain grain, in the absence of an applied field, can take up two orientations of equal minimum energy, $\theta = 0$ and $\theta = 180°$ from (2.12). The potential barrier between these two positions is represented by the positions of maximum energy which occur at $\theta = \pm 90°$, and from (2.12) has value E_r given by

$$E_r = vK. \tag{2.19}$$

The thermal fluctuations of energy E_t given by

$$E_t = kT, \tag{2.20}$$

where k is Boltzmann's constant and T is the absolute temperature, are capable of moving the magnetic moment from one position to the other if $E_t > E_r$ or if $kT > vK$. For a given value of T there must always be some grains of volume v for which the thermal fluctuations are large enough to cause the moment to change spontaneously from one position to the other. Thus if an assemblage of identical grains has an initial moment M_0, then under these conditions it will decay exponentially to zero according to the relation

$$M_r = M_0 \exp(-t/\tau), \tag{2.21}$$

where M_r is the moment remaining after time t and τ is the *relaxation time* of the grains. The initial moment M_0 will have decayed to one-half of its value ($M_r = \frac{1}{2}M_0$) after time $t = 0.693\tau$, which can be thought of as the 'half-life' of the initial remanence.

The relaxation time is related to the ratio of the two energies E_r and E_t by the equation

$$\tau = \frac{1}{C}\exp\left(\frac{E_r}{E_t}\right)$$

$$= \frac{1}{C}\exp\left(\frac{vK}{kT}\right), \tag{2.22}$$

where C is a frequency factor equal to about 10^{10} s^{-1}. From (2.13) the anisotropy constant K is related to the coercivity H_c, so that τ may alternatively be given as

$$\tau = \frac{1}{C}\exp\left(\frac{vH_cJ_s}{2kT}\right). \tag{2.23}$$

Note that τ is thus directly related to coercivity. This has important implications relating to the magnetic stability of rocks and forms the basis of the method of 'magnetic cleaning' (§3.4.1).

When the relaxation time is small, say equal to a typical laboratory experiment time of 100 to 1000 seconds, the magnetization acquired by an assemblage of grains will be lost almost as soon as it has been acquired. The grains are rendered unstable by thermal agitation and on application of a weak field, they very quickly reach equilibrium with this field. The moment so acquired is called the *equilibrium magnetization*, which, on removal of the applied field, quickly dies away at a rate determined by the relaxation time. Grains such as these are said to be *superparamagnetic*. The relaxation time according to (2.21) becomes small when T is large (that is at higher temperatures) and also when v is small (small grain size). For each grain of volume v, there is thus a *critical blocking temperature*, T_B, at which τ becomes small (say 100 to 1000 s), but which might also be below the Curie temperature. Similarly at any given temperature T, there is a *critical locking diameter*, d_B (corresponding to a sphere of volume v), at which τ becomes small.

Neglecting changes in the anisotropy constant K with temperature, the relaxation time τ_1 at temperature T_1, is simply related to the relaxation time τ_2 at temperature T_2 from (2.22)

$$T_1\ln C\tau_1 = T_2\ln C\tau_2. \tag{2.24}$$

Thus the same effect upon the remanence is obtained *either* by maintaining at temperature T_1 for sufficient time τ_1 *or* by raising to a higher temperature T_2 and maintaining this for some shorter time τ_2. Putting $C = 10^{10}$ s^{-1}, (2.24) shows that maintaining at a temperature of 150 °C for 10^6 years is equivalent to maintaining at 450 °C for only 1000 seconds. This important

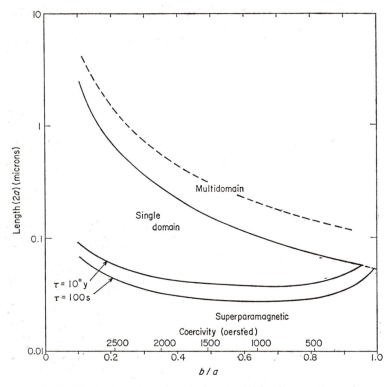

FIGURE 36   Single domain and multidomain fields of magnetite grains as a function of length and axial ratio of prolate spheroids. Coercivity arising from shape anisotropy is indicated. The approximate nature of the calculation of the critical size could enable the single domain field to extend up to the dashed line. From Evans & McElhinny (1969).

relationship is discussed more fully in §3.4.2 where it relates to the thermal demagnetization of rocks, and in §3.4.5 where it is used as a test for palaeomagnetic stability.

At room temperatures spherical grains of magnetite of size 0.03 μ are superparamagnetic, and this is why Stacey (1963) and Néel (1955) supposed that single domain magnetite was of no importance in palaeomagnetism. However, as discussed in §2.3.1, single domain grains of magnetite occur above the superparamagnetic size when they are elongated. The general case is for prolate ellipsoids of axial length $2a$, and minor axis $2b$. Evans & McElhinny (1969) have extended the calculation of Morrish & Yu (1955) throughout the size–shape range to produce the single domain and multidomain fields of magnetite illustrated in fig. 36. Strictly speaking this calculation is not valid for axial ratios greater than about 7:1. A more exact analysis for greatly elongated magnetite particles has been carried out by Murthy *et al.* (1971), but only minor differences are obtained. For elongated grains of magnetite, shape anisotropy predominates so that the coercivity is directly related to the axial ratio through (2.15). The region of super-paramagnetic behaviour at room temperature (300 °K) is also shown in fig. 36, being defined for relaxation times $\tau \leqslant$ 100 s, given by (2.22). Note that the boundary for grains with relaxation time of 10^{11} years lies very close to the superparamagnetic boundary, so that the transition region is very sharp. Fig. 36 indicates that the size range over which single domain behaviour can be expected in magnetite is fairly restricted varying from 0.03 to several microns. The approximate nature of the calculations may cause the critical size to be increased by a factor of about 2. The field given in fig. 36 is for the lower limits of the critical size. Allowance for this uncertainty could displace the single domain boundary up to the position of the dashed curve. Throughout almost the whole of the single domain field, the relaxation time at room temperature is very much greater than 10^{11} years, so that these grains are of considerable importance to palaeomagnetism, being capable of retaining their initial remanence throughout the whole of geological time. The presence of small magnetite grains in the appropriate size–shape range has been confirmed by electron microscopy (Evans & Wayman, 1970).

2.3.4 *Thermoremanent magnetization* (TRM)

The remanence acquired by a rock specimen during cooling from the Curie point to room temperature is called the *total* TRM. Upon cooling from a high temperature, spontaneous magnetization appears at the Curie point T_c, and this assumes an equilibrium magnetization in the presence of an applied field. Grains of different volumes v will each have differing blocking temperatures T_B. As the temperature cools below T_c and passes through each T_B, the relaxation time of each of these grains increases very rapidly. The equilibrium magnetization becomes 'frozen in' and subsequent changes in the field direction occurring at temperatures below T_B are ineffectual in changing the direction of magnetization. From (2.22), using a value of τ = 1000 s at the blocking temperature, a grain with T_B = 530 °C (800 °K) has relaxation time of 10^{17} years on cooling to room temperature (30 °C). Even grains with relatively low blocking temperature of 330 °C (600 °K) have a relaxation time of 10^9 years on cooling to room temperature. This is one of the basic appeals of palaeomagnetism, that TRM is essentially stable on the geological time scale.

The TRM is not all acquired at the Curie point T_c, but over a range of

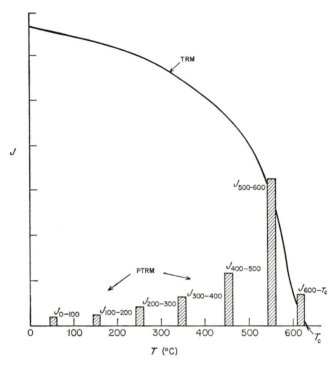

FIGURE 37   The acquisition of TRM. The PTRMs acquired over successive
temperature intervals add up to give the total TRM curve.

blocking temperatures from the Curie point down to room temperature
(fig. 37). The total TRM may be considered to have been acquired in steps
within successive temperature intervals $(T_c - T_1)$, $(T_1 - T_2)$ and so on. The
fraction of the total TRM acquired in these temperature intervals is called
the *partial* TRM for each temperature interval. The PTRM acquired in any
temperature interval depends only on the field applied during that
interval and is not affected by the field applied at subsequent intervals on
cooling. Thus the total TRM is equal to the sum of the PTRMs acquired in
each consecutive temperature interval between the Curie point and room
temperature (fig. 37). This is known as the *law of additivity of* PTRM
(Thellier, 1951). Conversely, on reheating to any temperature $T < T_c$, the
original magnetization of all grains with blocking temperatures less than
T is destroyed. This has important implications relating to the magnetic
stability of rocks and forms the basis of the method of 'thermal cleaning'
(§3.4.2).

   Néel (1955) considered the TRM of uniaxial grains of volume v aligned with the direction of the field H, in which case the intensity of TRM, J_{TRM} is given by

$$J_{\text{TRM}} = J_s \tanh\left(\frac{vJ_{sB}H}{kT_B}\right), \tag{2.25}$$

where J_s is the saturation magnetization at room temperature and J_{sB} that at the blocking temperature T_B. This is the basic equation of the theory of TRM. The general case of a randomly oriented assembly of grains can be solved for small fields (0–1 oersted) as in the case of the earth's field and from Stacey (1963) is given by

$$J_{\text{TRM}} = \frac{J_s}{3}\left(\frac{vJ_{sB}H}{kT_B}\right). \tag{2.26}$$

In this case TRM is proportional to the applied field. The theory of course applies strictly only to non-interacting single domain grains. It applies in this case to haematite grains, but only to magnetite grains of suitable shape and dimensions up to a few microns. For magnetite grains up to about 20 μ in the pseudo single domain region, the TRM is quite different and given by

$$J_{\text{TRM}} = \left(\frac{m}{v}\right)\cos\alpha \left[\frac{\ln\cosh(2\beta m)}{2\beta m}\right], \tag{2.27}$$

where

$$\beta = \frac{H\cos\alpha}{kT_B}\cdot\frac{J_{sB}}{J_s},$$

and $\cos\alpha = 0.85$ and $m = 4\times 10^{-4}d^2$ for grains of diameter d cm and volume v (Stacey, 1963). The factor in square brackets in (2.27) is approximately unity for the grain sizes being considered, from which $J_{\text{TRM}} \propto d^{-1}$ in the p.s.d. range.

For true multidomain grains, the TRM is roughly independent of grain size and from Dickson *et al.* (1966) is given by

$$J_{\text{TRM}} = 0.125H. \tag{2.28}$$

The relations of (2.27) and (2.28) are illustrated in fig. 38. Below 17 μ the TRM increases sharply due to p.s.d. behaviour and the data obtained by Parry (1965) on sized, dispersed magnetite powders show excellent agreement with this domain theory (Dickson *et al.*, 1966).

It is worth noting that (2.28) and (2.18) enable the Koenigsberger ratio Q_t of (2.4) to be calculated for multidomain grains of magnetite. For spheres $N \approx 4$, and thus $\chi_m = 0.25$, giving $Q_t = 0.5$. The susceptibility for single and pseudo single domain grains is very similar (2.17), but the TRM is very much higher, so that $Q_t > 1.0$ in general. Stacey (1967) has

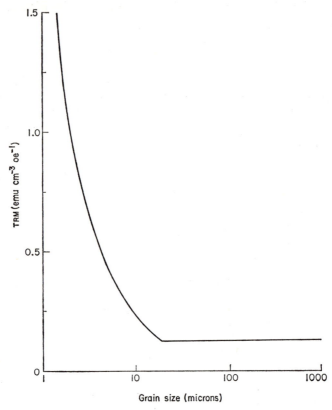

FIGURE 38 Theoretical variation with grain size of the TRM induced by low fields in approximately equidimensional magnetite grains from the multidomain and pseudo single domain theories of Stacey (1963). After Dickson *et al.* (1966).

pointed out that this indicates that the Koenigsberger ratio has a fundamental significance for palaeomagnetism, for most rocks used in palaeomagnetic work have $Q_t > 1.0$. He therefore concluded that pure multidomain TRM is probably rare.

In low fields the TRM is proportional to the applied field, but in stronger fields it eventually saturates. The efficiency of the process of acquisition of TRM is illustrated in fig. 39, where it is compared with IRM. The TRM of magnetite powder saturates in a few hundred oersteds, whereas the IRM approaches saturation more slowly needing about 1000 oersted or more. For haematite powder, the effects are comparable, except that the fields required for saturation are much greater; the TRM is not yet saturated in 800 oersted, nor the IRM in 30000 oersted.

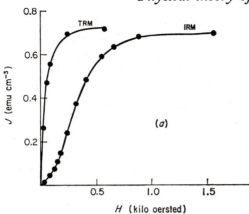

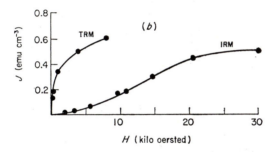

FIGURE 39. Variation of TRM and IRM with applied field for (*a*) dispersed magnetite powder, and (*b*) dispersed haematite powder. Data from Roquet (1954).

2.3.5   *Chemical remanent magnetization* (CRM)

CRM occurs during the formation of a magnetic mineral at low temperatures (below the Curie point) by chemical or phase change in the presence of an applied field. CRM is probably the most common origin of the magnetization of redbeds (Collinson, 1965*a*), and there are several quite different circumstances in which it can be produced. Haematite can be formed by chemical deposition during the consolidation of sediments. The dehydration of iron oxyhydroxide

$$2\text{FeOOH} \rightarrow \text{Fe}_2\text{O}_3 + \text{H}_2\text{O} \tag{2.29}$$

has been examined by Hedley (1968) and Strangway *et al.* (1968*b*). Other possibilities include the oxidation of titanomagnetites to titanomaghaemites or titanohaematites. Porath (1968) has investigated the oxidation of maghaemite to haematite, whilst Haigh (1958) and Kobayashi (1959) have examined the CRM produced by the reduction of haematite to magnetite at

59

temperatures of about 300 °C. Because of their greater magnetization, redbeds are the most frequently used sediments in palaeomagnetism. The reduction of haematite is unlikely to be of any significance due to the oxidizing environment present when redbeds are formed (Van Houten, 1961). The main point of interest in palaeomagnetism is the CRM of haematite acquired during the consolidation and compaction of sediments. The single domain theory (§2.3.3) therefore applies.

CRM can be imagined to arise from a process of nucleation. During the early stages the grains of haematite are small enough that their relaxation time is very short and the grains are superparamagnetic. The spontaneous magnetization assumes an equilibrium magnetization in the presence of an applied field. As the grain grows it passes through the critical blocking diameter d_B, and the relaxation time increases very rapidly. The equilibrium magnetization becomes 'frozen in' and subsequent changes in field direction as the grain grows further have no effect upon its direction of magnetization. The process is analogous to the acquisition of TRM.

From (2.22), as a single domain grain grows at temperature T, its magnetization becomes stabilized at some volume v_B when

$$\frac{v_B K}{kT} = \ln C\tau \approx 30, \tag{2.30}$$

where $\tau = 1000$ s and $C = 10^{10}$ s^{-1}. Putting $K = 1250$ erg cm^{-3}, which is reasonable for haematite (Wohlfarth, 1955; Strangway et al., 1968b), then for a temperature of 30 °C (300 °K) and assuming the grains are spheres, the critical blocking diameter $d_B \approx 0.1\,\mu$. This approximate result is confirmed by the experiments of Strangway et al. (1967) which show that the critical size is in the region of $\frac{1}{4}$ to $\frac{1}{2}\mu$. Experiments on the magnetic properties of synthetic samples of ultrafine haematite (Bannerjee, 1971) suggest, however, that the superparamagnetic threshold is lower and occurs at $0.02\,\mu$. From (2.30), when the grains have grown from a diameter of $0.1\,\mu$ to $0.126\,\mu$, the relaxation time has already increased to 10^9 years and at $0.14\,\mu$ becomes 10^{18} years! Once again this is one of the basic appeals of palaeomagnetism that the CRM of sediments can be stable over the geological time scale.

Stacey (1963) has considered the CRM of single domain grains, whose anisotropies are aligned with the direction of the applied field H, in which they are growing at temperature T. The intensity is then given by

$$J_{CRM} = J_s \tanh\left(\frac{v_B J_s H}{kT}\right). \tag{2.31}$$

In small fields (0–1 oersted), a randomly oriented assemblage of such grains produces a CRM given by

$$J_{\text{CRM}} = \frac{v_B J_s^2 H}{3kT} = \frac{10 J_s^2 H}{K},\tag{2.32}$$

the second expression resulting from the substitution of (2.30). Substituting appropriate values for haematite this gives

$$J_{\text{CRM}} \approx 0.04H.\tag{2.33}$$

This indicates that the CRM of non-interacting grains of haematite is independent of the size to which the grains have grown, providing the size exceeds the critical blocking diameter. Thus the CRM of a given rock specimen will not necessarily be directly related to the amount of haematite present, even if all the grains are above the critical size. The maximum CRM for a given amount of haematite will be observed if all the grains are just above the critical size. A rock specimen typically containing one per cent haematite by volume produced chemically at atmospheric temperatures in the earth's field (say 0.5 oersted) should thus be capable of acquiring a CRM of maximum value about 10^{-4} emu cm^{-3}. This is commonly the upper limit of intensities (10^{-7} to 10^{-4} emu cm^{-3}) observed is red sediments (Collinson, 1965a).

The magnetic characteristics of CRM are very similar to those of TRM but different from IRM (see §3.4.1), as would be expected from theory. Experimental work confirms that the intensity of CRM is proportional to the applied field (for small fields) as theory predicts, but Haigh (1958) and Porath (1968) note that it is only about one-tenth of the total TRM produced in the same specimens in the same field. Eqn. (2.26) for TRM can be written in the same form of (2.32) by noting that (2.30) applies in this case at the blocking temperature T_B, at which the anisotropy constant will have a new value K_B. The ratio of CRM to TRM induced in the same field H is then simply given by

$$\frac{J_{\text{CRM}}(H)}{J_{\text{TRM}}(H)} = \frac{J_s}{J_{sB}} \frac{K_B}{K}.\tag{2.34}$$

Since $J_s > J_{sB}$, this means that the anisotropy constant must be considerably reduced at higher temperatures as might be expected.

2.3.6 *Detrital remanent magnetization* (DRM)

Although the magnetization of most red sediments probably arises from CRM (Collinson, 1965a), it is known however that detrital magnetization is a naturally occurring process, capable of explaining the observed remanent magnetization of some sediments. DRM describes the process of alignment of magnetic particles by an applied field as they fall through water and the rotation of such particles into the field direction when they are in the water-filled interstitial holes of a wet sediment. The former, acquired at deposition due to particle alignment during sedimentation, is termed *depositional* DRM, whilst the latter, due to particle rotation after deposition but prior to consolidation, is termed *post-depositional* DRM. The remanent magnetization of varved clays and similar detrital sediments is thought to arise from the detrital process in some cases. The depositional process has been investigated under controlled laboratory conditions by Johnson *et al.* (1948), King (1955) and Griffiths *et al.* (1960) and the post-depositional process by Irving & Major (1964). Nagata (1961), Stacey (1963), Collinson (1965 *b*) and King & Rees (1966) have considered theoretically the alignment of magnetic particles as they fall through water.

Suppose a spherical grain of diameter d is falling through water of viscosity η, then there is a couple L turning the magnetic moment of the grain towards the field direction, given by

$$L = A\sin\theta = \tfrac{1}{6}\pi d^3 HJ\sin\theta, \tag{2.35}$$

where θ is the angle between the field H and the intensity of magnetization J of the grain. The motion of the detrital magnetic particles is likely to be rather highly damped (Collinson, 1965 *b*) so that inertia can be neglected. Under these conditions the couple can be equated to the viscous drag on the rotation of the grain and thus

$$L = -B\frac{d\theta}{dt} = -\frac{\pi^2}{4}\eta d^2 \cdot \frac{d\theta}{dt}. \tag{2.36}$$

If the angle θ is given by θ_0 at time $t = 0$, then for small angles

$$\theta = \theta_0 \exp(-t/t_0), \tag{2.37}$$

where $$t_0 = B/A,$$

and is the 'time-constant' of the rotation, or the time taken for the initial angle θ_0 to be reduced to $1/e$ of its value. Thus

$$t_0 = \frac{3\pi\eta}{2dHJ}. \tag{2.38}$$

If the alignment process is to be reasonably complete, the particle must fall through the water for at least time t_0. From Stokes' law, the time of fall t of a spherical grain through a depth h of water given by

$$t = \frac{18\eta h}{d^2(\rho - \rho_0)g}, \tag{2.39}$$

where $(\rho-\rho_0)$ is the density difference between the grain and the water and g is the acceleration due to gravity. There is thus some critical height h_0 through which the grain must fall for the time t_0 to have elapsed. Equating (2.38) and (2.39) this height is given by

$$h_0 = \frac{\pi d(\rho-\rho_0)g}{12HJ}.\qquad(2.40)$$

Typical conditions that might exist for varves are $H = 0.5$ oersted, $(\rho-\rho_0) = 4$ g cm^{-3} and for magnetite particles of size $d = 1\,\mu$ in the p.s.d. range, whose magnetization was originally TRM, $J = 1$ emu cm^{-3} say. This gives $h_0 \approx 0.2$ cm, so that complete alignment occurs rapidly especially under natural deposition where there might be considerable depth of water.

Whilst laboratory studies of post-depositional DRM show impressive agreement between the direction of the applied field and that of the magnetization of the sediment (Irving & Major, 1964), studies of depositional DRM tend to show a so-called 'inclination error' (Johnson et al., 1948; King, 1955). In the latter case, whereas the declination of the field was faithfully recorded by the sediment, the sediment inclination I_s was invariably less than the applied field inclination I_H, where

$$\tan I_s = f\tan I_H,\qquad(2.41)$$

and f is normally about 0.4. King (1955), Griffiths et al. (1960) and Rees (1961) have examined the effects of the shape of the grains deposited both on irregular and sloping surfaces, and also the effects of flowing water.

For most rocks of palaeomagnetic interest, these effects do not seem to be of any significance for whereas an inclination error might occur during the deposition of particles, either subsequent post-depositional DRM or CRM occurs and this corrects for it. The observation of uniform magnetization in slump beds (Irving, 1957a) and the consistency between igneous and sedimentary rocks of the same age (Opdyke, 1961; Irving, 1967) confirm this view.

2.3.7 Stress effects and anisotropy

Most rocks are subjected to stresses during their history, either from deep burial or from tectonism. Magnetocrystalline anisotropy is stress dependent so that the application of stress can result in a change in the magnetization of a grain, the effect being generally referred to as magnetostriction (see also §2.3.1). It should be noted that stress alone cannot induce magnetic moments; the application of stress to an initially isotropic material causes stress induced anisotropy (Stacey, 1960a, b) which may change the state of magnetization.

Graham (1956) first drew attention to the possible importance of magnetostriction in palaeomagnetism, by suggesting that the simple, reversible application of elastic stress to a rock would cause a substantial deflection of its remanent magnetization. Thus a rock which cooled under stress would acquire a magnetization in the direction of the field in which it cooled. Upon release of the stress before measurement, changes in magnetization would occur which would make the original field direction impossible to determine. Stott & Stacey (1959, 1960) and Kern (1961) have studied the effects of stress on TRM of igneous rocks. For isotropic rocks they found that the TRM was always acquired parallel to the applied field with or without being cooled under stresses of up to 1000 kg cm^{-2}. For anisotropic rocks, stress often had a positive effect on the TRM, which could be deflected through large angles from the direction of the applied field. Kern (1961), however, showed that this effect could be eliminated by partial demagnetization in alternating magnetic fields.

Stacey (1960a, b) presumes that when an intrinsically isotropic rock is subjected to stress while cooling in a field, it acquires a TRM in a direction

---

PLATE I   Progressive oxidation of magnetic minerals in rocks. Photomicrographs were taken using an oil immersion objective, and the scale line shows 25 μ. From Wilson & Haggerty (1966).

A–F. Progressive high-temperature oxidation of titanomagnetite in basalts.

A   A brown homogeneous titanomagnetite grain with sharp well-formed crystal faces. Titanomagnetite has a face-centred cubic structure.

B   A well-shaped titanomagnetite grain with coarse ilmenite lamellae along [111] planes. The fine needles are short lamellae of spinel exsolved along the [100] cube faces.

C   The original broad ilmenite lamellae have been oxidized to pseudobrookite (grey) and a submicroscopic intergrowth ('metailmenite') of rutile and titanohaematite (yellow). Brown titanomagnetite again contains dark spinel rods.

D   The smaller dark brown titanomagnetite areas have been enveloped by the streaky rutile–titanohaematite intergrowth. The larger titanomagnetite areas still survive.

E   Development of pseudobrookite (grey) and titanohaematite (bright yellowish) from the 'metailmenite' of C. Titanomagnetite contains well-formed and abundant dark exsolved spinel rods.

F   This represents the highest oxidation state in the titanomagnetite series. The pseudobrookite (grey) has become redistributed into fewer, larger, more irregular areas in a host of titanohaematite.

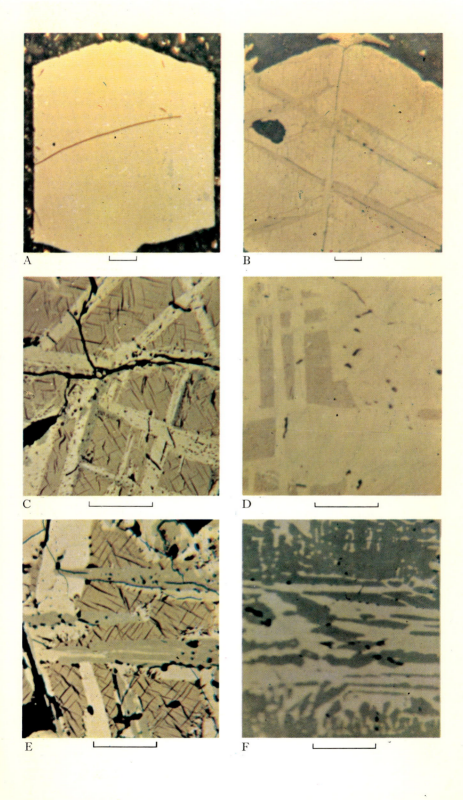

A

B

C

D

E

F

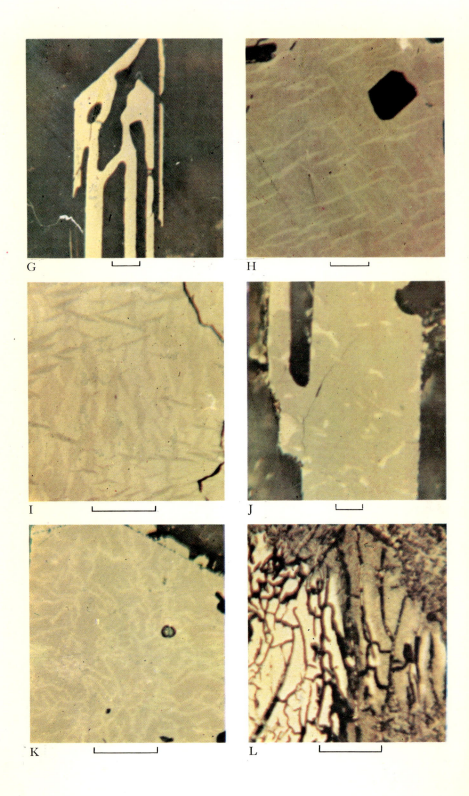

G

H

I

J

K

L

deflected away from the field by such an angle that it returns to the field direction precisely when unloaded. All experimental results presently available are consistent with this hypothesis.

If rocks have an intrinsic anisotropy, the TRM may be deflected away from the direction of the applied field towards the direction of easy magnetization. Anisotropy is measured by the variation in values of susceptibility, saturation magnetization, IRM or TRM in different directions in a specimen. Normally anisotropy of susceptibility is considered, since this is the easiest to measure and the procedure does not generally alter the state of magnetization of the specimen. The degree of anisotropy A_n is then expressed as the ratio of maximum to minimum susceptibility

$$A_n = \frac{\chi_{max}}{\chi_{min}}. \tag{2.42}$$

A value of A_n of 1.25 means that the maximum susceptibility exceeds the minimum by 25 per cent, and such a specimen is often referred to as having '25 per cent anisotropy'.

The main question that arises and which is of interest in palaeomagnetism is what degree of anisotropy can be tolerated in a specimen before the TRM is deflected through a significant angle. The problem has been considered in detail by Stacey (1960 a, b), Uyeda et al. (1963) and Fuller (1963). In the present discussion it is simplest to consider the most favourable case

---

G–J   Progressive high-temperature oxidation in discrete ilmenite grains.

G   Homogeneous skeletal ilmenite in the lowest oxidation state. Ilmenite crystallizes in the rhombohedral system and is typically long and lath-shaped.

H   Development of fine light-coloured ferri-rutile blades in ilmenite. The blades have grown in the [0112] and [0001] planes.

I   Sigmoidal (or leaf textured) rutile in titanohaematite, completely replacing the original ilmenite.

J   Pseudobrookite with relic undigested haematite (bright blebs) and rutile (less bright elongated blebs) completely replacing the original ilmenite. This represents the highest oxidation state in ilmenite.

K–L   Low-temperature oxidation (maghaematization) and replacement of titanomagnetite.

K   Fine vermicular replacement of titanomagnetite (brown) by titanomaghaemite (lighter coloured). Note renewed growth along sharp upper edge after original crystal had become well formed.

L   The bright area is totally maghaematized. The cracks are typical at this stage. In the darker cracked area, the titanomaghaemite has been replaced by a fine granular, amorphous iron–titanium oxide.

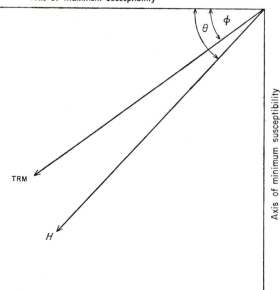

FIGURE 40   Deflection of TRM from the applied field due to anisotropy.

for the deflection of the TRM, which occurs when the direction of the applied field lies in the plane containing the maximum and minimum susceptibility directions. Suppose the applied field H makes an angle θ with the direction of maximum susceptibility (fig. 40). Since TRM is proportional to the applied field in low fields ($J = cH$, say), the specimen will acquire TRM components in the maximum and minimum directions given by

$$\left.\begin{array}{l} J_{\max} = c_{\max} H \cos \theta, \\ J_{\min} = c_{\min} H \sin \theta, \end{array}\right\} \qquad (2.43)$$

where $c_{\max}$ and $c_{\min}$ are the constants of proportionality in the maximum and minimum directions respectively. The angle ϕ which the resulting TRM will make with the direction of maximum susceptibility is then given by

$$\tan\phi = \frac{J_{\min}}{J_{\max}} = \frac{c_{\min} H \sin \theta}{c_{\max} H \cos \theta}$$

$$= \frac{1}{P} \tan \theta \qquad (2.44)$$

where $P = c_{\max}/c_{\min}$ and is the degree of anisotropy of TRM. The angle through which the TRM is deflected $(\theta - \phi)$ will have its maximum value when $\theta = \tan^{-1}\sqrt{P}$, so that

$$(\theta - \phi)_{\max} = \tan^{-1}\left(\frac{P-1}{2\sqrt{P}}\right). \qquad (2.45)$$

For an anisotropy of 10 per cent, $P = 1.10$ and the maximum deflection is $2.7°$, and for 20 per cent anisotropy it is $5.2°$. Even when $P = 1.50$ the maximum deflection is only $11.6°$. The rocks most commonly used in palaeomagnetism, basic igneous rocks and redbeds, rarely have anisotropies exceeding a few per cent. The above calculations indicate that ten or even twenty per cent anisotropy can be tolerated without significantly deflecting the TRM. The effect is likely to be significant only in some metamorphic and strongly foliated rocks.

3 Methods and techniques

3.1 COLLECTION AND MEASUREMENT OF SAMPLES

3.1.1 *Sample collection in the field*

In a palaeomagnetic study it is usual to collect oriented samples from a set of sedimentary beds, lava flows or intrusives, which have been recognized as units on geological grounds. The samples are collected from a number of separate exposures referred to as collecting sites or localities. In palaeomagnetic terms a 'site' is defined as a single point in time, that is a sufficient thickness (or lack of thickness) of exposure which will record only a spot reading of the palaeomagnetic field. A single lava flow is unlikely to represent more than an instant in time, since its cooling time is very short. In general, therefore, no matter how many different localities in a single flow are collected, they will all represent the same 'site' in palaeomagnetic terms. In sedimentary rocks, however, several metres thickness of sediment may represent a considerable length of time and spot readings can only really be made by sampling over a limited thickness. In this case several 'sites' may be obtained at a single locality by sampling different horizons several metres apart stratigraphically.

At each site one or more samples are collected. A *sample* is usually defined as a separately oriented piece of rock, which might be either an oriented block or an oriented core drilled from an outcrop. From each sample a number of *specimens* are then cut and/or drilled and sliced.

The collection of samples in the field together with accounts of various palaeomagnetic techniques used by different workers are given in Collinson *et al.* (1967). In palaeomagnetic investigations it is necessary to collect oriented samples from suitable rock exposures which are still *in situ*. Samples are collected either by hand sampling oriented blocks or by drilling cores from an outcrop by means of a portable rock drill. The simplest procedure used for hand sampling is to mark a strike and dip line on a suitably flat surface, using a Brunton compass and inclinometer. Convenient flat surfaces are often difficult to find and the remanent magnetization of the outcrop may cause errors in the compass readings, so that the accuracy of the method is at best only a few degrees. This problem is overcome by using a simple sun-compass device mounted on a tripod which forms an isosceles triangle (fig. 41). The plate mounted immediately on the legs creates an artificial flat surface and the positions of the legs are marked on the sample before removal. The upper plate is hinged to the lower along the

68

FIGURE 41   Sun compass for orienting hand samples in use at the Australian National University. The points of the three legs define a reference plane, whose tilt is measured using the vertical clinometer, and whose 'strike' is the line joining the upper pair of legs. The circular scale (10 cm diameter) is a sundial whose zero is aligned with the hinged edge and with the 'strike' of the reference plane.

'strike' line and the position of the legs is adjusted until the upper surface is horizontal. The shadow cast by the sun on the scale is noted together with the time of day, which should be known to the nearest minute. Details of the method of calculating the bearing of a strike line on a sun compass are given by Creer & Sanver (1967). In the laboratory the sample is set up with the flat surface horizontal, and cylindrical specimens are then cored and sliced from it.

The basic objection to block sampling is that the most convenient samples to collect are those associated with cracks or joints and therefore tend to be more weathered. This objection is overcome by using a portable coring drill such as that described by Doell & Cox (1965), in which case the freshest and least jointed part of an outcrop can be selected. A further problem arises in tropical countries where the incidence of lightning strikes to ground is high. Lightning strikes consist of currents of the order of 10^4 to 10^5 amps (Schonland, 1953) which travel along the surface of an outcrop upon striking the ground. In the immediate vicinity of the current the

69

magnetic field created is large and virtually remagnetizes the outcrop at these points.

Assuming a lightning strike is equivalent to an infinite conductor carrying a current I amps, then the magnetic field H at a distance d cm from it is given by

$$H = \frac{I}{5d}.$$ (3.1)

At a distance of 1 m, a current of 5×10^4 amps produces a field of 100 oersted. Experiments carried out on outcrops which have been struck by lightning (Graham, 1961; Cox, 1961) show that the IRM produced by fields of this order can be selectively removed by alternating field demagnetization (§3.4.1) in peak fields of the same order. If samples are always collected several metres apart, the chances of their having been taken too near the path of a strike are minimized. The occasional spurious sample can then be rejected. Since lightning travels along the surface of an outcrop, drilling to a depth of about 1 m is certain to avoid severe lightning effects. A portable coring rig for this purpose has been described by Graham & Keiller (1960). Techniques for coring and orienting at greater depths are described by Gough & Opdyke (1963).

3.1.2 Astatic magnetometer

Of the two main methods used for the measurement of the magnetization of rock samples, the astatic magnetometer is based upon the principle invented by Nobili over 100 years ago. In its present form it is largely due to Blackett (1952). Basically the instrument is very simple and consists of two magnets, each of equal moment P, mounted on the ends of a light stem (fig. 42). The two magnets are arranged so that their moments cancel one another, and the system has zero or a very small residual moment ΔP and thus has negligible response to a uniform magnetic field across the system. It is suspended by a torsion fibre of torsion constant c and a small mirror is mounted on the stem so that its movement can be noted through the deflection of a light beam. The *astaticism* (S) or degree of cancellation of the moments of the two oppositely directed magnets is defined as

$$S = \frac{P}{\Delta P}.$$ (3.2)

The system responds to a vertical gradient in the horizontal field. A rock sample brought up under the magnet system produces a horizontal field F at the lower magnet and $F - \Delta F$ at the upper magnet, so that the field difference ΔF produces a torque on the system $(\Delta F \times P)$ which causes it to

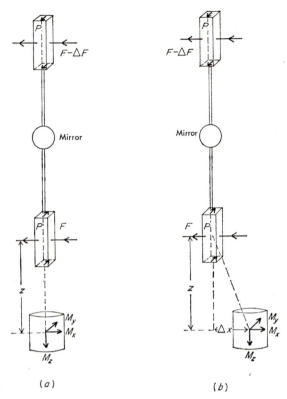

(a) (b)

FIGURE 42   Measurement of specimen magnetization with an astatic magneto-meter. (a) On-centre position, where only the component M_x produces a deflecting field F and $F-\Delta F$ at the lower and upper magnets respectively. (b) Off-centre position, where the component M_x and part of M_z combine to produce F and $F-\Delta F$.

deflect through an angle θ. This torque is balanced by the restoring couple due to the torsion fibre ($c\theta$) and the couple tending to turn the residual moment into the earth's field H, ($\Delta P \times H$), so that

$$\Delta F \times P = c\theta + \Delta P \times H. \tag{3.3}$$

Depending on the dominance of one or other of the two terms on the right hand side of this equation, two types of astatic magnetometers have been developed.

(a) *Under torsional control* ($\Delta P \times H \ll c\theta$). This is commonly the most used method. To achieve torsional control requires a high astaticism be attained and that the earth's field H be reduced through a set of cancelling coils. Usually F is arranged to act horizontally at right angles to the lower magnet so that

$$c\theta = P\Delta F. \tag{3.4}$$

Methods and techniques

The sensitivity, which is given by the deflection per unit field difference $(\theta/\Delta F)$ is thus given by

$$\frac{\theta}{\Delta F} = \frac{P}{c}. \tag{3.5}$$

(b) *Under magnetic control* $(c\theta \ll \Delta P \times H)$. This arrangement is used virtually exclusively by Dutch workers (As, 1960). In this case (3.3) becomes

$$\Delta F \times P = \Delta P \times H. \tag{3.6}$$

If F is arranged to be horizontal and at right angles to P, then this becomes

$$P\Delta F = H\sin(\Delta P, H)\,\Delta P, \tag{3.7}$$

where $(\Delta P, H)$ is the angle between ΔP and H. Under magnetic control the system naturally sets itself in the position of ΔP aligned along H so that this angle is zero. On applying a deflecting torque, the deflection θ is thus equal to the angle $(\Delta P, H)$ and for small deflections then

$$P\Delta F = H\theta\Delta P, \tag{3.8}$$

and the sensitivity is

$$\frac{\theta}{\Delta F} = \frac{P}{H\Delta P} = \frac{S}{H}. \tag{3.9}$$

A high sensitivity thus requires a high astaticism and reduction of the earth's magnetic field, the same conditions as required under torsional control.

The objection to this last arrangement is that the system is under the control of the earth's magnetic field or its residual, so that there are problems due to drift. This is not the case under torsional control and as Collinson (1967) has pointed out, in order to reduce the source of drift to a minimum it is desirable to keep the applied fields over the magnet system to as few as possible. Thus most workers tend to favour using torsional control. Details of design considerations for astatic magneto-meters used in palaeomagnetism are discussed by Roy (1963).

The magnetization of a specimen is approximated to a dipole at its centre. Corrections for the cylindrical shape of specimens, for uniformly magnetized cylinders, have been calculated by Papapetrou (Blackett, 1952). The effect of non-uniform magnetization can be corrected for by suitable measurement procedure and averaging (Collinson *et al.*, 1957). In the *on-centre method* (fig. 42 *a*), the specimen is always situated directly below the magnet system, so that only the component of magnetization in the horizontal plane at right angles to the lower magnet produces a deflection. The three components M_x, M_y, M_z can each be brought into position in turn by turning the specimen through successive right angles. The deflections observed will be proportional to the respective components. In the *off-centre method*, the specimen is displaced through a small distance Δx to the east or

72

west of the magnet system (fig. 42*b*). In this position, in addition to the component M_x or M_y acting horizontally at right angles to the lower magnet, there is a component of M_z which also produces a horizontal field and consequently a deflection of the system.

Providing $\Delta x \ll z$, the horizontal field F_x, produced by the component of magnetization M_x at the lower magnet (fig. 42*b*) is given by

$$F_x = \frac{M_x}{z^3}. \tag{3.10}$$

The vertical component of the specimen M_z produces a horizontal field F_z at right angles to the lower magnet given by

$$F_z = \frac{3M_z}{z^4}.\Delta x. \tag{3.11}$$

If the specimen is rotated in 90° steps about the vertical axis, then the component M_y will in turn produce a field F_y following (3.10). In both cases the field F_z adds to F_x and F_y. When the specimen is moved to the west position and the procedure repeated, the field F_z now subtracts from F_x and F_y. The three components M_x, M_y, M_z are then determined from (3.10) and (3.11). This procedure involves fewer measurements to obtain all three components compared with the on-centre method. The off-centre method is therefore the most commonly used with astatic magnetometers and is described in detail by Creer (1967*a*).

An ingenious method by which inhomogeneities are allowed for has been described by Collinson (1970). In this method the sample is rotated about the axis whose component is to be measured, at a frequency of 10 revs per second, sufficiently fast compared with the period of a typical magnet system (say 12 s) to prevent oscillations in the deflecting field during one revolution, due to inhomogeneities of magnetization in the sample, from being recorded.

The ordinary static deflection produced by the sample can be greatly amplified through resonant coupling of a rotating sample with an under-damped astatic system. The sample is revolved on a horizontal axis at constant frequency of a few cycles per second. The astatic magnet system, with the lower magnet parallel to the rotational axis, is suspended above the sample on a relatively stiff torsion fibre, which allows torsional oscillations at exactly the same frequency. Resonance amplification of the order of 150 times the deflections produced by the revolving sample can be accomplished by precise tuning. This type of resonance magnetometer has been evaluated by Graham (1967), who concludes that there are inherent difficulties in applying the instrument to the measurement of very weakly magnetized specimens.

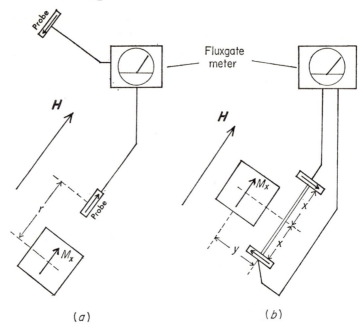

FIGURE 43 Two possible configurations for measurement of specimen magnetization with a fluxgate magnetometer, after Helbig (1965). The method in (b) is more effective when the specimen is inhomogeneously magnetized.

3.1.3 *Fluxgate magnetometer*

The development of the fluxgate magnetometer as an instrument for measuring very small magnetic fields has provided an alternative and simple method of measuring the magnetization of rock samples. By placing two probes in opposition, a fluxgate gradiometer is produced and this can be used in much the same way as an astatic magnetometer (Helbig, 1965). Two possible configurations for measuring the remanent magnetization of a rock sample are shown in fig. 43. In the first case (fig. 43 a) the axis of one of the fluxgate probes is set up parallel to the earth's magnetic field and the other antiparallel to the earth's field at some distance away. A sample brought near the first probe causes an imbalance between the two fluxgates caused by the field of the sample parallel to the earth's field. The sample is positioned so that one component (the x-component say, as in fig. 43 a) lies along the axis of the probe. The y- and z-components then act at right angles to the probe and have no effect. The x-component can be calculated from the normal dipole equation

$$F_x = \frac{2M_x}{r^3},$$

(3.12)

74

where F_x is the field intensity read on the fluxgate meter and r is the distance from the centre of the sample to the centre of the adjacent fluxgate probe. Eqn. (3.12) is correct for spherical samples, but for samples of different shapes corrections must be applied as with the astatic magnetometer. The effect of induced magnetization is eliminated by measuring with the x-component antiparallel to the earth's field and averaging the readings. The y- and z-components are similarly measured in turn, a total of six measurements being made.

If the sample is inhomogeneously magnetized the procedure does not eliminate the effects. The configuration of fig. 43b is then superior. Here the probes are set up antiparallel at a distance apart slightly larger than the size of the samples to be measured. The line through the probes at right angles to their axes is placed in the direction of the earth's field. In this case both fluxgates measure the magnetization of the sample parallel to the earth's field. Again, reversing the sample eliminates the induced magnetization by averaging. The effects of inhomogeneities can now be allowed for by rotating the sample about the x-axis and taking a measurement at 90° intervals, that is four measurements with the x-component parallel to the earth's field and four with it antiparallel. The average of these eight readings enables even the most strongly inhomogeneous sample to be measured correctly. In this configuration the corrections for samples of different shapes have been discussed by Helbig (1965).

A simple fluxgate magnetometer for the measurement of natural remanent magnetization at the outcrop has been made by Doell & Cox (1967a). The fluxgate probe is set up on a tripod so that its axis is exactly perpendicular to the earth's field, as determined by the centre zero reading on the scale at the highest sensitivity. An oriented sample is removed from the outcrop and then brought up to the probe in the position of its original orientation. The method is particularly useful for the quick determination of magnetic polarity in late Tertiary and younger volcanics, whose directions of NRM are roughly parallel or antiparallel to the geomagnetic field at the sampling site. By rotating the sample in front of the probe, approximate NRM directions can be determined to an accuracy of about 20°.

The fluxgate gradiometer is probably the fastest and most convenient instrument for measuring strongly magnetized specimens, but it is not sensitive enough to measure most sediments. Its use with a slow speed spinner magnetometer, however, increases its sensitivity to that achieved by most other types of magnetometer as described below.

Methods and techniques

3.1.4 *Spinner magnetometer*

The spinner magnetometer or rock generator operates on the principle that a magnetic moment rotating within a coil about an axis in the plane of the coil will produce an alternating e.m.f. whose amplitude and phase can be detected and measured. A block diagram showing the principal components is shown in fig. 44. A reference signal is generated by means of a light beam and photocell and the phase of the alternating e.m.f. is measured with respect to this signal by means of a phase sensitive detector. Alternatively, as in one commercial instrument, two reference signals separated by 90° in phase are generated, and two amplifiers then amplify the 'in phase' and 'quadrature' components separately. The ratio of the two amplitudes then gives the tangent of the phase angle.

The fundamental limitation of the sensitivity of spinner magnetometers is the thermal noise in the pick-up coil as has been discussed by Johnson (1938). The minimum detectable moment (M_0) for a spinner magnetometer is given by

$$M_0 = k \cdot \frac{R(\Delta f)^{\frac{1}{2}}}{f}, \tag{3.13}$$

where f is the frequency, Δf is the bandwidth, R is the signal to noise ratio and k is a constant depending upon the configuration of the pick-up coil. Narrow band-pass filtering and long integration times enhance the signal to noise ratio. Recently developments in commercially made narrow band-pass, phase sensitive electronic detection systems (lock-in amplifiers) have become available. The lock-in amplifiers with their associated preamplifiers are capable of recovering very weak signals in the nanovolt (10^{-9} volt) range, which may be buried in a large noise background. This type of system was used by Gough (1964) and is also made use of in commercially made spinners.

Most shaft spinners operate at frequencies between 5 and 100 cycles per second, whilst most air turbine spinners operate between 150 and 500 cycles per second. Higher frequencies increase the signal to noise ratio (R), but this increase is generally offset by other factors such as increased electrostatic effects. Furthermore, friable samples cannot be spun at high speeds. To overcome this problem, especially to measure the weak magnetization of ocean sediments, Foster (1966) has combined a fluxgate gradiometer with a slow speed spinner. The fluxgate magnetometer is capable of measuring varying magnetic field strengths from steady fields to a few tens of cycles per second. In Foster's magnetometer the five cycles per second signal generated by the fluxgate gradiometer is fed into a lock-in amplifier, whose reference signal is generated by the spinning shaft. Inte-

76

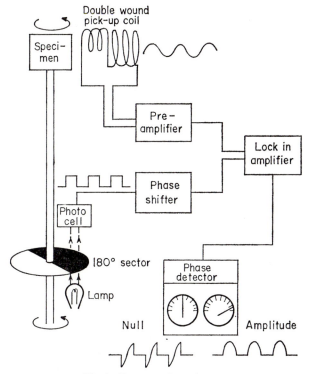

FIGURE 44   Block diagram of a spinner magnetometer.

gration times from one to one hundred seconds produce a high sensitivity to measure most sediments.

In all spinner type magnetometers, the component of magnetic moment in a plane at right angles to the spin axis is measured. By spinning about three axes, the three components are then measured. Generally a procedure involving six spins is used by most workers. Sensitivities of both fast and slow speed spinners are comparable with those obtained by sensitive astatic magnetometers. In both instruments it is extremely difficult to measure samples with intensities of less than 1×10^{-7} emu cm^{-3}, whereas the measurement of intensities greater than 1×10^{-6} emu cm^{-3} is relatively simple.

3.2   STATISTICAL METHODS

3.2.1   *Dispersion on a sphere*

The usual sampling scheme in palaeomagnetic studies is of the hierarchical type comprising several levels. The directions of magnetization of specimens are combined first to give a sample mean. Sample means are then in

77

turn combined to give site means, which are then combined to give the overall formation or unit mean. The averaging of the directions of magnetization at each level requires the development of a method for the statistical analysis of a set of vectors. Each direction of magnetization is represented by a vector of unit length and there is no weighting in favour of samples with greater intensity of magnetization.

Fisher (1953) has suggested that these vectors, when regarded as points on a unit sphere, will be distributed with probability density P, given by

$$P = \frac{\kappa}{4\pi \sinh \kappa} \exp(\kappa \cos \psi), \qquad (3.14)$$

where ψ is the angle between the direction of a sample and the true direction at which $\psi = 0$ and density is a maximum. This means that the proportion of samples expected to fall in a small area δA, the normal to which makes an angle ψ with the true mean direction will be given by $P\delta A$. The parameter κ is called the *precision parameter* and determines the dispersion of the points. If $\kappa = 0$ they are uniformly distributed (the directions are therefore random) and when κ is large the points cluster about the true mean direction. The constant factor in (3.14) ensures that the density adds up to unity over the whole sphere.

Given a sample of points dispersed from a common centre, the best estimate of the position of this centre (the mean direction) is that of the vector sum of the unit vectors having the directions of the several observations. In palaeomagnetic studies the direction of magnetization of a rock sample is specified by the declination D, measured clockwise from true north, and the inclination I, measured positively downwards from the horizontal. This direction may be specified by its three direction cosines, as follows:

$$
\left.
\begin{aligned}
\text{north component} \quad l &= \cos D \cos I, \\
\text{east component} \quad m &= \sin D \cos I, \\
\text{down component} \quad n &= \sin I.
\end{aligned}
\right\} \qquad (3.15)
$$

The direction cosines (X, Y, Z) of the resultant of N such directions of magnetization are proportional to the sum of the separate direction cosines, and are given by

$$X = \frac{1}{R}\sum_{i=1}^{N} l_i, \quad Y = \frac{1}{R}\sum_{i=1}^{N} m_i, \quad Z = \frac{1}{R}\sum_{i=1}^{N} n_i. \qquad (3.16)$$

The vector sum of these vectors will have length R (where $R \leqslant N$) given by

$$R^2 = (\Sigma l_i)^2 + (\Sigma m_i)^2 + (\Sigma n_i)^2, \qquad (3.17)$$

and the declination D_R and inclination I_R, of this mean direction are given by

$$\tan D_R = \frac{\Sigma m_i}{\Sigma l_i} \tag{3.18}$$

and

$$\sin I_R = \frac{1}{R} \Sigma n_i. \tag{3.19}$$

The best estimate k of the precision parameter κ is given by Fisher (1953) for $k > 3$ as

$$k = \frac{N-1}{N-R}. \tag{3.20}$$

The probability that a direction will be observed which makes an angle ψ with the true mean direction is given approximately by the following relations for various probabilities P.

(a) $P = 0.5$, $\quad \psi_{50} = \dfrac{67.5}{\sqrt{k}}$ degrees; $\tag{3.21}$

(b) $P = 0.37$, $\quad \psi_{63} = \dfrac{81}{\sqrt{k}}$ degrees; $\tag{3.22}$

(c) $P = 0.05$, $\quad \psi_{95} = \dfrac{140}{\sqrt{k}}$ degrees. $\tag{3.23}$

These are analogous to (a) the quartile distance, (b) the standard deviation, and (c) the ninety-five per cent deviation, for normal distributions. This last represents the angle from the mean direction beyond which only $\frac{1}{20}$th of the directions lie.

When κ is not too small the distribution is confined to a small portion of the sphere near the maximum, and tends to conform to a two-dimensional Gaussian distribution. In such cases the precision parameter κ is in effect the *invariance* or the reciprocal of the variance in all directions.

3.2.2 *Statistical tests*

(a) *Estimate of accuracy.* Fisher (1953) has shown that the true mean direction of the population of N directions lies within a circular cone about the resultant vector R with semi-angle α at probability level $(1-P)$, for $k > 3$, where

$$\cos \alpha_{(1-P)} = 1 - \frac{N-R}{R} \left\{ \left(\frac{1}{P}\right)^{1/N-1} - 1 \right\}. \tag{3.24}$$

Normally P is taken as 0.05 and values of the factor in brackets in (3.24) for various values of N are tabulated by Irving (1964). When α is small the approximate relations

$$\text{Standard error of the mean} \quad \alpha_{63} = \frac{81}{\sqrt{(kN)}}, \tag{3.25}$$

$$\text{Circle of 95 per cent confidence} \quad \alpha_{95} = \frac{140}{\sqrt{(kN)}} \tag{3.26}$$

may be used. In order to determine whether a palaeomagnetically determined direction differs significantly from some known direction such as the present earth's field at the sampling site, α_{95} may be used directly. The two directions are significantly different at the 95 per cent probability level if the angle between them is greater than α_{95}.

It is often necessary or desirable to compare one palaeomagnetically determined direction with another, rather than with a known direction. A criterion sometimes used is that the two mean directions are significantly different if their cones of confidence do not intersect. This criterion is certainly correct, but two mean directions may still be significantly different even when their cones of confidence intersect. In order to test these cases Watson (1956a) has devised a more exact significance test as follows:

(b) *Comparison of mean directions.* Suppose the two populations have samples N_1 and N_2 and the lengths of the resultant vectors in each case are R_1 and R_2 respectively. Assuming the populations have the same value of κ, the statistic

$$(N-2)\frac{(R_1+R_2-R)}{(N-R_1-R_2)} \tag{3.27}$$

may be referred to the F-ratio tables with 2 and $2(N-2)$ degrees of freedom. R is the length of the vector sum of the resultants of the separate populations, and $N(=N_1+N_2)$ is the total number of samples in the two populations. Large values of the statistic suggest that the assumption of identical true mean directions is false, because the algebraic sum of the sample resultants (R_1+R_2) will be very much greater than their vector sum R. Before carrying out this test it is necessary first to establish identity o precisions as will be described below.

(c) *Comparison of precisions.* If κ is not too small, the ordinary methods for the comparison of variances may be used to test whether the precisions observed in several populations differ from one another (Watson 1956a). If samples N_1 and N_2 gave precision estimates k_1 and k_2 respectively, then the ratio k_1/k_2 is given by

$$\frac{k_1}{k_2} = \frac{\text{variance with } 2(N_2-1) \text{ degrees of freedom}}{\text{variance with } 2(N_1-1) \text{ degrees of freedom}}. \tag{3.28}$$

The assumption that the two populations have the same value of κ may be tested since the right hand side has the variance ratio or F distribution. Values of $F = k_1/k_2$ far from unity strongly suggest the two populations do not have the same precision.

If more than two populations are involved, the ratio of the largest to the smallest k may be used to test the hypothesis that κ is constant over the populations. This ratio may be referred to the tables of maximum F-ratio.

(d) *Randomness test.* In some cases, directions of magnetization are widely scattered, and the question then arises as to whether or not these directions could arise from sampling a random population. In that case the mean direction would have no significance. For a truly random population κ is zero. In practice, however, the observed k, the best estimate of κ, is never zero. Watson (1956b) has devised the following test. For a sample of size N, the length of the resultant vector R will be large if a preferred direction exists, or small if it does not. Assuming no preferred direction exists, a value R_0 may be calculated which will be exceeded by R with any stated probability. Watson (1956b) has tabulated R_0 for various sample sizes for probabilities of 0.05 and 0.01. To carry out the test one merely enters Watson's table at the row corresponding to the sample size N in order to find the value of R_0 which will be exceeded with given probability. Significance points for R_0 up to $N = 100$ have also been listed by Irving (1964) for $P = 0.05$.

3.2.3 *Estimation of formation mean direction*

The final problem in the analysis of palaeomagnetic data is the estimation of the mean direction of a geological formation. Suppose that N_i samples have been collected from the ith of B sites, the sites having been spaced uniformly throughout the thickness and areal extent of the formation. Suppose also that the observations within the ith site obey Fisher's distribution with precision k_W, and the mean direction varies from site to site with precision k_B about the overall mean direction. This overall mean direction, the formation mean, may be estimated either (1) as the direction of the vector resultant of the B site mean directions or (2) by the direction of the vector resultant of all $N = \Sigma N_i$ observations. There are two cases to consider. In basic igneous rocks, notably basalt flows and dolerite sills and dykes, one might expect the within-site precision k_W to be constant at each site. However in sedimentary rocks, or in an igneous unit with a wide variety of rock types, one might expect the within-site precision to vary according to the rock type samples.

(a) *Case when k_W is constant.* The treatment is approximate and valid for

TABLE 4   *Analysis of dispersion*

Source	Degrees of freedom	Sum of squares	Mean square	Expectations of mean squares
Between sites	$2(B-1)$	$\Sigma R_i - R$	$\dfrac{\Sigma R_i - R}{2(B-1)}$	$\dfrac{1}{2}\left(\dfrac{1}{k_{\mathrm{W}}} + \dfrac{\overline{N_i}}{k_{\mathrm{B}}}\right)$
Within sites	$2\Sigma(N_i-1)$	$\Sigma(N_i - R_i)$	$\dfrac{\Sigma(N_i - R_i)}{2\Sigma(N_i-1)}$	$\dfrac{1}{2} \cdot \dfrac{1}{k_{\mathrm{W}}}$
Total	$2(N-1)$	$N-R$		

small dispersions (i.e. k_{W} and k_{B} large) and is due to Watson & Irving (1957). It is most appropriate in the case of basalt lava flows and dolerite sills and dykes.

The observations are first analysed to give the lengths of the vector resultants at each of the B sites, R_1, R_2, $R_3 \ldots R_{\mathrm{B}}$, and the length of the resultant of all $N = \Sigma N_i$ observations (R say). The analysis of dispersion of table 4 may then be used. Here

$$\overline{N}_i = \frac{1}{B-1}\left\{N - \frac{1}{N}\Sigma N_i^2\right\}, \tag{3.29}$$

and is the weighted average of the N_i samples.

The significance of the between-site variation may be judged by an F-test as follows:

$$\frac{\text{mean square between sites}}{\text{mean square within sites}} = F_2(B-1),\, 2\Sigma(N_i-1). \tag{3.30}$$

If the resultant is significant, estimates k_{W} and k_{B} may be found by equating the mean squares to their expectations, and solving the resulting equations. If the result is not significant the between-site variation may be ignored since $k_{\mathrm{B}} \approx \infty$.

If the direction of the resultant of all N observations is used as an estimate of the direction of magnetization of the rock formation, it will be distributed approximately in Fisher's distribution with precision k_0 given by

$$k_0^{-1} = (k_{\mathrm{W}} N)^{-1} + (k_{\mathrm{B}} B)^{-1}. \tag{3.31}$$

The radius of the circle of ninety-five per cent confidence may then be found by substituting the values of k_0 for k in (3.26).

(b) *Case when k_W is variable.* When k_{W} is variable from site to site, there is no satisfactory way known in which all the vector directions can be taken into account in estimating the overall formation mean. However, if the within-site and between-site dispersions follow Fisher's distribution, the site mean directions will also obey Fisher's distribution. Thus the product

$k_W N_i$ must be kept constant at each site. Under these conditions the mean formation direction and its error can be calculated in the usual way. Examples of the application of Fisher statistics are given in McElhinny (1967a).

3.2.4 *Analysis of poles*

The palaeomagnetic pole may be calculated from the formation mean direction of magnetization (D_m, I_m) according to (1.12), (1.13) and (1.14). The mean direction has its associated circle of confidence α_{95} and if dI_m and dD_m are the corresponding errors in inclination and declination, then

$$\alpha_{95} = dI_m = dD_m \cos I_m. \tag{3.32}$$

From (1.11) the error dI_m in the inclination corresponds to an error dp in the ancient colatitude p given by

$$dp = \tfrac{1}{2}\alpha_{95}(1 + 3\cos^2 p). \tag{3.33}$$

The error dp lies along the great circle passing through the sampling site S and the palaeomagnetic pole P, and is the error in determining the distance from S to P. The error in the declination corresponds to a displacement dm from P in the direction perpendicular to the great circle SP where

$$dm = \alpha_{95} \frac{\sin p}{\cos I_m}. \tag{3.34}$$

The polar error (dp, dm) is termed the oval of ninety-five per cent confidence about the pole position.

Alternatively each site direction may be specified by a virtual geomagnetic pole (VGP). The VGPs may themselves be analysed by Fisher's method, since they correspond to points on a sphere. The declination is replaced by longitude east and the inclination by the latitude of each VGP. The mean pole position will have its associated precision parameter (K) and circle of ninety-five per cent confidence (A_{95}). To distinguish the analysis of VGPs with that of directions, capital symbols are always used for the analysis of poles.

3.3 FIELD TESTS FOR STABILITY

It is important in palaeomagnetic studies, in attempting to reconstruct the past history of the geomagnetic field, to establish as far as is possible that the palaeomagnetic directions measured in rocks relates to their time of formation and not some later time. A number of tests have been applied as a means of establishing this important point, and they fall into two broad

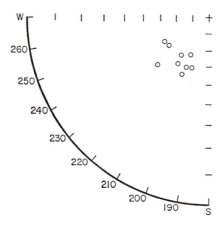

FIGURE 45 Stereographic projection of directions of magnetization observed at nine sites in the Great Dyke of Rhodesia. Open circles are on the upper hemisphere. Redrawn from McElhinny & Gough (1963).

categories – field tests and laboratory tests. Laboratory tests may in general establish the capability of all or some of the magnetic grains to remain stable over the necessary geological time, but they do not by themselves establish that this has definitely been the case. Field tests, on the other hand, can provide information of the stability of the magnetization over all or part of the geological time interval since the rock was formed. Usually most field tests are enhanced by laboratory studies and the use of cleaning techniques (see §3.4).

3.3.1 *Consistency tests*

If a single geological unit or formation can be sampled over a wide area and through a considerable thickness, in which are represented a variety of rock types, and if consistent directions of magnetization are observed differing from that of the present earth's field, then there is good reason to to believe this magnetization to be stable. The use of cleaning techniques is now generally used to enhance the self-consistency of the data. The study of McElhinny & Gough (1963) of the Great Dyke of Southern Rhodesia illustrates this test. Consistent directions of magnetization were obtained at nine sites after magnetic cleaning (fig. 45). The sampling sites were spread over a distance of more than 300 km, through a thickness of 700 m, and represented five different rock types. The directions were quite different from that of the present earth's field and stability since the time of intrusion 2.5×10^9 years ago is strongly indicated.

84

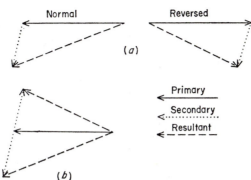

FIGURE 46   Consistency of reversals test. (*a*) A secondary component causes misalignment of the normal and reversed directions. (*b*) When one of the directions is reversed, the vector sum tends to cancel out the secondary component.

3.3.2   *Reversals test*

The parallelism between tightly grouped mean directions of magnetization in two groups of samples which are reversely magnetized with respect to each other is a much stronger test than simple consistency of directions without reversals. The test is equally applicable to reversals due either to field- or self-reversal, since in both cases the mean directions of magnetization are 180° apart. If subsequent to formation, both groups acquire a secondary component of magnetization, it is quite probable that they would both change in the same direction towards the ambient magnetic field. The two resultant groups would then not be 180° apart as shown in fig. 46*a* (Cox & Doell, 1960). However, in this particular case, if the reverse of one of the groups is taken and their mean is then calculated (fig. 46*b*), the secondary magnetizations tend to cancel one another out. The effect is, to a certain extent, the same as magnetic cleaning. Although this technique is useful in judging the reliability of some of the older data in the palaeomagnetic literature, it is now common practice to use magnetic cleaning or other cleaning techniques in all cases as a matter of routine.

3.3.3   *Baked contact test*

When an igneous rock intrudes a rock formation at a time subsequent to the formation of the latter, the intrusion heats the surrounding rock, which upon cooling will acquire a remanence in the same magnetic field as that in which the intrusive rock becomes magnetized. Since the country rock and the igneous intrusion are generally very different materials, agreement between the direction of the intrusion and of the country rock provides

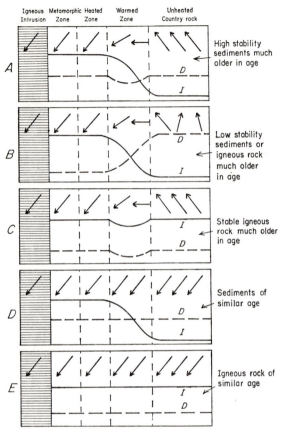

FIGURE 47 Baked contact test. The variation of direction of magnetization (arrows), intensity (*I*) and dispersion (*D*) with distance from an igneous intrusion in five possible situations is indicated schematically. Modified from Irving, *Paleomagnetism and its application to geological and geophysical problems.* © John Wiley & Sons. Inc. 1964.

evidence for the stability of the magnetization of the intrusion. This situation also applies to the baked rock underlying an extruded lava flow. The test is much enhanced if it can be demonstrated that there are changes in properties with distance from the baked contact that correspond to the diminishing heating effects of the intrusion (Everitt & Clegg, 1962).

It is useful to define three gradational zones outside the igneous body (fig. 47). In the *metamorphic zone* there will be extensive changes in the magnetic minerals, whilst in the adjacent *heated zone* these changes will be comparatively small, but in both cases the rock will have acquired a total TRM. In the *warmed zone*, the temperature never rose to the Curie temperature of the magnetic minerals and a partial TRM is induced, which combines vectorially with any high temperature stable component in the country rock. Outside the warmed zone is the unheated country rock. Five possible situations are illustrated in fig. 47.

In the first case (A), the country rock is much older than the intrusion and consists of sedimentary material of high stability. The occurrence of high intensity at the contact falling to low values at a distance, coupled with agreement between the direction of magnetization of the intrusion and contact rock, indicates that both are stable and that their magnetization was acquired at the time the igneous intrusion cooled. The change in intensity is of the order of ten times, since the magnetization of the sediment is usually due to CRM, whilst the baked rock now has a TRM. The dispersion may or may not change except in the warmed zone due to the addition of two differing vector directions. Consistent directions in the unheated sedimentary rock, which differ both from that of the present earth's field and that of the intrusion, provide further evidence for the stability of the unheated sedimentary rock (other than evidence from consistency or laboratory tests etc.). The baked contact test now provides indirect evidence that the magnetization of the unheated sediment has been stable at least since the time of the intrusion.

In case B, the country rock is still much older than the intrusion but consists of sedimentary or igneous material of low magnetic stability. The intensity decreases from the contact into the unheated zone, whilst the dispersion increases and scattered directions are observed. In case C, the country rock is again much older than the intrusion but consists of stable igneous material. This situation is similar to that in A but in this case no change in intensity need be observed between the baked zone and the unheated zone. When the country rock is only a little older or of comparable age to that of the intrusion the situation is less favourable. If the country rock is sedimentary material (D) the main variation will be in the intensity between the baked zone and the unheated zone, which should change by a factor of ten, assuming the sedimentary rock originally acquired a CRM. The most unfavourable situation (E) is when the country rock is stable igneous rock of the same age as the intrusion. No variations will be seen from the baked zone to the unheated zone, a situation that could also arise from general heating due to a period of regional metamorphism.

3.3.4 *Fold test*

In all the previously described field tests the assumption has been made that the rocks have not been tilted or folded. The classical *fold test* of Graham (1949) uses the folding to establish the stability of magnetization.

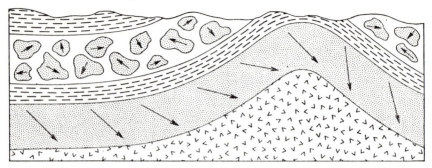

FIGURE 48   Field relationships in Graham's Fold Test and Conglomerate Test. From Cox & Doell (1960), *Geol. Soc. Amer. Bull.* **71**, 647–768. Redrawn with permission of the Geological Society of America.

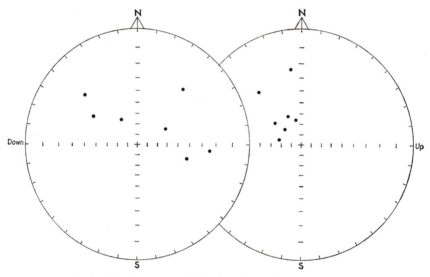

FIGURE 49   Reduction in scatter of the directions of magnetization in seven sites collected from a fold of Caledonian age in the Torridonian sandstone. (*a*) Before correction for geological dip ($k = 4.4$), and (*b*) after correction for dip ($k = 25.5$). Equatorial stereographic projection; pole of projection west, horizontal; plane of projection, vertical N–S. Redrawn from Irving & Runcorn (1957).

The situation is illustrated in fig. 48. If the directions of magnetization of samples collected from different limbs of a fold differ with the beds in their present attitude, but are brought into agreement after 'unfolding' the beds and rotating the directions of magnetization along with them, the following may be concluded. The direction of magnetization of the beds was acquired prior to their folding and has remained stable since that time. The test may be applied to a single structure, such as a fold, within an otherwise

88

undeformed sequence, or more generally over a wide area where samples are collected at sites with differing bedding tilts.

An example of the application of the fold test is given in fig. 49. The originally divergent directions converge upon applying the tilt corrections. The improvement in grouping can be tested statistically as shown by McElhinny (1964). The procedure is essentially an application of the test for comparison of precision (§3.2.2) except that in this case instead of comparing the precisions of two separate groups of samples, the precision of the directions *in situ* (k_b) is compared with their precision after unwinding (k_a). Their ratio k_a/k_b may be compared with F-ratio tables with equal degrees of freedom ($2N-1$), where N is the number of directions involved. McElhinny (1964) has tabulated the ninety-five and ninety-nine per cent confidence limits for the application of the fold test for various values of N.

3.3.5   Conglomerate test

To use this test it is necessary to identify conglomerate pebbles in one formation derived from beds in another formation whose stability of magnetization is being investigated. The situation is illustrated in fig. 48. If the directions of magnetization of the conglomerate pebbles are random then this suggests the magnetization of the parent formation has been stable since the deposition of the conglomerate (Graham, 1949). The pebbles will be large enough that mechanical forces will far exceed the magnetic aligning forces during deposition. The test for randomness of directions may be used in this case (§3.2.2).

This test is probably the least definitive of the various field tests outlined. Uniform directions observed in pebbles, whilst implying the magnetization of the parent formation is secondary, does not necessarily mean that this is so, since the process of conglomerate formation may itself have affected the magnetization. Furthermore random magnetizations at outcrops can also be produced by lightning strikes (Cox, 1961; Graham, 1961), although these can usually be distinguished by abnormally high intensities.

3.4   LABORATORY TESTS AND CLEANING TECHNIQUES

3.4.1   *Alternating field demagnetization – magnetic cleaning*

If a rock specimen is placed in an alternating magnetic field with peak value H, then all domains with coercive force less than $H\cos\theta$ (where θ is the angle between the domain coercive force and H) will follow the field as it alternates. If the alternating magnetic field can be applied to all domain orientations then as the alternating field is slowly decreased to zero, domains with progressively lower coercive force become fixed in different orientations, and hence ultimately all domains with coercive forces less than H will have random orientations. The procedure involves taking the specimen domains around successively larger hysteresis loops (see fig. 24, §2.1.4) and then around successively smaller ones.

For single domain grains (the most important for palaeomagnetic purposes) the relaxation time τ, is directly related to the coercive force H_c through (2.23) viz:

$$\tau = \frac{1}{C} \exp\left(\frac{vH_cJ_s}{2kT}\right). \tag{3.35}$$

Thus progressive demagnetization of a rock specimen randomizes those domains with low coercive force, which are for any given grain volume also those with the shortest relaxation time. If it can be demonstrated that the NRM of a rock specimen is capable of withstanding high alternating fields then this also demonstrates that these grains have long relaxation times, capable of retaining their magnetization for considerably longer than the age of the rock.

Stacey (1961) has discussed the theory of the magnetic properties of multidomain grains in alternating magnetic fields, but the coercive force of such grains is unlikely to exceed a few hundred oersted. Apart from their lack of importance to palaeomagnetism, they are easily demagnetized and even in the most favourable situations the maximum coercive force is 900 oersted (Evans & McElhinny, 1969). Thus the ability of rocks to withstand fields of this order is a clear demonstration of the presence of single domain grains (§2.3.2). For the most stable single domain magnetite grains, the coercive force is essentially controlled by their shape anisotropy (§2.3.2) so that the coercivity is directly related to the length to breadth ratio of the grains (see fig. 36). In general the effect of thermal agitation in very small grains reduces their effective grain coercivity (Néel, 1949; Dunlop, 1965) so that the peak alternating field required to 'unblock' them (H_B) is less than the theoretical coercive force (H_c) according to the relation

$$H_B = H_c - [(Q + \ln t)\, 2kTH_c/J_s v]^{\frac{1}{2}}, \tag{3.36}$$

where t is a typical experiment time (100 s say), Q has a numerical value of about 22 (Néel, 1949) and the other symbols are as in (3.35). Solution of this equation for grains throughout the single domain region for magnetite, produces the curves of fig. 50 (Evans & McElhinny, 1969). With grain sizes nearer the superparamagnetic size, the effective coercivity (given by the unblocking field) may be less than half the theoretical coercive force (compare with fig. 36).

Whilst the maximum theoretical coercivity for single domain magnetite

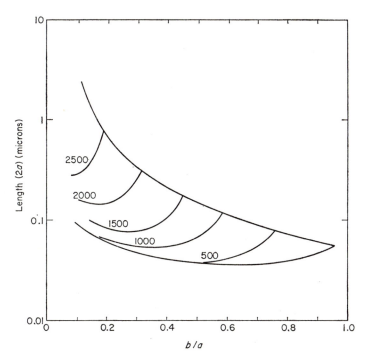

FIGURE 50   Unblocking field (H_B) for single domain magnetite grains at 300 °K, calculated from (3.36), as a function of length (2a) and axial ratio (b/a) of prolate spheroids. The curves for the various unblocking fields, given in oersted, are drawn within the single domain region for magnetite shown in fig. 36 (§2.3.3). After Evans & McElhinny (1969).

is 3000 oersted (for infinitely long needles, §2.3.2), that for haematite is considerably greater. From a practical standpoint it is fairly simple to construct alternating field demagnetization apparatus to produce peak fields up to 2000 oersted, but the very much larger fields required for dealing with haematite create problems. For this reason, it has been found that alternating field demagnetization is very appropriate when dealing with magnetite bearing rocks, but that thermal demagnetization is more useful when dealing with haematite bearing rocks (see §3.4.2). Discussions of alternating field demagnetization methods are therefore largely related to magnetite bearing rocks.

The technique of alternating field demagnetization was first carried out on a routine basis by As & Zijderveld (1958). In their apparatus, they demagnetized the specimen three times along three axes at progressively higher peak alternating fields. Objections to this technique are that all directions in the specimen are not exposed to the same peak field. Creer (1959)

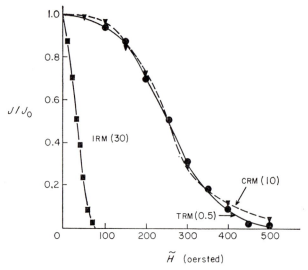

J/J_0

IRM (30)

CRM (10)

TRM(0.5) —

$\tilde{H}$ (oersted)

FIGURE 51   Alternating field demagnetization of IRM, CRM and TRM in magnetite samples. The field in oersted used in each process is indicated in brackets in each case. Redrawn from Kobayashi (1959).

introduced the technique of tumbling the specimen about two axes at right angles and perpendicular to the axis of the demagnetizing coil. Doell & Cox (1967b) have employed a more elaborate three axis tumbler. A simple analysis of the two axis tumbler is given by McElhinny (1966). The procedure is to place the specimen in an alternating field (generally produced by a solenoid), which is decreased continuously to zero. For this purpose full symmetry of the field is necessary; that is no steady field should be present nor even harmonics of the alternating field. The latter are easily eliminated by tuning the demagnetizing coil and the former by cancelling the geomagnetic field at the demagnetizing coil. Alternating field demagnetization in the presence of a steady field produces an *anhysteretic remanent magnetization* (ARM), which must be avoided. The use of a two axis tumbler also assists in removing this effect and is commonly used in most alternating field demagnetization equipment.

   Apart from an interest in the stability of rocks as judged by the ability of specimens to withstand alternating magnetic fields, it is important to be able to judge the possible origin of the magnetization being studied. Studies of the demagnetization of TRM and CRM produced in weak fields show that they have similar demagnetization characteristics, which are quite different from that of IRM even in high fields (Kobayashi, 1959) as illustrated in fig. 51. The IRM acquired by magnetite in a field of 100 oersted is effectively destroyed in a peak alternating field of the same value, but the TRM acquired in 0.5 oersted has decreased only slightly at 100 oersted and a measurable part still remains at 500 oersted. Thus it is relatively simple to distinguish IRM from TRM or CRM, but not to distinguish CRM from TRM.

Viscous magnetization (VRM) differs from IRM in requiring an alternating field larger than the field in which it was produced, for its destruction. VRM acquired in a field of five oersted over five minutes required a thirty-seven oersted peak field for its removal, whilst that acquired over two months in the same field required a field of 180 oersted (Rimbert, 1956). On the geological time scale VRM acquired in a field of say one oersted over a million years would probably be destroyed in alternating fields of only a few hundred oersteds (Brynjolfsson, 1957).

In general, therefore, both IRM and VRM tend to be less resistant to alternating field demagnetization than TRM or CRM, and may thus be selectively demagnetized, whilst preserving an original TRM or CRM. This procedure forms the basis of the technique of *magnetic cleaning*. An empirical approach is usually employed as it is not known in advance what value of alternating field will be required to effect the cleaning process. Some workers use the criterion that the stable component is obtained after vector rotation stops and changes occur only in intensity (As & Zijderveld, 1958; McElhinny & Gough, 1963). An alternative approach makes use of the change in dispersion of directions for several test specimens from the same site. The treatment necessary to produce minimum dispersion is then selected and applied to all specimens from the same site (Irving *et al.*, 1961 a).

Two examples of magnetic cleaning are given in fig. 52. In the first case the NRM directions from samples at a site lie in a plane containing the present earth's field. VRM of differing relative magnitude are present in each sample, directed along the present field direction. Cleaning in a field of 200 oersted peak value causes the sample directions to cluster about a mean direction which is presumed to be that of the primary magnetization acquired when the rock was formed. In the second case the NRM directions are widely scattered and can be shown to be random on application of the test for randomness of directions (§3.2.2). Samples from this locality have probably been struck by lightning and a randomly directed IRM has been added to the primary magnetization. After cleaning in 400 oersted, the IRM is removed and the directions cluster about the primary direction. Normalized demagnetization curves for the intensities of two samples from this group are shown in fig. 53. One sample has a large IRM component and the other a very small component, both samples being from the same site. When the relative intensities are plotted on a linear scale (fig. 53 a), it appears that the one sample consists entirely of IRM (and might thus be rejected), whilst the other is pure TRM or CRM. When

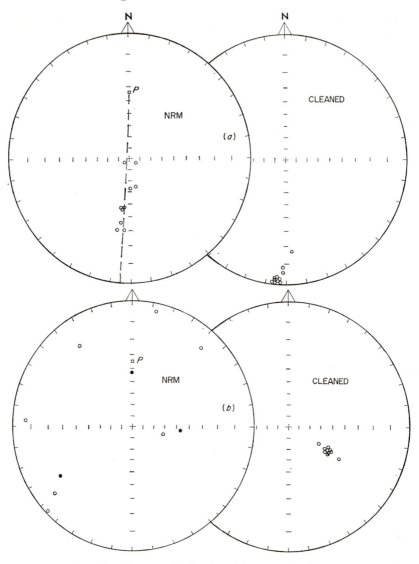

FIGURE 52 Examples of magnetic cleaning. (*a*) NRM directions are strung along a great circle from the direction of the present axial dipole field (*P*). (*b*) NRM directions are randomly magnetized, probably due to lightning strikes. Stereographic projection, solid circles on the lower hemisphere and open circles on the upper hemisphere. Redrawn after McElhinny & Opdyke (1964).

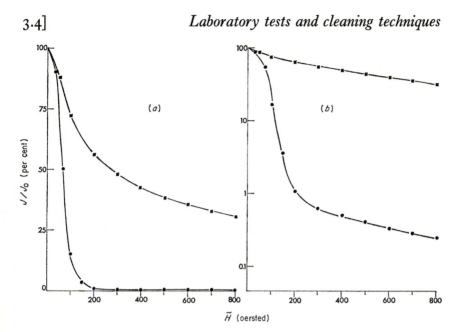

FIGURE 53 Normalized a.f. demagnetization curves of two samples from the group shown in fig. 52 *b*. (*a*) Plotted on a linear scale, (*b*) plotted on a log scale. Squares are for a sample containing little or no secondary component, circles are for a sample containing a very large IRM component, probably due to lightning. Modified from McElhinny & Opdyke (1964) with the addition of unpublished data.

plotted on a log scale (fig. 53 *b*), however, it becomes apparent that the IRM type curve in the first sample flattens out above 200 oersted and becomes parallel to the curve of the other sample. This occurs in spite of only 0.3 per cent of the NRM remaining above 400 oersted (the cleaning field). The important point is that in spite of removing 99.7 per cent of the NRM, the remainder not only has the same demagnetization characteristics of TRM or CRM as its sister sample, but also has the same direction of magnetization. This last example demonstrates the power of the magnetic cleaning technique in preserving a small measureable primary component even against the background of a secondary component nearly 1000 times as large.

The choice of optimum treatment for magnetic cleaning is often based upon a combination of criteria which, although objective in themselves, have commonly been combined in a subjective manner by palaeomagnetists. Attempts to overcome this problem and place the procedure on a more rigorous basis have been made by defining a *stability index*, a property related to the remanent magnetization and which changes during a.f. demagnetization. Tarling & Symons (1967) first attempted to define a stability index, but related it to directional changes only. Wilson *et al.*

95

(1968) succeeded in balancing the effects of intensity and direction change, but the resulting formula had certain drawbacks. These are that the index is predominantly a measure of the dominance of low coercivity components and that rocks with such components are condemned with low stability indices even though they may carry high coercivity components of high stability. The most useful approach appears to be through a stability index proposed by Briden (1972) which overcomes these drawbacks. The index (*SI*) is defined for comparison of two vectors J_1, J_2 by the equation

$$SI_{1-2} = 1 - \frac{J_1 - J_2}{J_1}.$$
(3.37)

Two identical vectors have $SI = 1.0$ and two equal antiparallel vectors have $SI = -1.0$. To use the index, SI is defined for successive equal increments of alternating field, say 100 oersted, which Briden (1972) suggests be used as a standard. During successive demagnetization the field at which SI peaks is then chosen as the most suitable cleaning field.

3.4.2 *Thermal demagnetization – thermal cleaning*

Eqn. (3.35) shows that the logarithm of the relaxation time of single domain grains is inversely proportional to temperature. The blocking temperature T_B for a grain of volume v has already been defined as that temperature at which the relaxation time τ becomes small, say 100 s (§2.3.3). A high blocking temperature indicates the grains will have a long relaxation time at room temperatures. The blocking temperature spectrum can be investigated through thermal demagnetization studies. Two methods have been used. In the *continuous method*, the magnetization of the sample is measured whilst still hot by using a small furnace under an astatic magnetometer (Wilson, 1962c). In the *progressive method*, the sample is heated to successively higher temperatures and cooled in field-free space after which it is measured before the next heating (Irving *et al.*, 1961b). Irving & Opdyke (1965) describe the blocking temperature spectrum as consisting of two basic types: *thermally discrete components* are those of very great stability which remain unchanged up to temperatures near the Curie point, whilst *thermally distributed components* are those whose properties are defined by a series of blocking temperatures. These are the less stable components and the ones which are the more capable of acquiring secondary components of magnetization. Examples of the thermal demagnetization curves for these types of components are illustrated in fig. 54.

Because the low stability, low relaxation time components also have the lower blocking temperatures, partial thermal demagnetization lends itself

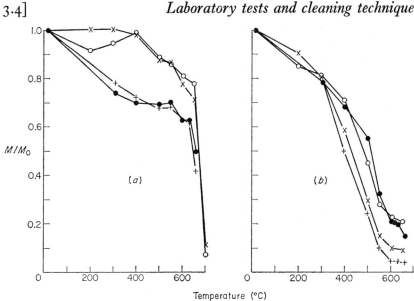

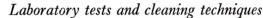

FIGURE 54 Normalized thermal demagnetization curves of samples from the Waterberg redbeds of South Africa. (*a*) Samples having both thermally discrete and thermally distributed components; (*b*) samples having only thermally distributed components. From Jones & McElhinny (1967), *J. Geophys. Res.* **72**, 4173.

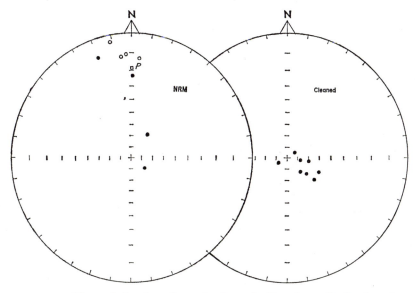

FIGURE 55 Thermal cleaning of samples from the Permian redbeds of southwest Tanganyika. (*a*) NRM directions tending to lie in a plane containing the axial dipole field direction (*P*). (*b*) After thermal cleaning at 300° C. Redrawn from Opdyke (1964), *J. Geophys. Res.* **69**, 2483. Projection and symbols as in fig. 52.

to use as a method of cleaning secondary components from samples. The technique is referred to as *thermal cleaning*, and is largely used when dealing with sediments in which haematite is the main carrier of the remanence. Whilst thermal demagnetization has been effective in removing low stability viscous components directed along the present geomagnetic field, it appears to have little use in removing large IRM components, such as those produced by lightning strikes. The latter are more readily removed by a.f. demagnetization. An example of thermal cleaning is given in fig. 55 in which it can be seen that viscous components aligned along the direction of the present field were easily removed after thermal cleaning at 300 °C. It should be noted that the stability index *SI* of (3.37) may also be used to select the optimum temperature for thermal cleaning. Briden (1972) proposes that to use the index, it should be defined for successive equal temperature increments of 50 °C as standard.

There are examples in which the NRM of rock formations which show no obvious petrological signs of baking, possess a secondary component that is stable and of considerable geological antiquity (Chamalaun & Creer, 1964; Briden, 1965). In the Old Red Sandstone (Chamalaun & Creer, 1964) the secondary component (Permo-Carboniferous) was found to be dominant, and a much smaller, more stable component was detected only after thermal demagnetization to at least 600 °C. Chamalaun (1964) ascribed the secondary component to a rapid increase in the relaxation time of the grains as a consequence of fall in temperature during uplift and folding, and refers to it as 'viscous PTRM' (see also §2.3.4). In the Bloomsburg redbeds, Irving & Opdyke (1965) identified a secondary component (Permian) which was of comparable magnitude to a more stable component that is believed to be of the same age (Silurian) as the Bloomsburg formation itself. They suggested that the secondary magnetization was a 'moderate temperature viscous remanent magnetization' acquired when the beds were buried beneath a thick sedimentary pile or that it could be a chemical effect due to increase in grain size. Thermal cleaning at 550 °C was required to remove this ancient secondary magnetization as illustrated in fig. 56. This effect is consistent with the theories of magnetic viscosity (§2.3.3) and has been investigated in the laboratory by Briden (1965).

The simple minded approach is to suppose that if rocks suffer deep burial and elevation to temperatures of say 100 to 200 °C, then they will acquire a PTRM on cooling from this temperature to 0 °C say. Whilst this is true, the situation is much more complicated since there are relationships between relaxation time and temperature as given in (2.24) viz:

$$T_1 \ln C\tau_1 = T_2 \ln C\tau_2. \tag{3.38}$$

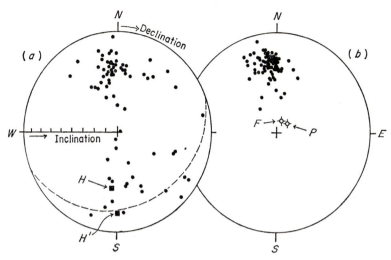

FIGURE 56   Thermal cleaning of samples from the Bloomsburg redbeds (Silurian) of Pennsylvania. (*a*) NRM directions; (*b*) after cleaning at 550 °C. Equal area projection, dots (circles) indicate negative (positive) inclination. The dashed line is the trace of the present horizontal plane on the upper hemisphere. Stars give the direction of the present field (*F*) and axial dipole field (*P*), both downwards. Squares give the direction of the Permian field assuming beds folded prior to (*H*) and after (*H'*) the Permian (upwards). From Irving & Opdyke (1965).

At the blocking temperatures observed in thermal demagnetization experiments in the laboratory, the relaxation time τ_2 may be assumed to be 5 min (300 s), since this is commonly the time for which the specimens are kept at the maximum temperature before cooling. A family of curves can be plotted for different blocking temperatures for $\tau_2 = 5$ min as shown in fig. 57 (Chamalaun, 1964; Briden, 1965). An analysis of these curves has important implications for thermal demagnetization experiments and the thermal cleaning of rocks. Each curve represents a set of grains having the same relaxation time–temperature relationship. After being heated to T_1 for 5 min during a laboratory thermal demagnetization experiment, all grains represented by curves to the left of *A* (fig. 57) will contribute to the remanence (this is the PTRM component). On the other hand *either* heating to T_1 for time τ_1 *or* maintaining at T_2 for time τ_2 will result in all grains represented by curves to the left of *B* contributing to the remanence. This last situation is that typically encountered after burial for a length of time at elevated temperatures (say 150 °C as shown in fig. 57). The diagram shows that burial at 150 °C for 10^4 years would require 450 °C over laboratory experiment times ($\tau_2 = 5$ min) to demagnetize the effect, while burial for 10^3 years requires 420 °C and 10^5 years about 480 °C. The 'viscous PTRM'

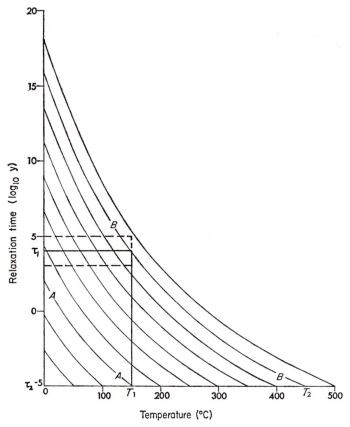

FIGURE 57   Variation of relaxation time with blocking temperature.
Modified from Chamalaun (1964) and Briden (1965).

produced after uplift and cooling is extremely stable having relaxation time of 10^{16} years at o °C for the first of the cases above (the point where the curve B cuts the vertical axis). These examples demonstrate the necessity of investigating the complete blocking temperature spectrum up to the Curie point during thermal demagnetization experiments. They show that even mild and very ancient orogenic effects may still be recognized by finding viscous PTRMs.

The theory of the origin of the ferromagnetism of haematite (§2.2.3) and the evidence put forward by Dunlop (1970) favours the view that in the case of red sediments the ferromagnetism consists of a very hard fundamental spin-canted moment and a softer structure-sensitive moment. Thermal demagnetization then serves the purpose of preferentially destroying the defect moment, both because the average blocking temperature of its NRM is lower than that of the spin-canted NRM and because the moment itself will partially anneal out. Thermal cleaning then becomes very effective in eliminating magnetic noise from red sediments.

3.4.3 *Chemical demagnetization*

The technique has been developed by Collinson (1965c) for investigating the magnetization of red sediments, whilst Carmichael (1961) has described its application to the NRM of haematite–ilmenite crystals. The iron oxide in red sediments, which are those largely used for palaeomagnetic purposes, occurs in two distinct forms. These are the red coating on the grains and in the interstices of the rock (the red cement), and the black particles, which usually occur to the extent of a few per cent by weight. The red form of the iron is believed to be anhydrous or hydrated ferric oxide, very finely divided, and the black form is generally identified as haematite. Small quantities of magnetite and haematite–ilmenite mixtures may also be present. In the experiments carried out by Collinson (1965c, 1966) the red cement is extracted chemically by dissolving it in cold concentrated hydrochloric acid. Specimens are immersed in acid for a suitable length of time, after which they are washed and dried and their magnetization is remeasured. The process is repeated whilst the red cement is preferentially dissolved.

Application of the technique to samples of the Triassic Chugwater formation showed clearly that the bulk of the NRM resides in the black particles (Collinson, 1965c). Seventy to eighty per cent of the original NRM remains at the point where the red colour has just disappeared (fig. 58). When applied to samples of the Taiguati formation of Bolivia (Collinson, 1966), chemical demagnetization showed that the natural remanence is carried by the red pigment, which is the dominant form of iron oxide in the samples. Chemical treatment of samples from the Supai formation (Permian) showed interesting effects (Collinson, 1965c). The rocks appear to have been remagnetized in the Quaternary as a result of a large uprising of volcanic material which formed the San Francisco mountains. During the chemical treatment, the directions of magnetization of two specimens move progressively towards that of the Permian field in North America (fig. 59). The process has thus revealed an original component of magnetization and suggests that *chemical cleaning* might be a useful procedure in palaeomagnetic investigations of sediments. In this case it is of interest that thermal cleaning did not reveal the original component, the direction of magnetization remaining unchanged up to the Curie point. Apart from the work of Collinson, experiments by Park (1970) also indicate that the red cement has much greater magnetic stability than the black particles.

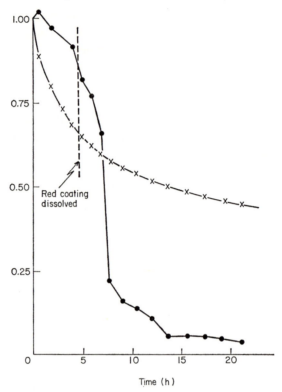

FIGURE 58   Decay of magnetization (circles) and decrease in iron content (crosses) during chemical demagnetization of a sample from the Chugwater formation. From Collinson (1965 *c*).

3.4.4   *Low temperature treatment*

Ozima *et al.* (1964) have noted that at low temperatures the anisotropy energy of magnetite decreases, causing a partial demagnetization of the NRM, especially of the unstable (low coercive force) components. They have suggested therefore, that the low temperature treatment or rocks (cooling to liquid nitrogen temperatures and heating to room temperature in field free space) may provide a quick and effective way of magnetic cleaning. In one of their experiments, a basalt with Curie point 510 °C was given a TRM by cooling in a geomagnetic field from the Curie point. An IRM at right angles to this TRM was then produced in the same sample in a field of 100 oersted. Successive cooling to liquid nitrogen temperatures (−196 °C) showed that the IRM was progressively removed during the treatment. This IRM component was removed by a.f. demagnetization in a field of 100 oersted peak value. Although the experiments of Ozima *et al.* (1964)

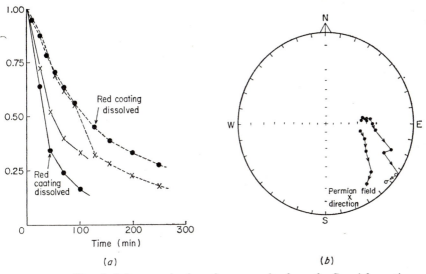

(a)   *(b)*

FIGURE 59   Chemical demagnetization of two samples from the Supai formation. *(a)* Decay of magnetization (circles) and decrease in iron content (crosses). *(b)* Corresponding change in directions of magnetization, projection and symbols as in fig. 52. From Collinson (1965c).

are an impressive demonstration of the method as a possible means of cleaning, virtually no use has been made as yet of the technique.

Merrill (1970) has compared low-temperature treatments and alternating field demagnetization of magnetite samples. These treatments affect different parts of the coercive force spectrum of magnetite, and it appears that some secondary magnetization can survive low-temperature cleaning. However, the combination of low-temperature treatment with a.f. demagnetization may be most useful in separating behaviour in multidomain grains from that in single domain grains. The latter are apparently unaffected magnetically by low-temperature cycling.

3.4.5   *Study of various magnetic properties*

Measurements of the coercivity of NRM (H_{cr}') has been used by some workers as an indicator of palaeomagnetic stability (see §2.1.4). The procedure is often referred to as steady field (or d.c.) demagnetization. The specimen is placed in a small magnetic field (say 10 oersted) with its remanence opposed to the direction of the field. On removal from the field the magnetization is remeasured, and the process repeated for progressively higher fields until the intensity becomes zero. This occurs at the field H_{cr}'. The method was introduced into palaeomagnetic work by Johnson *et al.* (1948) and has been used extensively by Russian workers (Petrova, 1961). High values of H_{cr}' indicate high stability but low values do not necessarily mean that stable components are absent; it might only mean that a high proportion of low coercivity grains are present. Petrova (1961) has suggested that values of H_{cr}' in excess of 40 oersted indicate the presence of a useful stable NRM.

The measurement of the coercivity of maximum IRM (H_{cr}) may also be used (§2.1.4) rather than H_{cr}'. In igneous rocks values of H_{cr} range from about 200 oersted to over 1000 oersted, whilst values in excess of 400 oersted are generally associated with a stable NRM, and for values less than 200 oersted it can be anticipated that unstable components are present.

The Koenigsberger ratio (Q) may indicate the presence of substantial amounts of multidomain grains for which $Q = 0.5$ (§2.3.4). For single domain grains $Q > 1.0$ in general, and this is sometimes used as a measure of stability. However large values can arise in rocks affected by lightning, and low values could arise if there has been viscous decay of NRM without the build-up of unstable secondary components.

The presence of Curie temperatures only a little above room temperature indicates that even lower blocking temperatures are present and with them unstable magnetizations. The occurrence of high Curie temperatures near the ideal value (585 °C in the case of magnetite), which are repeatable on cooling are sometimes associated with stable magnetization. Rocks with low Curie temperatures which are not reproducible on heating are unstable.

Finally, Stacey (1963) has devised a simple laboratory test for palaeo-magnetic stability based upon the relationship between relaxation time τ and temperature T of (2.24), which may be rewritten

$$\frac{T_2}{T_1} = \frac{\ln C\tau_1}{\ln C\tau_2}. \tag{3.39}$$

The temperature T_2 which the rock can withstand for $\tau_2 = 1000$ s, say, in the laboratory will be related to the temperature T_1 which the rock has withstood continually over geological time, say $\tau_1 = 10^9$ years $= 3 \times 10^{16}$ s. Putting $C = 10^{10}$ s^{-1} as before, then

$$\frac{T_2}{T_1} \simeq 2. \tag{3.40}$$

The simple test thus consists of heating the rock specimen for 1000 seconds to an absolute temperature equal to twice the temperature to which the rock may have been subjected subsequent to the acquisition of its primary remanence. For example, exposure to a maximum temperature of 100 °C (370 °K) over geological time requires that the specimen be heated to 470 °C (740 °K) for 1000 seconds. The test should preferably be carried out with the earth's field at right angles to the NRM for maximum effect. The acquisition of a large partial TRM does not necessarily mean the specimen be rejected, because the successful magnetic or thermal cleaning of this component will demonstrate that it is possible to distinguish between the primary and secondary components. There are two implications which result from the successful application of Stacey's test. The first is that there is a primary remanence present that can retain its direction of magnetization for at least 10^9 years. The second is that any effects on the primary remanence resulting from the thermal history of the rock can be preferentially eliminated by appropriate cleaning techniques.

3.5 CRITERIA FOR RELIABILITY

In making an analysis of palaeomagnetic data, it is necessary to determine what reliance can be placed on them as indicators of the palaeomagnetic field. Early work in palaeomagnetism relied on limited field tests and consistency criteria and were subject to difficulties then unknown. Later, new methods were developed, the most important of which are the cleaning techniques just outlined. Most modern investigations include some form of cleaning as a matter of routine and thorough investigations of magnetic stability are undertaken. Irving (1964) first listed a set of *minimum* criteria for reliability of palaeomagnetic results. These were appropriate to the period in question and related to all results up to the end of 1963. At that time cleaning techniques had only just been developed and were not yet in universal use. Irving himself pointed out that as the subject develops, then more stringent criteria will have to be applied. It now seems an appropriate moment to review the situation.

Annual compilations of palaeomagnetic results are published by the Geophysical Journal of the Royal Astronomical Society. For the period up to 1963 these were listed by Irving (1960–5). 554 results are listed and of these only twelve per cent were derived through cleaning techniques. From 1964 results have been listed by McElhinny (1968–72) and for the period up to the end of 1970 a further 812 results are listed of which eighty-one per cent were derived through the use of cleaning techniques. This comparison shows the dramatic change in the standard of palaeomagnetic data over the past seven years. It is obviously an appropriate moment to review Irving's criteria in the light of developments in the subject. A more stringent set of criteria are now listed, and again it should be stressed that these are to be regarded as *minimum* criteria. There will certainly be purists who will regard them as being too lenient.

(1) *Number of samples.* No result is considered adequate unless it is based on consistent observations from eight (or very rarely seven) or more samples. Results based on fewer than this number of samples are regarded as inadequate whatever their stratigraphic distribution and however many specimens were cut from them.

(2) *Time of origin of magnetization.* In some cases the authors have stated that they do not consider their results as indicative of the geomagnetic field direction at the time of formation of the rock unit being studied. These results and those from rocks which are known to have acquired their magnetization long after the period of formation are also not considered.

(3) *Age of the rock.* The geological age of the rock, or in the case of baked sediments, the age of the igneous body, should be known within the limits lower Palaeozoic, upper Palaeozoic, Mesozoic and Cainozoic. In general the ages of most of the beds studied are known very much better than this. Precambrian rocks are only considered if they have a radiometric age associated with them, or if their stratigraphic position is known in relation to formations with radiometric ages.

(4) *Stability tests.* All results should conform with at least one of the tests outlined in §3.3 or §3.4. This is an absolute minimum requirement and is lenient enough to allow consideration of some important early work in the subject. In general however it is preferable that *all* studies be accompanied by some laboratory tests for stability with the use of cleaning techniques where necessary. In some cases the same formation has been independently studied by two or more workers. In these cases those results derived from cleaning are regarded as superseding the others even if they conform with the minimum criteria.

The problem as to whether the result in question represents a palaeo-magnetic pole or only a virtual geomagnetic pole is a matter of some debate. In obvious cases where the result is a VGP (e.g. a single lava flow or dyke), it is useful to combine several results together to produce a single palaeomagnetic pole from a number of VGPs. Also a number of small studies, each of which does not conform with the sampling minimum, can be combined together to produce a useful result. It is unrealistic to consider a requirement that the circle of confidence be less than any given value. Statistical procedures vary so widely that even the most scattered results can be made to produce a mean with small confidence limits by suitable manipulation of the data (e.g. by using large numbers of specimens as unit vectors). The simplest method is to regard all results having better than five per cent probability of being drawn from a random population, according to the test outlined in §3.2.2, to have little physical meaning. It must be emphasized that these are *minimum* criteria and, as Irving (1964) has stressed, it is not intended to imply that any results which satisfy them necessarily give the direction of the geomagnetic field at the time of formation of the rock being studied. The appendix at the end of this book lists all results up to the end of 1970 that conform with these criteria.

4 Reversals of the earth's magnetic field

4.1 FIELD-REVERSAL OR SELF-REVERSAL?

The investigations of David (1904) and Brunhes (1906) into the magnetization of lava flows and their underlying baked clay led to the first observation
of directions of magnetization in rocks roughly opposed to that of the present
field. This led to speculation that the earth's magnetic field had reversed
itself in the past. Since then, the study of many rock formations around
the world has revealed directions of magnetization roughly opposed to one
another, which occur in rocks throughout the geological column. Nagata
et al. (1952) reported a number of detailed studies on the Haruna dacite
from Japan and showed that on cooling through its Curie point it acquired
a TRM directed antiparallel to the applied field. This was the first demonstration of the self-reversing property of the magnetization in a rock. The
problem then arose as to whether the observed reversals of magnetization
in rocks were due to self-reversal or reversals of the earth's magnetic field.

4.1.1 *Possible modes of self-reversal in rocks*

Néel (1955) has considered theoretically several mechanisms by which rocks
would undergo self-reversal of their remanence. Self-reversal requires the
coexistence and interaction of two ferromagnetic constituents. The simplest
case involves the magnetostatic interaction between two closely intergrown ferromagnetic minerals, A and B, with different Curie points. If the
geometry of the mineral intergrowths is suitable the process is as follows.
If the Curie point of A is higher than that of B, then on cooling through
the Curie point of B, the magnetization of A will produce a magnetization
in B in an opposing direction. An example is given in fig. 60a where layers
of two materials A and B are sandwiched together. If on cooling the magnetization of B is greater than A, self-reversal will have been produced.
The two constituents A and B need not be different ferromagnetic materials,
they may represent the two interwoven sub-lattices of a ferrimagnetic. In
this case the opposing magnetizations A and B are both acquired at the
same Curie temperature. If, however, the spontaneous magnetizations of
A and B vary with temperature in such a way that the net magnetization
reverses on cooling as shown in fig. 60b, then self-reversal will have been
produced.

The first self-reversal in natural minerals, produced in the laboratory by
Nagata *et al.* (1952) on the magnetic extract from the Haruna dacite, was

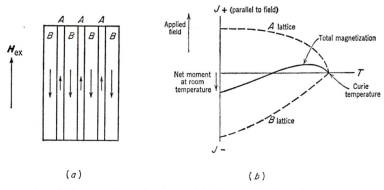

FIGURE 60   Self-reversal mechanisms. (*a*) Alternating lamellae of *A* and *B*, the magnetization of *B* being greater than that of *A*. (*b*) Variation of spontaneous magnetization with temperature is different for the two sub-lattices of a ferrimagnetic. From Irving, *Paleomagnetism and its application to geological and geophysical problems.* © John Wiley & Sons, Inc. 1964.

pursued by Uyeda (1958) using synthetic materials. He found that self-reversal is an intrinsic property of the ilmenite–haematite solid solution series in the region of 0.45 to 0.60 ilmenite (see §2.2.3). Ishikawa & Syono (1963) have shown that the reversal is connected with the ordering and disordering of Fe and Ti ions in the lattice. Self-reversal occurs only when ordered and disordered patches are simultaneously present in metastable equilibrium, and seems to be the result of an antiparallel superexchange coupling between the patches. The mechanism is essentially different from either of Néel's mechanisms given above.

Carmichael (1959, 1961) has produced a self-reversal in an ilmenite–haematite solid solution but this time with only 15–25 per cent ilmenite. He suggests the self-reversal may concern the exchange of electrons between Fe^{2+} and Fe^{3+} ions on the oppositely directed magnetic sub-lattices of the solid solution. Everitt (1962) first demonstrated the self-reversal behaviour of pyrrhotite in the laboratory, whilst Bhimasankaram (1964) has shown that a reversal in natural and synthetic pyrrhotites takes place between two components with 'Curie' temperatures 560 °C and 310 °C coupled antiparallel to each other. Robertson (1963) has also discovered self-reversal in pyrrhotite when investigating the NRM of a monzonite from Australia. The most obvious test for self-reversal in rocks might seem to be to heat and cool them in known magnetic fields and determine whether they are self-reversing. The studies of Uyeda (1958) and Ishikawa & Syono (1963) on minerals of the haematite–ilmenite series showed that the self-reversing properties in the limited composition range were a function of cooling rate. Since the rate of cooling in laboratory experiments is inevitably many

TABLE 5 *Comparison of polarities observed in igneous rocks and their baked contacts. Analysis of Wilson (1962b) updated. N, normal; R, reversed; I, intermediate*

Igneous	Baked contact	No. of observations
N	N	47
R	R	104
I	I	3
N	R	3
R	N	0

orders of magnitude greater than that at which the rocks originally cooled, the test is inconclusive. Furthermore, laboratory heating of rocks to their Curie points often brings about mineralogical changes, so that the self-reversing property, which was present when the rock originally cooled, may now have been obliterated.

4.1.2 *Evidence for field-reversal*

The arguments concerning the hypothesis that the reversed directions of magnetization observed in rocks reflect the fact that the earth's magnetic field has reversed many times in the past must depend upon the weight of circumstantial evidence, rather than on any one decisive experiment. For the hypothesis to be generally valid the following conditions need to be met:

(*a*) Studies of baked contacts adjacent to igneous intrusions or under-lying lava flows should in general show agreement in polarity between the intrusion and the baked rock.

(*b*) There should be world-wide simultaneous zones of one polarity.

(*c*) There should be records of the magnetic field observed in a rock sequence which show the polarity changing continuously from one to the other.

The first two conditions rely on the difference in rock type. In (*a*) the igneous rock and the baked rock will in general contain quite different kinds of magnetic minerals. Similarly, the world-wide occurrence of zones of one polarity as required in (*b*) must also be observed in many different rock types of differing magnetic mineralogy.

The evidence from the studies of baked contacts was first compiled by Wilson (1962b) and later updated by Irving (1964). In table 5 a further updated analysis is given. Rocks whose magnetization is in the same sense as the present field are termed normal (N), whilst in the opposite sense are termed reversed (R). In rare cases intermediate (I) directions are observed during the transition from one polarity to the other. Of the 157 cases now reported in the literature, agreement of the two polarities occurs in all but three cases. If the fraction of self-reversing ferromagnetic minerals in all rocks is x and if these self-reversing minerals are randomly distributed between igneous and baked rocks, then the fraction of pairs of igneous rock and baked rock with opposing polarities will be $2(x - x^2)$ according to Cox et al. (1964). On the basis of the present data, it appears that, at most, one per cent of the reversals in igneous and baked rocks are due to mineralogically controlled self-reversals. The evidence in favour of the field-reversal hypothesis is overwhelming.

Wilson (1962a) has produced a particularly subtle and convincing piece of evidence in favour of field-reversals from an investigation of a doubly-baked rock. A band of laterite had been baked by an overlying lava, the direction of magnetization of the lava and the baked laterite both agreeing and having reversed polarity. Subsequently both the laterite and the lava have been intruded by a basic dyke. The direction of magnetization of this dyke is also reversed and differs by 25° from that of the baked laterite. An investigation was made of the laterite baked by the lava, which was subsequently reheated by the intrusion of the dyke. As the dyke is approached the effect of increasing temperature has been to remove more and more of the initial thermoremanence given by the initial baking from the lava, and to substitute a new partial thermoremanence parallel to the new ambient field given by the dyke direction. The second heating by the dyke did not however exceed the Curie temperature, so that the heating effect could be removed by partial thermal demagnetization. As a result each laterite sample contains two independent magnetizations, one for each of two different temperature ranges (the first is the TRM from the lava, the second the PTRM from the dyke). In the same sample both these superimposed magnetizations were of reversed polarity. It seems almost impossible to explain this fact by any known or theoretical self-reversal mechanisms.

The second test for the field-reversal hypothesis is to date contemporaneous zones of normal and reversed rocks all over the world. For the past few million years this has been undertaken using the Potassium–Argon dating method and is discussed fully in §4.2.1 below. An impressive demonstration of the field-reversal hypothesis is given by the world-wide occurrence of rocks with reversed polarity from the Upper Carboniferous and Permian. Irving & Parry (1963) have termed this the Kiaman Magnetic Interval, and have demonstrated the general occurrence of reversed Kiaman rocks from all over the world. This period will be discussed more fully in §4.2.3. The observation of transition zones during the time when the field is in the process of changing polarity is discussed below in §4.3.1.

4.1.3 *Correlation of petrology and reversals*

The evidence of baked rocks, world-wide simultaneous zones of one polarity and the observation of transition zones, all suggest unanimously and independently that the earth's magnetic field has inverted its polarity many times in the past. It is however worth enquiring whether any other interpretation is possible. A number of workers (e.g. Balsley & Buddington, 1958; Ade-Hall & Wilson, 1963) have made extensive petrological investigations of normal and reversed rocks in an attempt to discover statistical differences between them. Balsley & Buddington (1958) discovered a correlation between the magnetic polarity and state of oxidation of metamorphic rocks in the Adirondack Mountains. Rocks of reversed polarity invariably contained ilmeno-haematite, whilst normal rocks invariably contained magnetite.

Ade-Hall & Wilson (1963) and Ade-Hall (1964*a*) have reported descriptions of a detailed petrological examination of normally and reversely magnetized samples from Tertiary basalt lava from the Isle of Mull. They reported that there were obvious chemical and structural differences between the normally and reversely magnetized lavas; the reversely magnetized lavas are more highly oxidized than the normal ones. The normally magnetized samples were found to contain olivine replaced only by serpentinous minerals together with relatively well-crystallized, optically homogeneous (unexsolved) primary magnetite grains. On the other hand, reversely magnetized samples were found to contain olivine replaced by serpentine, secondary iron oxides and a red hydrated iron oxide (iddingsite). In addition, the primary magnetite grains in this group are relatively large, show poor crystal form and are typified by varying amounts of ilmenite exsolved as lamellae in the octohedral planes of the magnetite. At the same time an investigation of the magnetic properties of these lavas by Wilson (1964) failed to reveal any definite corresponding differences between the normal and reversed groups.

Larson & Strangway (1966) have argued that the correlation between magnetic polarity and petrology is probably fortuitous, and cite extensive investigations of basaltic samples from Oregon, New Mexico, California and Japan in which there was no such correlation. R. L. Wilson (1966), however, continued to find distinct petrological differences between normal and reversed Tertiary basalts from Japan and Iceland and Carboniferous lavas from Scotland. Subsequently Wilson & Watkins (1967) presented an impressive, well-documented correlation between the oxidation state of the opaque grains and the magnetic polarity of a sequence of basaltic lavas in south-central Oregon. Samples showing normal polarity tended

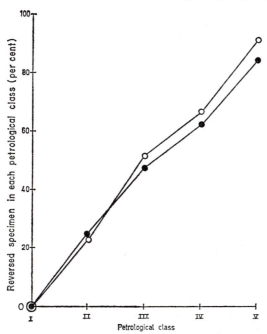

FIGURE 61   Percentage of reversely magnetized specimens as a function of oxidation state in the Columbia Plateau Basalts. Specimens are placed in five oxidation classes (I–V in terms of increasing oxidation). From Wilson & Watkins (1967).

to have ferrimagnetic oxides (chiefly titanomagnetite) with low oxidational states, whilst reverse polarity samples tended to have high oxidational states. The result is summarized in fig. 61. Larson & Strangway (1968), however, have disputed this correlation and their own re-analysis of the same data produced no such correlation.

The general picture concerning the relationship between petrology and magnetic polarity in basaltic rocks has yet to be satisfactorily defined. Further examples of significant associations of relatively high magnetic deuteric oxidation and/or ilmenite percentage with reverse polarity have been reported in lavas from Iceland (Watkins & Haggerty, 1968) and from Mull dykes (Ade-Hall & Wilson, 1969). However, collections of Icelandic dykes showed no significant correlation (Watkins & Haggerty, 1968), whilst a particularly detailed investigation of specimens from 168 Miocene to Pliocene lava flows from the Canary Islands (Ade-Hall & Watkins, 1970) showed no correlation between opaque petrological parameters and magnetic polarity regardless of the way in which the data for the lava collection were subdivided.

As Bullard (1968) has pointed out, the existence of the observed associations, usually a feature suggesting higher oxygen fugacity during lava cooling in a reverse polarity field, appears to be one of the major unsolved problems in geophysics, since no satisfactory explanation is as yet forthcoming. Several suggestions have been made in attempts to answer this difficult point. Wilson & Watkins (1967) state that the evidence is such that a causal connection, however indirect, between field polarity and oxidation state would permit the evidence to be reconciled. Watkins & Haggerty (1968) mentioned that until a meaningful model can be proposed, the observations they described could only be understood to be a function of sampling. A search for correlations in the basaltic lavas that contribute to the polarity time-scale (see §4.2.1 below) and in sediments of deep sea cores (§4.2.2) may well help in deciding whether or not these correlations are either fortuitous or perhaps a function of sampling.

4.2 POLARITY TIME-SCALE

Although the first discovery of reversals of the earth's magnetic field is attributed to Brunhes (1906), the first attempts to delineate the times when the field was reversed were made by Mercanton (1926) and Matuyama (1929). Matuyama's study of volcanic rocks from Japan and Korea is especially important because it was the first attempt to date the most recent change from a reversed polarity epoch to the present normal polarity epoch. He proposed that during the early part of the Quaternary Period the earth's magnetic field was reversed and that this gradually changed over to the normal state. This result, like that of Brunhes, has been amply confirmed by subsequent investigations.

If the field-reversal theory is correct, there must be a precise stratigraphic correlation of normally and reversely magnetized strata from all over the world. This approach is contingent on the condition that the duration of epochs during which the magnetic polarity remains unchanged, must be sufficiently long to be resolved by the available geological techniques. For example, the classical techniques of palaeontology can hardly be used if polarity epochs lasted only 50000 years. The development of the Potassium–Argon (K–Ar) dating technique has made it possible to use the method to date very precisely volcanic rocks whose magnetic polarities have been determined. From these measurements made at a number of laboratories around the world, a geomagnetic polarity time-scale covering the past few million years has been built up over the past eight years.

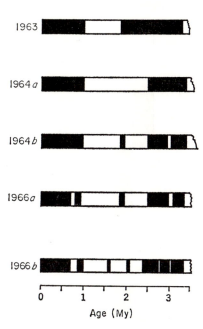

FIGURE 62   Successive versions of the geomagnetic polarity time-scale for the past
3.5 My. From Cox (1969), *Science*, **163**, 237–45. © 1969 by the American Association for the Advancement of Science.

4.2.1   *Polarity dating of lava flows*

The first quantitative time-scale for geomagnetic reversals using the K–Ar
technique was proposed by Cox *et al.* (1963 *a*, *b*). It appeared that polarity
intervals were of nearly equal length of about 1 My (fig. 62). As more data
were obtained by McDougall & Tarling (1964), Doell & Dalrymple (1966)
and McDougall & Chamalaum (1966), some of the ages and polarities were
found to be inconsistent with the simple pattern of intervals. This inconsistency led to the discovery of short polarity events of duration about 10^5
years within the major polarity intervals of about 10^6 years duration (Cox
et al., 1964). These polarity intervals were termed *polarity epochs* and
given the names of early workers in the field of geomagnetism (Brunhes,
Matuyama, Gauss, Gilbert), whilst the short duration *polarity events* were
named after the sites of their discovery (Cox *et al.*, 1964). As more data
were produced, more of these events have been discovered. A brief history
of the development of the geomagnetic polarity time-scale is given in
fig. 62 and illustrates this point.

The most recent summary of the geomagnetic polarity time-scale for the
past 4.5 My based on 150 radiometric ages and polarity determinations
has been given by Cox (1969) and Cox *et al.* (1968). The criteria used to

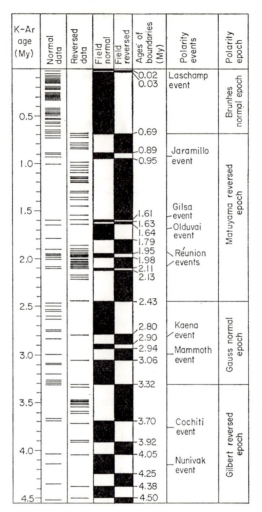

FIGURE 63 Geomagnetic polarity time-scale for the past 4.5 My based on polarity
and K–Ar measurements on lava flows throughout the world. Each data point is
represented by a horizontal bar in the second and third columns. From Cox (1969),
Science, **163**, 237–45. © 1969 by the American Association for the Advancement
of Science. But the identification of the Gilsa, Olduvai and Réunion events follows
Grommé & Hay (1971).

judge the reliability and precision of each data point as meeting reasonable
standards are as follows:

(i) The palaeomagnetic study involves laboratory measurements of
magnetic stability.

(ii) The precision of the Potassium–Argon age determinations is less

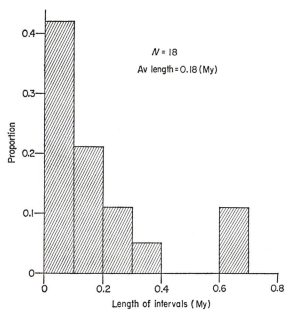

FIGURE 64   Histogram of the lengths of geomagnetic polarity intervals based on the time-scale of fig. 63. After Cox (1969), *Science*, **163**, 237–45. © 1969 by the American Association for the Advancement of Science.

than 0.1 My for ages between 0 and 2.0 My, and less than five per cent for ages greater than 2.0 My.

(iii) The magnetic and age measurements were made on rocks and minerals of types known to yield reliable results, and the samples for both measurements were collected from the same volcanic unit.

The time-scale and data points are summarized in fig. 63, in which the identification of the Gilsa, Olduvai and Réunion events follows the more recent work of Grommé & Hay (1971). Attempts to extend the radiometric time-scale for reversals back beyond 4.5 My have been largely unsuccessful because the errors in the radiometric ages of the older rocks are too large (Cox & Dalrymple, 1967). A dating error of five per cent in a 5 My old sample is 2.5×10^5 years. This is larger than many of the lengths of the polarity events shown in fig. 63. In the case of these short events, their precise length can only be deduced from other information to be described later, including palaeomagnetic data from ocean sediments (§4.2.2) and magnetic profiles at sea analysed on the basis of the sea-floor spreading hypothesis (see chapter 5).

The lengths of the polarity intervals are varied and based on the information summarized in fig. 63, the mean length for the past 4.5 My is 0.18 My. The distribution of the lengths of these polarity intervals is illustrated in fig. 64. Whereas all the longer intervals seem to have been discovered, there are gaps of 10^5 years or more in the data so that it is possible that several very short events might have been missed. The discovery by Bonhommet & Babkine (1967) of the Laschamp event whose duration is probably only about 10^4 years demonstrates that additional events may exist in gaps even shorter than 10^5 years. The presence of some of these short events appears to be indicated in ocean cores (§4.2.2).

For rocks older than 4.5 My the use of well-defined stratigraphic successions is the only really suitable approach. A number of such investigations have been reported by Dalrymple *et al.* (1967), Dagley *et al.* (1967), McDougall & Chamalaum (1969), Wellman *et al.* (1969), for lava successions of Tertiary age.

4.2.2   *Geochronometry of ocean sediment cores*

The study of lava successions does not produce a continuous sequence of polarity information, since the volcanic activity is essentially an intermittent process. Investigations of deep-sea sediments however, have provided a quite independent method of determining the polarity time-scale since a continuous sequence can be obtained. For typical oceanic sediment deposition rates of say 1 to 10 mm per 1000 years, the Brunhes–Matuyama boundary at 0.7 My will be found at a depth of between 0.7 and 7 m, whilst the Gauss–Gilbert boundary at 3.3 My will be found between 3.3 and 33 m depth.

The earliest investigations of marine sediments (McNish & Johnson, 1938) showed no reversals, probably because the cores were too short. The first observations of a reversal in a core were made by Harrison & Funnell (1964) who studied some sediments from the equatorial Pacific. More detailed work on longer cores has been reported by Dickson & Foster (1966), Harrison (1966), Ninkovich *et al.* (1966), Opdyke *et al.* (1966), Hays *et al.* (1969), Opdyke & Glass (1969) and Foster & Opdyke (1970). The cores are unoriented but are cored vertically from the ocean floor so that the magnetic inclination measured in the core simply indicates the polarity. Two examples from the North Pacific are given in fig. 65, where the sedimentation rate in one of the cores (fig. 65*b*) is about fifty per cent greater than in the other. In the first example the inclination is normal (positive) for the upper 525 cm and fluctuates between 40° and 60° about a mean of 52°, which is not far from the present dip of 57° at the sampling locality. At 525 cm the inclination changes sign, remaining negative until 675 cm where it again reverses and remains for the next 35 cm. It reverses yet again and remains so until near the bottom of the core where it changes sign again. In the second example, the first major change of sign from the

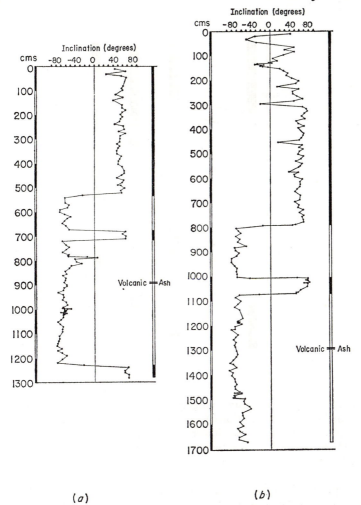

(a)   (b)

FIGURE 65   Variation of magnetic inclination with depth in two deep-sea sediment
cores from the North Pacific, (a) V–20–107 and (b) V–20–108. On the right of each
log, the shaded strip indicates normal polarity, unshaded reversed. From Ninkovich *et al.* (1966), *Earth Planet. Sci. Letters*, **1**, 476–92.

present positive inclination occurs at 800 cm depth, the negative inclination
persisting for the next 200 cm whereupon the sign changes back to normal
for only 70 cm. The negative inclination then persists to the bottom of the
core. The upper normal zone of positive inclination in each case is correlated with the Brunhes normal epoch, with the first polarity change marking
the Brunhes–Matuyama boundary. The brief return to normal polarity
further down the cores is correlated with the Jaramillo event and so on. Note

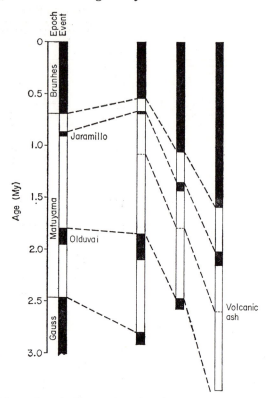

FIGURE 66   Magnetic polarity stratigraphy observed in three deep-sea sediment cores from the North Pacific. Cores v–20–107 and v–20–108 in the centre and on the right respectively are those also illustrated in fig. 65. Shaded regions are normal, unshaded are reversed. Figure redrawn by Bullard (1968) from Ninkovich *et al.* (1966), *Earth Planet. Sci. Letters*, **1**, 476–92.

that in fig. 65 *a* there is an anomalous change in inclination at the 800 cm mark, whilst in fig. 65 *b* three such changes are observed at 30, 130 and 300 cm. These changes resemble short events, and whilst originally Ninkovich *et al.* (1966) attributed them to disturbances which occurred during coring operations, they might also be correlated with short events or excursions of the geomagnetic field as is discussed below.

The correlation of the polarity changes down the cores with the geomagnetic polarity time-scale enables various horizons to be dated precisely so that the rate of sedimentation can be determined. A summary of the magnetic stratigraphy of three cores from the North Pacific including the two shown in fig. 65 is illustrated in fig. 66 and the rates of sedimentation deduced from these cores is given in fig. 67. In one of the cores a change in sedimentation rate can be deduced to have taken place about 0.9 My ago.

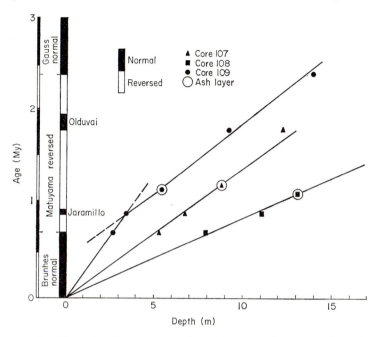

FIGURE 67   Rate of sedimentation deduced from the three cores shown in fig. 66.
From Ninkovich *et al.* (1966), *Earth Planet. Sci. Letters*, **1**, 476–92.

The continuous nature of the record from ocean cores enables the lengths of the various polarity events to be determined with some precision. An example is given in fig. 68 for the Jaramillo event, the first event to be seen in most cores. Opdyke (1969) has analysed the Jaramillo event as observed in 15 cores whose rate of sedimentation varies from 5 to 12 mm per 1000 years. Assuming that the Brunhes–Matuyama boundary occurs at 0.7 My the average rate of sedimentation can be calculated and extrapolated to the time of the Jaramillo event at about 0.9 My as shown in fig. 68. The parameters for the Jaramillo event estimated from each core in this way are given in table 6. The average duration of the event is 56050 years with standard deviation 14600 years.

If very short polarity events of duration 10^4 years or less are present, these can only be detected in cores with very fast sedimentation rates. An examination of seven cores from the Caribbean and the North Atlantic and Indian Oceans with sedimentation rates of 24 to 62 mm per 1000 years has revealed the existence of a short polarity event (reversed) within the Brunhes normal epoch (Smith & Foster, 1969). Ericson *et al.* (1961) have established an X-zone in deep-sea sediments based on the presence or absence of certain warm water species of planktonic foraminifera. Broecker

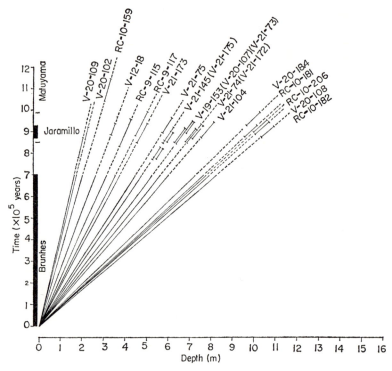

FIGURE 68   Plot of the time of the Brunhes–Matuyama boundary against depth of its occurrence in fifteen deep-sea sediment cores. Extrapolation provides an estimate of the beginning, end and duration of the Jaramillo event as listed in table 6. From Opdyke, *The application of modern physics to the Earth and planetary interiors*. © Interscience Publishers, Inc. 1969.

et al. (1968) place the lower boundary of this X-zone at 126 000 ± 6000 years on the basis of radiometric dating. By identifying the boundaries of the X-zone in each core, the rate of sedimentation with the zone could be determined. The short polarity event occurs within this zone and its boundaries were then estimated to be 108 000 and 114 000 years ago ± 10 per cent. This very short event of only about 6000 years in length has been named the Blake event. Unpublished data mentioned in Stacey (1969) suggest there is another short reversed event within the Brunhes epoch occurring in the V-zone of Ericson *et al.* (1961). Watkins (1968) has also suggested the existence of two short-term polarity events (normal) within Matuyama reversed epoch at 0.82 and 1.07 My. Their verification however awaits more detailed analyses of cores with unusually high sedimentation rates as undertaken by Smith & Foster (1969) in identifying the Blake event. Whereas it is possible to correlate the anomalous changes in

TABLE 6   *Parameters for the Jaramillo event as observed
in oceanic cores (Opdyke, 1969)*

Core	Sedimentation rate (mm/1000 years)	End of event (× 10³ years)	Beginning of event (× 10³ years)	Length of core (cm)	Duration of event (× 10³ years)
RC–10–182	11.8	877	928	60	50.8
V–20–108	11.3	889	947	65	57.5
RC–10–206	11.1	898	953	61	54.9
RC–10–181	10.9	876	913	40	36.7
V–20–184	10.35	932	990	60	57.9
V–20–204	8.65	884	942	50	57.8
V–21–74	8.10	846	907	50	61.7
V–21–172	8.10	870	932	50	61.7
V–21–73	7.50	847	913	50	66.6
V–20–107	7.50	900	950	38	50.6
V–19–153	7.50	849	884	26	34.6
V–21–145	6.93	859	902	30	43.3
V–21–175	6.93	786	830	30	43.3
V–21–75	6.43	895	964	45	70.0
V–21–173	5.35	850	943	50	93.4
Average		870.5	926	—	56.05
Standard deviation		32.56			14.60

inclination observed in fig. 65 with short term events, an alternative explanation is that they represent large excursions of the geomagnetic field. A feature of these deviations is that they are never observed to reach 180° and 'lock-in' to the opposite polarity. The Laschamp and Blake events could thus be explained as geomagnetic excursions.

Although the precision of K–Ar dating makes it difficult to extend the land based geomagnetic polarity time-scale beyond 4.5 My, the magnetic stratigraphy back to about 9 My has been determined from very long piston cores of deep-sea sediment. Two of the longest piston cores of deep-sea sediment, 24 and 28 metres in length, were taken in the equatorial Pacific and their magnetic stratigraphy was investigated by Foster & Opdyke (1970). In the equatorial region the magnetic inclination is near zero so that it is not very easy to distinguish polarity changes from changes in the sign of the inclination. It is easier to measure changes in declination with respect to an internally consistent reference mark on the cores. In the recovery procedure a reference mark is painted at each end of every core pipe near the pipe couplings. As the sediment is being extruded a scribe mark is made in the sediment along the reference azimuth. A string is placed in the scribe mark to serve as a reference guideline for splitting the core in the laboratory. The split face is then available for an azimuthal

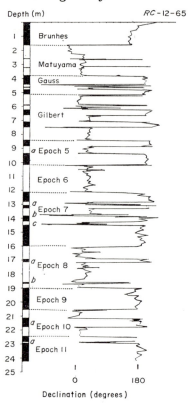

FIGURE 69 Magnetic stratigraphy in a very long (24 m) deep-sea sediment core (Conrad 12–65) from the equatorial Pacific. Changes in declination with respect to a scribe mark on the core are plotted against depth. The black (normal) and white (reversed) bar diagram on the left is a proposed extension of the geomagnetic polarity time-scale beyond 4.5 My. Redrawn from Foster & Opdyke (1970), *J. Geophys. Res.* **15**, 4468.

reference. Conrad 12–65 core 24 metres long is the first reported measurement of magnetic changes in clear stratigraphic continuity with the established magnetic stratigraphy extending down through eleven polarity epochs. Changes in magnetic declination in this core with respect to the split face of the core are shown in fig. 69. Prior to the Gilbert reversed epoch, the epochs have been numbered as 5 to 11 in sequence. The as yet unnamed third event in the Gilbert epoch is termed event '*c*' and a number of events discernible in each of the preceding epochs have been labelled '*a*', '*b*', '*c*' etc. The base of the core penetrates calcareous and siliceous ooze to sediments of Middle Upper Miocene Age, about 9 My or more.

4.2.3   Kiaman Magnetic Interval

If the frequency of reversals observed over the past few million years has persisted throughout geological time, it would be extremely difficult to determine whether or not there exists a precise stratigraphic correlation of reversals on a world-wide scale. If however the field maintained the same polarity for a considerable length of time, such as several tens of millions of years, the possibility of correlating this long epoch on a world-wide scale becomes feasible. Graham (1955) first noted that most rocks of Permian age from the United States were reversely magnetized. Irving & Parry (1963), in studying Permian volcanics from New South Wales, noted that there were twenty-three rock units from other continents that indicated that the time range Late Carboniferous to Late Permian was a predominantly long reversed period. They proposed this time interval be termed the *Kiaman Magnetic Interval*, since it was in a sample from the Upper Marine Latites near Kiama, New South Wales that Mercanton (1926) first reported the occurrence of reversely magnetized rocks for this period. The corresponding rock term proposed is *Kiaman Magnetic Division*.

The Kiaman Magnetic Interval is defined from the section in south-east Australia. Irving (1966) places the base of the interval at the level of the Paterson Toscanite in the Upper Carboniferous section of the Hunter Valley, New South Wales. The Paterson Toscanite is regarded as late Westphalian in age (Brown *et al.*, 1968) and has normal polarity whereas the overlying Seaham Formation has reversed polarity. The lower boundary of the Kiaman Magnetic Interval has therefore been termed the *Paterson Reversal*. The upper boundary in Australia occurs within 90 m of un-sampled section between the latites of the Upper Marine Series (showing reversed polarity) and the Lower Triassic Narrabeen sediments, which are normally magnetized (Irving, 1963). Irving & Parry (1963) termed this the *Illawarra Reversal*. An analysis of the Kiaman Interval in the u.s.s.r., u.s.a., western Europe and Australia has been given by McElhinny (1969 *b*).

Khramov (1967) has produced an extensive summary of all late Palaeozoic palaeomagnetic data from the u.s.s.r. The remarkable consistency between the six sections summarized for different parts of Russia demonstrates the reality of the Kiaman Magnetic Interval, which on the Russian nomenclature extends from the Baskhirian–Moscovian to the upper Tartarian, an estimated time interval of 60 My. McMahon & Strangway (1968 *a*, *b*) have made a detailed study of sediments on the eastern and western slopes of the Rocky Mountains which cover the period of the Kiaman Interval. On the basis of all the data from the u.s.a., McMahon & Strangway concluded that the Kiaman Division extends from the Upper

TABLE 7   *Polarities of Carboniferous, Permian and Lower Triassic rocks from western Europe. N, normal; R, reversed; M, mixed*

Age	Rock unit	Polarity	
Trl	Bunter Sandstone (Germany)	M	
Trl	Vosges Sandstone (France)	M	
Pu	Dome de Barrot redbeds (France)	R	
Pm	Montcenis Sandstone (France)	R	
P(m?)	Nideck Porphyry (France)	R	
P	Nahe igneous rocks (Germany)	R	
P	Mauchline beds (Scotland)	R	
P	Ayrshire kylites (Scotland)	R	
P	Esterel Suite (France)	R	
P	Vallée de Guil Volcanics (France)	R	
P	Malmedy conglomerate (Belgium)	R	
Pl	Exeter traps (England)	R	
Pl	Oslo igneous complex (Norway)	R	Kiaman
Pl	St Wendel Sandstone (Germany)	R	Magnetic
Pl	Rotliegende beds (Germany)	R	Interval
Pl	Krakow Volcanics (Poland)	R	
Cu–Pl	Whin Sill (England)	R	
Cu–Pl	Intrusives (Sweden)	R	
Cu–Pl	Lower Silesian Volcanics (Poland)	R	
Cu–Pl	Ny-Hellesund diabase (Norway)	R	
Cu	Inner Sudetic Basin (Poland/Czechoslovakia)	R	
Cu	Plzen Basin (Czechoslovakia)	R	
Cu	Kladuo-Rakovink Basin (Czechoslovakia)	R	
Cu	Stephanian Sandstone (France)	R	
Cu	Midland sills (England)	R	
Cu	Pennant Sandstone (England)	N	
C	British Carboniferous (England)	M	
C	Southdean basanite (Scotland)	R	
Cm	Tideswelldale rocks (England)	R	
Cl	Derbyshire lavas, Kinghorn lavas, etc. (U.K.)	M	

Pennsylvanian (Desmoinesian), at the base, to the Lower Triassic (lower Scythian) at the top, although the latter is not so well defined. The close of the Kiaman appears to be located in the lower part of the Chugwater Formation (Picard, 1964), but although this formation is commonly regarded as extending down into the Permian the precise stratigraphic position remains uncertain.

Palaeomagnetic data from western Europe during the late Palaeozoic are derived mainly from igneous rocks. In many cases the rocks are intrusive so that their precise stratigraphic positions are not known or are based upon K–Ar age determinations. Continuous stratigraphic sequences have not been sampled with the exception of the Lower Triassic Bunter Sandstone, where Burek (1964) identified a large number of alternating layers of normal and reversely magnetized rocks. These are clearly post-Kiaman. Table 7

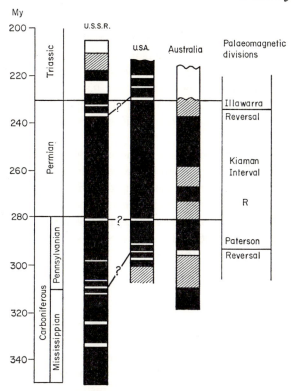

FIGURE 70 Summary of polarity information for the Kiaman Magnetic Interval from three continents. Shaded zones, reversed polarity; white zones, normal polarity; striped zones, unsampled. After McElhinny (1969 b).

lists the results from western Europe, which are not intended to be in precise stratigraphic order, but they do demonstrate the persistent reverse magnetization observed in all Permian rocks. Although the exact horizon has not yet been located on the ground, the lower boundary of the Kiaman appears to occur within the Westphalian or between the Westphalian and Stephanian. The upper boundary cannot be defined better than within the time Late Permian to middle Scythian.

Composite sections from the U.S.A., U.S.S.R., western Europe and Australia are compared in fig. 70. They clearly demonstrate the reality of the Kiaman Magnetic Interval as a world-wide phenomenon. The upper and lower boundaries then serve as precise marker horizons. For example, it appears reasonable to suppose that the first occurrence of alternating layers of normal and reversed polarity above the base of the Chugwater Formation is to be correlated with the upper Tartarian of the Russian sequence. It seems that the Russians place the lower boundary of the Kiaman

127

too early in relation to the other continents. The thin layer of normal polarity observed in the late Moscovian more realistically defines the lower boundary. On this basis the Paterson Reversal (late Westphalian in Australia) is correlated with the Desmoinesian (Upper Pennsylvanian) in the U.S.A., with the upper Moscovian in the U.S.S.R., and lies somewhere between the Westphalian and Stephanian of Europe. The upper boundary of the Kiaman is best defined from the Russian sequence as occurring in the upper Tartarian.

The existence of a narrow zone of normal polarity near the Permo-Carboniferous boundary appears to be well established by both the U.S.S.R. and the U.S.A. data. This horizon has been identified in at least two widely separated sections in the U.S.S.R. and appears to be correlated with a band of normal polarity observed in the lower part of the Dunkard Series of West Virginia (Helsley, 1965). This zone could serve as a useful marker horizon within the Kiaman Interval.

4.2.4 *Reversals during the Phanerozoic*

Khramov *et al.* (1965) have attempted to define a complete polarity time-scale for the Palaeozoic based on extensive investigations of Palaeozoic sections in Siberia. Their work suggests another predominantly reversed interval existed prior to the Kiaman Interval extending over roughly 50 My from Middle Devonian to Middle Visean. This interval contains only one small normal zone at the Frasnian–Famenian boundary. Helsley & Steiner (1969) have noted that the majority of Cretaceous rocks studied in various parts of the world are normally magnetized and produced a Cretaceous polarity time-scale whose main feature is a long normal interval of about 50 My containing two narrow reversed zones. The existence of these long intervals of predominantly one polarity confirms the suggestion of McMahon & Strangway (1967) that the time-scale of polarity intervals of the geo-magnetic field are 5×10^7 years, 10^6 years and 10^5 years or less.

In order to define a complete polarity time-scale for the Phanerozoic it would be necessary to make very careful and complete palaeomagnetic analyses of continuous sedimentary profiles of known stratigraphy. This would involve an enormous amount of work. The existence of rather long polarity intervals makes the problem much easier because it should be possible to correlate the data from all over the world to produce a relatively simple time-scale for these long intervals. To find out during which periods one might identify these long intervals in the Phanerozoic, it is a simple matter to analyse world-wide palaeomagnetic polarity measurements period by period. This was undertaken recently by McElhinny (1971), who

analysed 1094 palaeomagnetic results for the Phanerozoic making use of
the tables of results given by Irving (1964), McElhinny (1968*a*, *b*, 1969*a*,
1970*a*) and Khramov & Sholpo (1967). Polarity is often defined according
to which pole falls in the northern hemisphere. When the north pole falls
in the northern (southern) hemisphere, the sample has normal (reversed)
polarity. This definition is correct for late Palaeozoic and younger rocks,
but in the early Palaeozoic some apparent polar-wander paths cross the
equator on the present grid, so that strictly polarity should be defined with
respect to the apparent polar-wander paths for the region under consideration (see §6.1.3). The correct definition has been used here. Because it is
only polarity information that is being analysed, it is possible to include
a very much greater body of data than listed in the appendix.

It is unlikely that each palaeomagnetic investigation will represent
observations of polarity over equal lengths of time. The investigations will
represent the polarity over a range of intervals and it might reasonably be
assumed that this range will be approximately similar from one geological
period to the next. McElhinny (1971) analysed the proportion of normal
and reversed results period by period. The mixed results (when both
polarities occur in the same unit under investigation) were apportioned to
the normal and reversed figures as a fraction of the occurrence of each. An
updated version of the original figures including results from a later compilation (McElhinny, 1972) is summarized by geological periods in table 8.
These updated figures involve 1231 palaeomagnetic investigations for the
Phanerozoic. In detail they involve virtually no change from the original
analysis, which is illustrated in fig. 71. The percentage of normal polarity
measurements is plotted as a function of time and is a measure of the proportion of the time the geomagnetic field has normal polarity for the period
in question. For the Upper Tertiary (Miocene to Pliocene) fifty per cent of
measurements have each polarity, whilst in the Lower Tertiary forty-five
per cent are normal. There is a marked difference between the Lower and
Middle Devonian, which have roughly equal occurrences of normal and
reversed data, and the Upper Devonian which is largely reversed and marks
the beginning of a predominantly reversed era until the close of the
Palaeozoic. For the mid-Carboniferous to Lower Permian only a few results
have normal polarity, the predominance of reversed measurements being
related to the Kiaman Magnetic Interval. Apart from the Lower Triassic,
where frequent reversals have been observed in several formations (Picard,
1964; Burek, 1964; Helsley, 1969), the Mesozoic has predominantly normal
polarity with seventy-five per cent of results in this category. Of particular
interest is the Upper Triassic with eighty-three per cent of results showing
normal polarity. This predominantly normal period probably contains

TABLE 8   *Polarity observed in 1231 palaeomagnetic investigations for the Phanerozoic. The division of the occurrences of mixed polarity between the normal and reversed is indicated by the numbers with the plus sign in parentheses. N, normal; R, reversed; M, mixed*

Geological period (no.)	Number of occurrences			Overall percentage	
	N	R	M	N	R
Tertiary (312)	97 (+65)	84 (+66)	131 (42%)	52	48
Cretaceous (129)	90 (+12)	14 (+13)	25 (19%)	79	21
Jurassic (70)	38 (+11)	12 (+9)	20 (29%)	70	30
Triassic (140)	77 (+21)	27 (+15)	36 (26%)	70	30
Permian (147)	16 (+9)	110 (+12)	21 (14%)	17	83
Carboniferous (193)	28 (+13)	134 (+18)	31 (16%)	21	79
Devonian (88)	12 (+18)	41 (+17)	35 (40%)	34	66
Silurian (28)	14 (+1)	12 (+1)	2 (7%)	54	46
Ordovician (57)	10 (+10)	26 (+11)	21 (37%)	35	65
Cambrian (67)	8 (+15)	29 (+15)	30 (45%)	34	66
All Phanerozoic (1231)	390 (+175)	489 (+177)	352 (29%)	45.9	54.1
All Phanerozoic exc. Tertiary (919)	293 (+110)	405 (+111)	221 (24%)	43.8	56.2
Precambrian (126)	—	—	30 (25%)	—	—

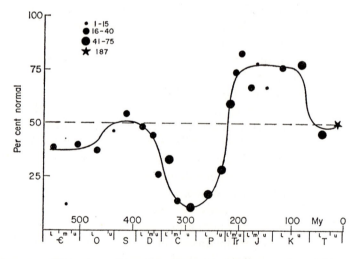

FIGURE 71   Percentage of normal polarity measurements observed in world-wide palaeomagnetic data for the Phanerozoic. The number of measurements for each period is indicated by the size of the point. From McElhinny (1971), *Science*, **172**, 157–9. © 1971 by the American Association for the Advancement of Science.

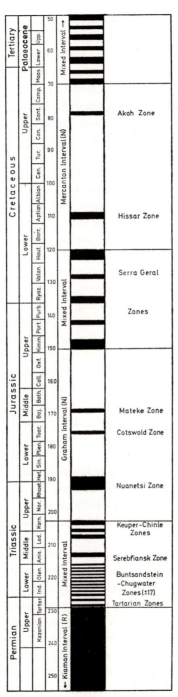

FIGURE 72 Mesozoic palaeomagnetic polarity stratigraphy. Shaded regions reversed, unshaded normal. From McElhinny & Burek (1971), *Nature*, **232**, 98–102.

a single reversed epoch which occurred close to the Triassic–Jurassic boundary, and which represents a horizon that could be of use in geological correlation (Opdyke & McElhinny, 1965; Brock, 1968). The general observation that the Mesozoic was predominantly normal led McElhinny & Burek (1971) to propose a simple polarity time-scale for the whole of the Mesozoic. They identified two long normal intervals, the one ranging from Upper Triassic to Upper Jurassic and the other during the Cretaceous following Helsley & Steiner (1969). The time-scale is shown in fig. 72.

McElhinny & Burek (1971) proposed that it should be possible to divide geological time into a number of *Magnetic Intervals*, whose durations are 10^7 to 10^8 years. These intervals might be periods of time during which either very occasional or no polarity changes took place (such as the Kiaman Magnetic Interval) or they might be long periods of time during which polarity changes occurred quite regularly. Two such intervals shown in fig. 72 have been named the Graham Interval and Mercanton Interval. The shorter periods of 10^6 years or less when the field temporarily changed polarity, have been termed *magnetic zones*. The term was chosen because in practice it will not be possible to distinguish between epochs and events, the term zone is intended to cover both possibilities. These zones are named after the place or rock formation of their discovery or association. Several such zones have been named in fig. 72. The unnamed zones in the mixed intervals are intended to be representative only and not to define precise horizons. Pecherskiy (1970b) has independently drawn up a Mesozoic polarity time-scale from palaeomagnetic measurements in the U.S.S.R. (Creer, 1971). His time-scale is remarkably similar to that proposed by McElhinny & Burek (1971), but he does not recognize the existence of magnetic intervals nor the occurrence of frequent reversals during the Early Triassic. With one exception he has given the various magnetic zones Russian names. Unfortunately his 'Maseru' zone (South Africa) cannot be placed as accurately stratigraphically as one is led to believe. In fact McElhinny & Burek (1971) relied only on well-dated horizons in establishing their magnetic stratigraphy.

McElhinny (1971) observed that the results shown in fig. 71 suggest a periodicity of about 350 My in the polarity of the geomagnetic field during the Phanerozoic. However Ulrych (1972) has determined a maximum entropy power spectrum of long period geomagnetic reversals from the data of fig. 71. This analysis gives two pronounced peaks with periods of 700 ± 100 My and 250 ± 50 My showing that the long period spectrum is actually a combination of two periods. In fig. 73 the percentage of mixed measurements is plotted as a function of time. The variation gives an indication of the frequency of reversals, and the pattern produced is not

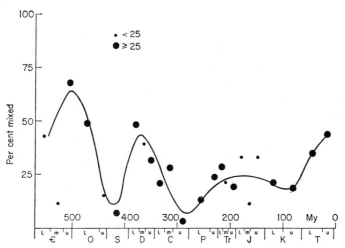

FIGURE 73 Percentage of mixed measurements observed in world-wide palaeomagnetic data, indicating the frequency of reversals during the Phanerozoic. From McElhinny (1971), *Science*, **172**, 157–9. © 1971 by the American Association for the Advancement of Science.

as simple as in the previous case, only one-quarter to one-third of the data being used in this case. Several general conclusions can however be drawn. Reversals were infrequent during the Upper Carboniferous and Permian, that is during the Kiaman Magnetic Interval, as is already well known. Reversals were also infrequent during the Upper Ordovician and Silurian. Frequent reversals seem to have taken place during the Upper Cambrian and Lower Ordovician, the Lower and Middle Devonian, and the Upper Tertiary, the last being specially well established.

It is not possible to define polarity for Precambrian rocks until the apparent polar-wander paths are known with certainty. This is not the case at present (with the possible exception of North America), but the percentage of mixed measurements observed in Precambrian investigations gives an idea of the frequency of reversals throughout the Precambrian compared with the Phanerozoic (table 8). The results are similar, twenty-five per cent of Precambrian investigations have mixed polarity compared with twenty-nine per cent of Phanerozoic results. This would indicate that the frequency of reversals throughout the Precambrian has been similar to that observed during the Phanerozoic.

4.3 REVERSAL MECHANISM

4.3.1 *Records of polarity changes*

Studies of baked contacts (§4.1) and the demonstration that there exist world-wide simultaneous zones of one polarity (§4.2) provide the most compelling evidence that the observations of reversed magnetization in rocks are due to reversals of the earth's magnetic field. The final piece of evidence comes from observations of the polarity changing continuously from one to the other in a rock sequence. For the sake of clarity the term reversal is always taken to indicate either the state of magnetization or polarity of the earth's magnetic field in the sense when they are opposed to the present-day situation. The actual period of change from one polarity to the other is termed a *polarity change* or *polarity transition*. Directions of magnetization observed in rocks during such changes are termed *intermediate* (I) or sometimes *transitional*.

Although the investigations of deep-sea sediments provide a continuous record of the field, it is generally noted that polarity changes occur very abruptly within a few cm at most. The details of the actual polarity transition are often lost because the sedimentation rate is generally too slow. The behaviour of the field during a polarity change in deep-sea sediments has been investigated by Harrison & Somayajulu (1966), whilst Vlassov & Kovalenko (1963) have studied several transitions in a sequence of Lower Devonian sediments from Siberia. The problem that arises is that a sample of sediment measured for palaeomagnetic purposes is necessarily an average over too long a period. A sequence of lava flows provides a series of instantaneous measurements of the field, but the requirement is that there should be a sufficiently rapid rate of extrusion at just the moment a polarity transition is taking place. These are stringent requirements and for this reason the observation of a polarity transition over more than just one or two lava flows is rare. Van Zijl *et al.* (1962 *a*, *b*) observed a polarity transition over ten successive flows in the Triassic–Jurassic Stormberg lavas of South Africa (fig. 74). This study is particularly impressive in that not only was the same transition zone observed in two sequences separated horizontally by 130 km, but also agreement between the magnetization of the lavas and the underlying baked sandstones was observed, both for the basal lavas and for one of the lavas in the transition zone. Normally magnetized dolerite dykes intruding reversed lavas also produced a positive baked contact test. Watkins (1965, 1969) has observed a polarity transition which occurred in the upper Miocene over twenty-nine successive lava flows in the Steens Mountain, Oregon. Goldstein *et al.* (1969) have also studied this transition in two sequences separated horizontally by about 700 m and a third sequence

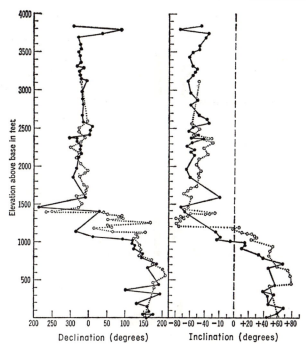

FIGURE 74   Polarity change over successive lava flows observed in the Stormberg lavas of South Africa. Variation of declination and inclination with height is given for two sections about 130 km apart. Dots from the Sani Pass section and open circles from the Maseru area. Redrawn from Van Zijl *et al.* (1962*a*, *b*). From Irving, *Paleomagnetism and its application to geological and geophysical problems.* © John Wiley & Sons, Inc. 1964.

90 km away. Wellman *et al.* (1969) report a polarity transition covering nine lava flows from a lower Oligocene volcanic sequence in southeast Australia.

   Attempts to estimate the time involved in a polarity transition by K–Ar dating techniques on lava sequences inevitably fail since it is never known what is the rate of extrusion of the flows and a continuous record does not exist. Ninkovich *et al.* (1966) have made estimates from three North Pacific deep-sea cores. The log of core v–20–108 given in fig. 67*b* is given in more detail in fig. 75 for the region near the Brunhes–Matuyama boundary and at both boundaries of the Jaramillo event. The rate of sedimentation is about 1 cm per 1000 years and the average intensity of magnetization, although variable, is between 3 and 5×10^{-6} emu g^{-1}. Sampling of the core is approximately every 1 cm and it is noted that in the vicinity of the polarity changes, the inclination changes abruptly between adjacent samples. This suggests that the polarity change is completed in a time of the order of 1000 years. The intensity of magnetization does not, however,

135

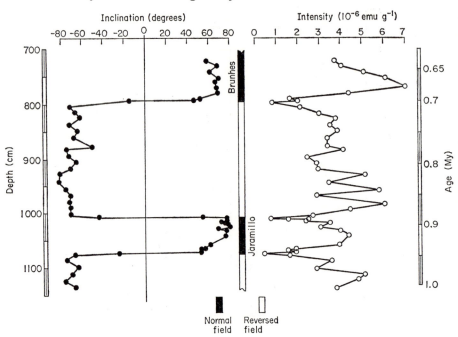

FIGURE 75 Variation of intensity of magnetization and inclination with depth of a portion of core v–20–108 from the North Pacific. From Ninkovich *et al.* (1966), *Earth Planet. Sci. Letters*, **1**, 476–92.

change so abruptly but progressively decreases to zero. The period of time represented can be estimated as 1000 to 10000 years. During this time the inclination remains practically unchanged. The samples closest to the point of transition have the lowest intensity. After the transition the intensity increases progressively for 1000 to 10000 years or more. Thus although the polarity change itself took only 1000 years, it appears that the time during which the dipole field of the earth was reduced was approximately 20000 years.

An estimate of the time required for a polarity transition can also be made from the proportion of observed intermediate directions of magnetization. Cox & Dalrymple (1967) have used the data upon which the geomagnetic polarity time-scale is based to make such an estimate. Assume that over some time interval θ the field has changed polarity M times and that the time required for the ith change is $\Delta\theta_i$, so that the total time $\Delta\theta_T$ during which the field was transitional is given by:

$$\Delta\theta_T = \sum_{i=1}^{M} \Delta\theta_i. \qquad (4.1)$$

If N age determinations have been made in the interval θ and if they are randomly distributed with respect to the transition, the probability that exactly R of the age determinations will occur in transitional intervals is given by the binomial distribution. The best estimate of $\Delta\theta_T$ is then given by:

$$\Delta\theta_T = \frac{\theta R}{N}. \qquad (4.2)$$

Cox & Dalrymple (1967) used the following two criteria to identify those data points that are from transitional intervals.

(*a*) The average direction of magnetization of the volcanic unit differs from the direction of the dipole field at the sampling locality by at least 2δ, where δ is the angular standard deviation of geomagnetic field directions produced by long-period geomagnetic secular variation.

(*b*) The radiometric age of the volcanic unit differs by no more than 2σ from the age of the nearest epoch or event, where σ is the standard error of the age determination.

Cox & Dalrymple (1967) analysed eighty-eight data points for the past 3.6 My only one of which could be classified as transitional. This gave $\Delta\theta_T = 41\,000$ years and assuming nine transitions had occurred, this gave an average time of 4600 years for each. This estimate needs revision in the light of the new time-scale of Cox (1969) (see fig. 63). There are now $N = 150$ data points for the past 4.5 My with $R = 2$ points satisfying the criteria as being transitional. These two are unit 4D057, a rhyolite dome in New Mexico of age 0.88 My (Doell & Dalrymple, 1966), and a lava flow on St Paul Island, Alaska, of age less than 10^5 years, which, although originally rejected by Cox & Dalrymple (1967) as not conforming with criterion (*b*), now seems to be related to the Laschamps event. Putting $\theta = 4.5$ My, this gives the best estimate of $\Delta\theta_T$ as 60000 years. Fig. 63 gives twenty-five polarity changes over the past 4.5 My, and if the additional short events observed in deep-sea cores, as suggested in §4.2.2, are also taken into account, there are at least thirty polarity changes during this time. Thus the best estimate of the average time for each transition is about 2000 years. Considering the uncertainties involved, this figure agrees very well with that of 1000 years deduced from deep-sea sediment cores. Such rapid polarity changes are consistent with the time-constants of up to about 10^3 years for magnetohydrodynamic processes in the core (Hide & Roberts, 1961).

Ninkovich *et al.* (1966) observed that the minimum intensity of samples in the transition zone was about one-fifth of the average magnetic intensity in either the normal or reversed zones. McElhinny (1970*b*) also observed a decrease of intensity to one-quarter or one-fifth of the usual value in two transitions observed in some Lower Cambrian sediments. Estimates of the palaeointensity of the field during polarity transitions from lava flows give similar, but more precise figures. Van Zijl *et al.* (1962*b*) showed that the field was decreased by a factor of four to five during the transition. A similar conclusion was reached by Momose (1963). Goldstein *et al.* (1969) found that the palaeointensities in the Steens Mountain polarity transition ranged from 0.5 oersted outside the transition to a minimum of 0.025 oersted inside the transition. However, Prevot & Watkins (1969) in studying the same transition concluded that it was not necessarily accompanied by a regular decrease in intensity of the associated geomagnetic field, but may have included rapid changes of intensity. An examination of the data of Goldstein *et al.* (1969) shows that the *average* palaeointensity within the two transition zones studied was about twenty per cent of the *average* palaeointensity outside the zones.

Recently Lawley (1970) has made a considerable number of palaeointensity determinations from at least six transition zones observed in a 900 lava sequence in Eastern Iceland. The values obtained from seventeen transitional flows range from 0.022 to 0.127 oersted compared with an average value of 0.34 oersted obtained previously by Smith (1967) for the normal and reversed lavas. An interesting aspect of Lawley's investigation was a study of the palaeointensities of flows whose directions of magnetization differed markedly from the axial geocentric dipole direction for Iceland, but which do not lie in a transition from one polarity to the other. These departures, defined as having divergence angles of between 40° and 140° from the dipole field direction, are apparently followed by a return to the pre-existing polarity. The term *systematic deviation* is used to describe these flows. Palaeointensities determined from ten such flows ranged in value from 0.014 to 0.122 oersted, similar to those observed for the transition flows.

An examination of all transitional palaeointensity data as a function of the divergence from the axial geocentric dipole field direction (Lawley, 1970) for the Steens Mountain, Japanese and Eastern Iceland lavas showed a minimum of average value 0.04 oersted when the divergence was 90°. The analyses of the intensities from sediments and palaeointensities from lava flows both suggest, therefore, that the field is reduced to a very low value during a polarity transition, decreasing by a factor of five or more.

4.3.2   *Theoretical aspects*

An important conclusion may be drawn from the observation that the geomagnetic field intensity decreases during a polarity transition. There are two main processes by which the transition takes place. Either the main dipole field rotates through 180°, with or without change in moment, or the dipole field reduces to zero followed by an increase in the opposite sense. A further question arises as to what happens to the non-dipole field intensity during a polarity change of the main dipole. Any given dipole produces its minimum field intensity in the equatorial plane. It follows therefore that if the dipole rotates through 180° with no change in moment during a polarity change, the transition zone field intensity cannot be less than the equatorial field, that is one-half of the axial field. The data therefore indicate that simple rotation of a constant strength dipole has not occurred.

The observation that the palaeointensities in systematic deviation flows are similar to those in transition zones, suggests that the dipole field is capable of decrease to a very low value or even to zero, followed by growth in the original direction. This type of behaviour might be expected from dynamo theory, for there appears to be no reason why the decay of the main dipole field should necessarily be followed by its growth in the opposite direction.

Irving & Ward (1964) have produced a statistical model of the geomagnetic field in which the geocentric dipole is axial, producing an equatorial field H_0, and in which the non-dipole fluctuations are simulated by a component h which has constant magnitude but whose direction is randomly distributed in time. This has been referred to as model A of the geomagnetic field in §1.2.5, and the recent analysis of palaeomagnetic results in terms of this model (Brock, 1971) (see fig. 19) shows that it is a remarkably good approximation to the average palaeomagnetic field. For an assumed probability distribution of angles between h and H (where H is the field intensity produced by the geocentric axial dipole at latitude λ) Irving & Ward (1964) showed that

$$f^2 \approx 3k^{-1}, \tag{4.3}$$

where
$$f = hH^{-1}, \tag{4.4}$$

and k is the Fisher (1953) precision parameter estimate. Defining $f_0 = hH_0^{-1}$, Irving & Ward showed that

$$f_0 = f(1 + 3\sin^2\lambda)^{\frac{1}{2}}. \tag{4.5}$$

This model was tested by comparison with the 1945 geomagnetic field (by analysing the dispersion of directions around lines of geographic latitude) and by comparison with palaeomagnetic directional data for the Phanerozoic (using observed values of k). Irving & Ward (1964) obtained the following results:

analysis of 1945 field $\qquad\qquad f_0 = 0.40 \pm 0.01,$
average of palaeomagnetic values $\quad f_0 = 0.38 \pm 0.13,$

where the errors are the standard deviations.

If the polarity changes in the earth's magnetic field have occurred by the decrease of the dipole moment to zero followed by build-up in the opposite sense, then the field intensities observed during polarity transitions will relate to the non-dipole field only. The ratio of this residual field to that observed before and after the polarity change is then a measure of f from (4.4). The value of f_0 is found by relating it to the equatorial field (H_0) outside the transition zone. Smith (1967) has evaluated the palaeo-intensity data of Van Zijl *et al.* (1962 b) and Momose (1963) in this way. In the Stormberg lavas, the mean equatorial field intensity for the fully normal and reversed fields was 0.13 ± 0.07 oersted. The mean transition zone field intensity was 0.05 ± 0.03 oersted. For the Japanese lavas of Momose (1963) the mean equatorial normal and reversed field was 0.22 ± 0.10 oersted, and the mean transition zone field was 0.09 ± 0.02 oersted. From these figures the values of f_0 are then:

$$\text{Stormberg lavas} \quad f_0 = 0.39 \pm 0.35,$$
$$\text{Japanese lavas} \quad f_0 = 0.41 \pm 0.19,$$

where the errors are the standard deviations. The agreement between these two estimates of f_0 with those from the 1945 field and palaeomagnetic data suggests that the dipole field reduces to zero during a transition but that the non-dipole field remains. Watkins (1969) has analysed the behaviour of the non-dipole field in some detail during the Steens Mountain polarity transition. Application of Irving & Ward's model A to the residual field at various stages suggests that the maximum dipole intensity was as low as two per cent of the present value for a period of not less than 100 years during the transition, outside of which the data are consistent with some present-day ratios of non-dipole to dipole field intensity.

Creer & Ispir (1970) have proposed an entirely different interpretation of the behaviour of the geomagnetic field during polarity transitions. They argue that if the non-dipole part of the field retained its present strength, it could be expected that the geomagnetic field direction would vary in an erratic manner during polarity transitions. If the succession of field changes are represented by the path traced out by the virtual geomagnetic poles computed from palaeomagnetic measurements made on successive lava flows extruded during a polarity transition, then the following should be observed:

(*a*) The path traced out by the virtual geomagnetic poles computed for sequences of lava flows collected at different places on the earth's surface but for the same polarity transition should not be the same.

(*b*) The path of virtual geomagnetic poles computed for sequences of lava flows representing different polarity transitions but from the same geographical area should not repeat itself.

From an analysis of seven polarity changes ranging in age from 0.02 to 20 My, Creer & Ispir (1970) argue that the paths traced out by the virtual geomagnetic poles are essentially similar. It is thus not easy to interpret this behaviour in terms of the decay of the main dipole field through zero intensity, leaving a dominantly non-dipole field during the polarity transitions. Instead, the three dipole model of the geomagnetic field of Bochev (1969) is used (see §1.1.4). In their interpretation Creer & Ispir propose that when the stronger axial dipole 1 (moment 7.05×10^{25} gauss cm^3) changes its polarity, the main field follows suit. Dipole 3 (moment 1.94×10^{25} gauss cm^3) is approximately parallel to dipole 1, whereas dipole 2 (moment 1.05×10^{25} gauss cm^3) is aligned obliquely to the earth's axis. Thus when dipole 1 undergoes a polarity change, dipole 2 might retain its strength and polarity. This would account for the similarity in paths observed for the virtual geomagnetic poles, suggesting that the field is still dipolar during polarity transitions. The possibility that dipole 2 might change polarity independently of dipole 1 and therefore at different times would explain why the usual path traced out is either the sequence of virtual north poles or the sequence of virtual south poles.

Cox (1968) has produced a probabilistic model for reversals of the earth's magnetic field in which it is assumed that polarity changes occur as the result of an interaction between steady oscillations and random processes. The steady oscillator is the dipole component of the field and the random variations are the components of the non-dipole field. Variations in the dipole moment of the geomagnetic field from archaeomagnetic palaeointensity measurements for the past 8000 years suggest a periodicity of the order of 10^4 years and amplitude about fifty per cent of the present value (Cox, 1968; see §1.2.2, fig. 15). Palaeointensity data from Tertiary and Quaternary rocks suggest a mean intensity of the dipole moment of about 6×10^{25} gauss cm^3, which is slightly smaller than the present value. The standard deviation of dipole intensity fluctuations is about fifty per cent (Smith, 1967). If the fluctuations are sinusoidal, the corresponding peak amplitude is about 4×10^{25} gauss cm^3. The dipole and non-dipole fields can be represented by a central dipole with an array of n smaller eccentric dipoles m_i at radius r (Lowes, 1955; Alldredge & Hurwitz, 1964; see §1.1.4, fig. 12). The dipole term found from the spherical harmonic analysis of this system, corresponds to a geocentric dipole of moment $M + M'$, where M' is the vector sum of the n dipoles m_i. On Cox's model the condition necessary for a polarity change is derived from the fundamental property of the magnetohydrodynamic dynamo that, given an initial weak axial poloidal field of either normal or reversed polarity, Coriolis forces act to reinforce the field, maintaining the initial polarity.

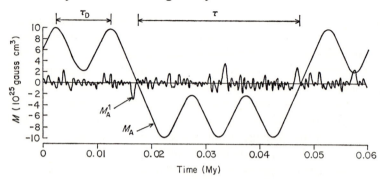

FIGURE 76 Probabilistic model for reversals. τ_D is the period of the dipole field and τ the length of a polarity interval. A polarity change occurs whenever the quantity $(M_A + M_A')$ changes sign, where M_A is the axial moment of the dipole field and M_A' is a measure of the non-dipole field (see §1.1.4, fig. 12). From Cox (1968), *J. Geophys. Res.* **73**, 3252.

Using the two-disk dynamo model of the field (Rikitake, 1958; see §1.1.4 fig. 9), a change in the poloidal component of the field is thus analogous to the change in current direction that characterizes a polarity change in this model. Only the dipole components of the poloidal field are considered for the sake of simplicity. The condition necessary for a polarity change is then that the quantity $M_A + M_A'$ change sign, where M_A and M_A' are the axial components of M and M' respectively (see §1.1.4, fig. 12). Cox's model is illustrated in fig. 76. An analysis of the model suggests that the probability that a polarity change will occur during one cycle of the dipole moment is 0.05. This means that a polarity change is expected on average every 20 cycles. If the period of the dipole moment is 10^4 years, then the average length of a polarity interval is $\tau = 0.2$ My. This is close to the observed value of 0.18 My over the past 4.5 My, but less than the average value of 0.23 My which Cox (1969) estimates for the past 10.6 My from sea-floor spreading (see chapter 5). Cox (1968) has thus suggested that additional short term polarity intervals with durations $0 < \tau \leqslant 0.05$ My, remain to be discovered in the time interval from 10 My ago to the present.

4.4 REVERSALS AND EVOLUTION

Uffen (1963) has proposed that if the earth's magnetic field was reduced to a very low value during a polarity change, the solar wind and a large proportion of the cosmic rays would be able to reach the surface of the earth. Furthermore the particles trapped in the radiation belts would also be dumped on the earth. Uffen supposed that the biological effects of the increase in radiation might affect the course of evolution and bring about

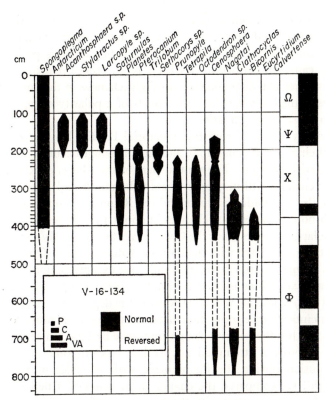

FIGURE 77   Extinction and appearance of radiolaria near the time of the Brunhes–
Matuyama boundary, as observed in a deep-sea sediment core from the Antarctic.
The width of the vertical bars indicates the abundance of species on a scale:
P, present; C, common; A, abundant; VA, very abundant. The faunal zones of
Hays (1965) and the polarity time-scale (black, normal; white, reversed) are
given on the right. From Opdyke *et al.* (1966), *Science*, **154**, 349–57. © 1966 by the
American Association for the Advancement of Science.

the extinction and transformation of species at the time of a polarity change.
The occasional sudden extinctions of large parts of the earth's fauna is one
of the great mysteries of biological evolution, resulting in a host of un-
proved explanations (Newell, 1963).

Several independent investigations have shown a close correlation between the extinctions of planktonic organisms and the times of polarity
changes over the past few million years (Harrison & Funnell, 1964; Opdyke
et al., 1966; Hays & Opdyke, 1967; Watkins & Goodell, 1967; Hays *et al.*,
1969). Fig. 77 shows a result for an Antarctic core. Six species of radiolaria
are found to become extinct and three species to make their first appearance
within 30 cm of the time of the magnetic transition from the Matuyama

to the Brunhes epoch. Since only a few organisms become extinct near a polarity change, and some have survived many such changes before becoming extinct, it is clear that a polarity change alone is probably insufficient to cause an extinction. As Hays *et al.* (1969) point out, highly specialized species may be vulnerable to slight changes in the environment and for these, small direct or indirect effects of a polarity change may be sufficient to cause their extinction.

In their study of cores from the equatorial Pacific, Hays *et al.* (1969) noted that their zones of major floral and faunal change occurred at times of frequent polarity changes. The central part of the Gauss epoch contains at least four changes within 300 000 years (Mammoth and Kaena events), whilst the Gilsa and Olduvai events and the Réunion events in the Matuyama epoch both produce four changes in about 150 000 years. They note that most polarity changes since the Cochiti event nearly 4 My ago that now have extinctions associated with them in the Antarctic cores (Opdyke *et al.*, 1966; Hays & Opdyke, 1967) are, with only one exception, not the same as those that have extinctions associated with them in the equatorial cores. Thus Hays *et al.* (1969) conclude that it is probably the degree of specialization of individual members of the population that make some species prone to extinction in any case, but the polarity change may have hastened their extinction.

Simpson (1966) has attempted to correlate accelerations in the rate of organic evolution during the Phanerozoic with reversals of the earth's magnetic field. He compared the change in percentage of new species as a function of existing species with the occurrence of reversely magnetized rocks. This is not a particularly useful comparison to make because the hypothesis being considered relates to polarity changes. The occurrence of reversely magnetized rocks during a particular period does not necessarily mean that any polarity changes occurred (e.g. the Kiaman Magnetic Interval). The frequency of reversals is a much more useful parameter, for obviously the effect will be enhanced when reversals occur frequently, as indeed Hays *et al.* (1969) have suggested. An attempt at delineating the variation in the reversal frequency during the Phanerozoic was made in §4.2.4 and the results were given in fig. 73. A comparison of this estimate of the variation in reversal frequency with Newell's (1963) estimates of the rate of organic evolution at various times is given in fig. 78. Two comparisons are made, the one with the number of extinctions and the other with the number of new families, in each case expressed as a percentage of the number of existing families. The maxima and minima in reversal frequency must *always* coincide with the maxima and minima respectively in either of the evolutionary curves for there to be a valid correlation. Accelerations in evolution seem to occur at times when the reversal rate has either

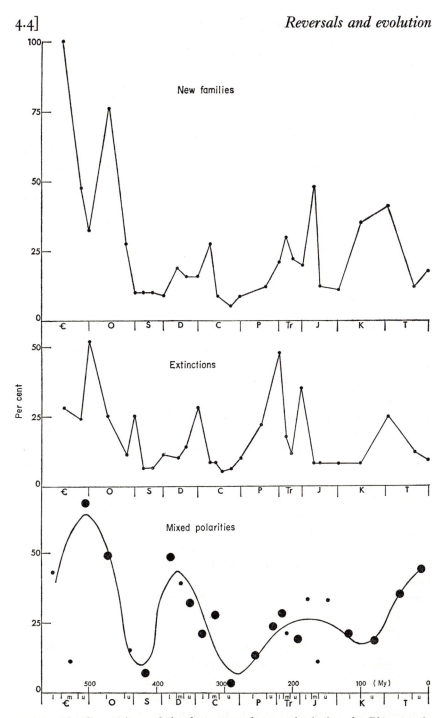

FIGURE 78   Comparison of the frequency of reversals during the Phanerozoic (from fig. 73) with the number of new families and extinctions, both expressed as a percentage of the number of existing families and as compiled by Newell (1963).

maximum or minimum values, so that there appears to be no significant correlation.

Quantitative estimates of the effects on the earth and the atmosphere at a polarity change have been given by Black (1967), Waddington (1967) and Harrison (1968). The surface of the earth is shielded from cosmic rays by the geomagnetic field and the atmosphere. Along the field axis (i.e. at the poles) all particles can reach the atmosphere. At the equator only particles greater than 15 GeV can reach the atmosphere. If the present geomagnetic field were reduced to zero during a polarity transition the increased radiation dose would amount to ten per cent at the equator and zero at the poles. If the decrease took place over 1000 years, this amounts to an increase in dose of 0.01 per cent per year. The rate of change of dose with sunspot cycle is 0.4 per cent per year, so the possible effects are negligible.

The protons that are trapped in the Van Allen Radiation Belts are slowly lost by atmospheric absorption and hydromagnetic wave scattering. The source of the trapped protons is the solar wind and in the steady state the lifetime of protons is probably less than ten years (Hess, 1964). If the source were cut off, this implies that the steady state flux into the atmosphere could not be sustained by the trapped protons for more than ten years. Thus at a polarity change, the equivalent of ten years steady state flux would be dumped into the atmosphere over a period of 1000 years say, as the geomagnetic field decays. This amounts to one per cent of the steady state flux being dumped per annum, a not very significant amount. The solar wind is excluded from the magnetosphere at the magnetopause, about ten earth radii distance. The atmosphere will be exposed to the solar wind at a polarity change if the reduction in the earth's field allows the magnetopause to come down to the earth's surface. Black (1967) has shown that this requires a reduction in field intensity to at least 10^{-3} of its normal value, a condition that is unlikely to be realized. If, however, the solar wind is not excluded from the atmosphere, the solar wind protons will cause ionization in the high atmosphere, the electrons being of too low energy to ionize air molecules. All would be absorbed in the high levels of the atmosphere. Thus it seems that only the high energy cosmic rays can reach the surface, most of the flux being secondary particles generated in the atmosphere.

Cosmic rays create radioactive isotopes in the atmosphere which lead to three radiation hazards. Radioisotopes contribute to the background radiation, ingested radioisotopes give an internal radiation dose, and radioisotopes incorporated in the genes may destroy the gene from the violence of their decay. At a polarity change, the production rate, concentration and

TABLE 9   *Radiation dose rate increase at zero geomagnetic field*
(*Black*, 1967)

(*a*) 2 mm below sea surface: Total cosmic ray dose + total background β-ray dose + ½ background γ-ray dose

Place	Direct cosmic ray (mrad y^{-1})	Back-ground (mrad y^{-1})	Internal (mrad y^{-1})	Total dose (mrad y^{-1})
Poles ⎱ present	25	3.2	1.5	29.7
Equator ⎰	22.5	3.2	1.5	27.2
World wide at zero field	25	3.2	2.3	30.5

Ratio of doses: $\dfrac{\text{at zero field}}{\text{at poles (present)}} = 1.03$; $\dfrac{\text{at zero field}}{\text{at equator (present)}} = 1.12.$

(*b*) 1 m below sea surface: Eighty-six per cent cosmic ray dose + total background β-ray dose + total background γ-ray dose

Place	Direct cosmic ray (mrad y^{-1})	Back-ground (mrad y^{-1})	Internal (mrad y^{-1})	Total dose (mrad y^{-1})
Poles ⎱ present	21.5	3.6	1.5	26.6
Equator ⎰	19.4	3.6	1.5	24.5
World wide at zero field	21.5	3.6	2.3	27.4

Ratio of doses: $\dfrac{\text{at zero field}}{\text{at poles (present)}} = 1.03$; $\dfrac{\text{at zero field}}{\text{at equator (present)}} = 1.12.$

the hazard of these isotopes will increase. Of particular interest is the background dose from radioisotopes in sea water in relation to planktonic organisms in deep-sea cores (Black, 1967). A sea organism will receive its maximum radiation dose from radioisotopes in solution in sea water when it is completely surrounded by a thickness of water greater than the range of α, β and γ rays emitted by the radioisotopes. This occurs at about 1 m depth whilst near the surface the γ-ray dose approaches about half its maximum value due to their very much longer range. Table 9 summarizes all these effects for a depth of 2 mm below the sea surface and 1 m below after Black (1967). It seems impossible that the total radiation dose within an organism at sea level can be increased by more than twelve per cent when the earth's magnetic field is reduced to zero.

Finally, it is possible that the solar wind might produce ozone in the high atmosphere which would absorb radiation and produce large changes in climate. Black (1967) shows that the ozone produced in this way is a very small fraction of that produced by ultraviolet light. Bullard (1968) points out that the only possibility of a causal connection between magnetic reversals and the extinction of species is that ionization in the high atmosphere has an unexpected effect on climate. In spite of our lack of knowledge of many aspects of the general circulation in the atmosphere, he feels that this is probably unlikely.

5 Sea-floor spreading and plate tectonics

5.1 SEA-FLOOR SPREADING AND PLATES

5.1.1 *The basic hypothesis*

The sea-floor spreading hypothesis was formulated by Hess (1960, 1962) and Dietz (1961, 1962). Hess attempted to provide a generalized history of the ocean basins which satisfied all the results then available from studies in marine geology and geophysics. Although at the time the hypothesis appeared to be a gross simplification, it is remarkable that during the past decade a number of additional concepts have been formulated within its framework and these have been subsequently substantiated by new data. In fact the more information that is produced about the ocean basins and their margins, the more the data seem to fit the simple unifying concept. Whilst this basic concept has been reformulated in terms of the new evidence which has become available (Vine & Hess, 1970), it is of interest first to consider the original basic hypothesis following Hess (1960, 1962).

The mid-ocean ridges are the largest topographic features on the surface of the earth. They are characterized by unusually high heat flow (Bullard *et al.*, 1956) along their crests. Heezen (1960) demonstrated that a median graben exists along the crests of the Atlantic, Arctic and Indian ocean ridges and that shallow depth earthquake foci are concentrated under the graben. He therefore postulated extension of the crust at right angles to the trend of the ridges. Hess (1960, 1962) noted that seismic profiles at sea show the striking uniformity in thickness of layer 3 of the oceanic crust. He considered that the only likely manner in which a layer of uniform thickness could be formed would be if its base represented a present or past isotherm, at which temperature and pressure some reaction occurred. Two possibilities are (*a*) the basalt to eclogite transformation (Sumner, 1954; Lovering, 1958; Kennedy, 1959), and (*b*) the hydration of olivine to serpentine at about 500 °C (Hess, 1954). Because of the common occurrence of peridotitic inclusions in oceanic basalts and the absence of eclogite inclusions, Hess accepted the hydration of olivine to serpentine, a view which was supported by the dredging of serpentinized peridotites from fault scarps in the oceans, where the displacement on the faults may have been sufficient to expose layer 3. The seismic velocity of layer 3, although variable, averages near 6.7 km s^{-1}, which would represent peridotite

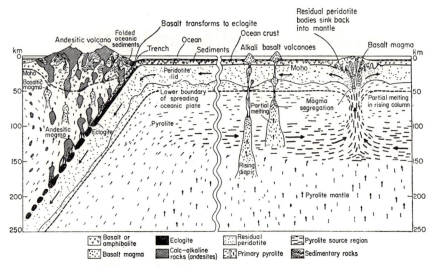

FIGURE 79   Modified version of the sea-floor spreading hypothesis of Hess (1960, 1962) from Ringwood (1969), *The earth's crust and upper mantle*, Geophysical Monograph no. 13, p. 12.

seventy per cent serpentinized. A significant feature of the seismic velocities of layer 3 is their reduction below the ridge crest.

On Hess' model the mid-ocean ridges are interpreted as representing the rising limbs of mantle convection cells. The topographic elevation is related to thermal expansion, and the lower seismic velocities both to higher than normal temperatures and microfracturing. Convective flow comes right through to the surface, and the oceanic crust is formed by hydration of mantle starting at a level of 5 km below the sea floor. The production of layer 3 by a convective system and serpentinization must be reversed over the downward limbs of the convection cells. As layer 3 is depressed into the downward limb it will deserpentinize at 500 °C and release its water upward. Thus the whole ocean is virtually swept clean (replaced by new mantle material) every few hundred million years. This accounts for the relatively thin veneer of sediments on the ocean floor, the relatively small number of volcanic seamounts and the absence of evidence of rocks much older than Cretaceous in the oceans. On the model, the water to produce the serpentine of the oceanic crust comes from the mantle and is released on deserpentinization at a rate consistent with a gradual evolution of ocean water throughout the earth's history.

The alternative model (fig. 79), making use of the basalt–eclogite transformation, has been proposed by Ringwood & Green (1966) and further developed by Ringwood (1969). A density difference develops in the mantle beneath mid-ocean ridges and pyrolite (a model mantle material) flows from the low velocity zone (the immediate source region) towards the ridge axis and rises upwards. At this stage fractional melting of pyrolite occurs, leading to the generation of basaltic magmas with residual unmelted peridotite, following Green & Ringwood (1967). At the axis of the ridge, the subsurface temperatures are high enough to maintain the stability of the basaltic mineral assemblage. The mid-ocean ridges thus develop as expanding features composed of heterogeneous mixtures of gabbro, dolerite, peridotite and pyrolite with surficial basalts. During the horizontal expansion, differentiation occurs so that an almost uniform mafic crust is formed overlying a depleted ultramafic mantle. The basaltic upper layers cool quickly and may become partly oxidized by interaction with sea water. On the other hand the deeper layers are protected from the sea water and remain relatively dry and unoxidized.

The crust and underlying plate of mantle cool as they move away from the ridge, sliding over the weak low-velocity zone. At the same time the dry basalt passes into the eclogite stability field. The transformation does not take place immediately, but when the crust approaches a continental margin the sedimentation on the oceanic crust causes an increase in pressure and temperature. This increase allows the lower dry mafic crust to transform to eclogite, whilst the upper wet regions transform to amphibolite. The higher density of the dry eclogite compared with that of mantle peridotite results in crustal subsidence. Ultimately all the basalt in the descending column must transform to dense eclogite, as will the amphibolite but at somewhat greater depths. Once the process starts it becomes self-sustaining the driving force being supplied by the high density of the cool descending column.

5.1.2   *Transform faults*

The conventional discussions of fault mechanisms (e.g. Anderson, 1951) assume that the faulted medium is continuous and conserved. If, however, new oceanic crust is created at mid-ocean ridges and subsequently destroyed in trenches or mountains, as proposed on the sea-floor spreading hypothesis, then this basic assumption is no longer true. The classical geological view has been that movements of the crust are concentrated in mobile belts, taking the form of mountains, mid-ocean ridges or major faults with large horizontal movements. A puzzling aspect of these features and the seismic activity along them is that they often appear to end abruptly. Wilson (1965 a, b) proposed that these features are not isolated but are connected into a continuous network of mobile belts around the earth. Any feature at its apparent termination may be transformed into another feature of one

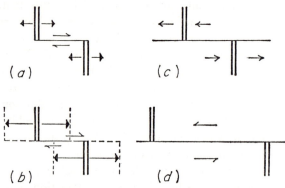

FIGURE 80   (*a*) Dextral ridge–ridge type transform fault connecting two expanding ridges. (*b*) The same fault after a period of motion. The motion has not changed the apparent offset. (*c*) Sinistral transcurrent fault offsetting a ridge, with offset in the same sense, but motion in the opposite sense to the transform fault in (*a*). (*d*) The same fault after a period of motion. Note that the offset has increased. Open-headed arrows indicate components of shearing motion. Solid-headed arrows indicate sea-floor spreading from the ridge axis. From Wilson (1965*b*). © 1965 by the American Association for the Advancement of Science.

of the other two types. Thus faults in which the displacement suddenly stops or changes form and direction are not true transcurrent faults. Wilson (1965*a*) therefore proposed that in addition to normal, thrust and transcurrent faults, a new class of strike-slip faults exist called transform faults.

On transcurrent faults, shear motion continues indefinitely, but in transform faults it ends abruptly by transformation into extension across a ridge or compression across a mountain or island arc. Thus a fault which terminates abruptly with mid-ocean ridges at both ends is called a ridge–ridge type of transform fault. A comparison between the behaviour of transform and transcurrent faults is given in fig. 80. In fig. 80*a* a dextral ridge–ridge type transform fault connects two expanding ridges. The motion does not change the apparent offset (fig. 80*b*). Classically the left-lateral offset of the ridge crest shown in fig. 80*a* would be interpreted as resulting from left-lateral movement along a transcurrent fault as shown in fig. 80*c*, and is thus in the opposite sense to the transform fault in fig. 80*a*. In this case the motion increases the offset as shown in fig. 80*d*.

Confirmation of the existence of transform faults of the ridge–ridge type has been made from studies of the locations and mechanisms of earthquakes on the mid-ocean ridges (Sykes, 1967). Use of data from the World-Wide Seismograph Network of the U.S. Coast and Geodetic Survey and from other long period seismographs enabled mechanism solutions of high precision to be obtained from seventeen earthquakes located on the mid-

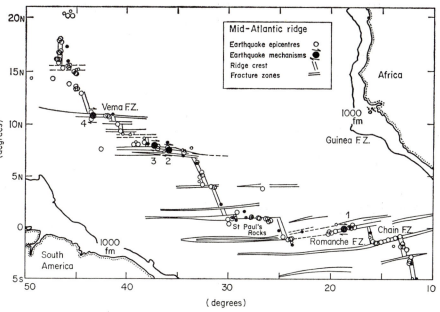

FIGURE 81   Epicentres of earthquakes (1955–65) and mechanism solutions for four earthquakes along the equatorial portion of the mid-Atlantic ridge. Sense of shear displacement and strike of inferred fault plane are indicated by the orientation of the set of arrows beside each mechanism. Large circles denote more precise epicentral determinations; smaller circles, poorer determinations. From Sykes (1967), *J. Geophys. Res.* **72**, 2137.

ocean ridges. The distinction between transcurrent and transform faulting can be clearly made. For transform faulting the location of earthquakes on the fracture zones should be confined to the region between ridge crests, whereas for transcurrent faulting it should extend beyond the crests. The strike–slip motion for the two types is in the opposite sense and clearly distinguishable as shown in fig. 80 *a* and *c*. Ten of the earthquakes that were studied occurred on fracture zones that intersect the crest of the mid-ocean ridge. The mechanism of each of these shocks was characterized by a predominance of strike–slip motion on a steeply dipping plane. The sense of the strike–slip motion in each of the ten solutions was in agreement with that predicted for transform faults, as was the spatial distribution of a large number of earthquakes across various fracture zones. The results obtained along the equatorial portion of the mid-Atlantic ridge are illustrated in fig. 81. They convincingly confirm the existence and predominance of transform faults.

Transform faults may be described in terms of the features which they connect. The six possible types of dextral transform faults are illustrated

153

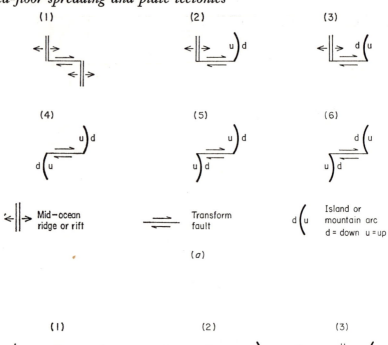

FIGURE 82   (*a*) The six possible types of dextral transform faults. (*b*) The same after a period of growth. Traces of former positions now inactive, but still expressed in the topography, are shown by dashed lines. (1) Ridge to ridge type; (2) Ridge to concave arc; (3) Ridge to convex arc; (4) Concave arc to concave arc; (5) Concave arc to convex arc; (6) Convex arc to convex arc. From Wilson (1965*a*), *Nature*, **207**, 343–7.

in fig. 82*a*. Another six sinistral types can also exist. In fig. 82 mountain or island arcs are described as being convex or concave depending on which face is first reached when proceeding in the direction indicated by an arrow depicting relative motion. The distinctions between the types can be made from the manner of their growth shown in fig. 82*b*. Ridges expand

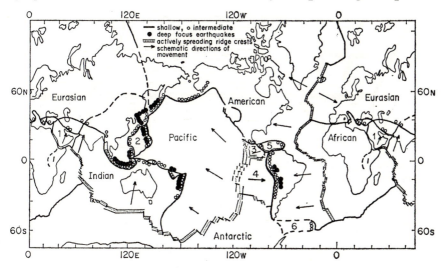

FIGURE 83 Summary of the seismicity of the earth after Gutenberg & Richter (1954) and Barazangi & Dorman (1969). The six major aseismic crustal plates of Le Pichon (1968) are named. Some minor plates are numbered: (1) Arabian; (2) Phillipine; (3) Cocos; (4) Nasca; (5) Caribbean; (6) Scotia. Spreading rates at ridge crests are indicated schematically and vary from 1 cm y^{-1} in the vicinity of Iceland to 6 cm y^{-1} in the equatorial Pacific (see table 10). From Vine (1970), *Nature*, **227**, 1013–17.

to produce new crust, leaving residual inactive traces in the topography of their former positions, whilst oceanic crust moves down under island arcs absorbing old crust and leaving no traces of past positions. The convex side of arcs thus advance.

5.1.3   *The concept of plates*

Earthquake activity throughout the world is largely confined to the young fold mountains and trench systems of the Alpine–Himalayan and circum-Pacific belts and to the crests of the mid-ocean ridges (Gutenberg & Richter, 1954). By far the greatest number of earthquakes occur in the circum-Pacific belt in association with the trench systems. They are less frequent in the Alpine–Himalayan belt and less frequent again in the mid-ocean ridges. Thus most of the current seismicity of the earth occurs in very restricted linear zones.

The activity associated with ridge crests and strike–slip faults is confined to very shallow depths, probably not exceeding 10 or 20 km (Isacks *et al.*, 1968). However, in the young fold mountains and trench systems, shallow focus earthquakes are present, but intermediate and deep focus earthquakes

155

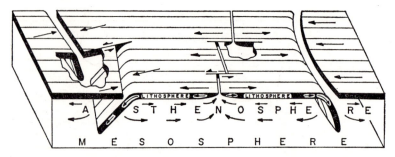

FIGURE 84 Model illustration of plate tectonics showing the roles of the litho-sphere, asthenosphere and mesosphere. Arrows on lithosphere indicate relative movements of adjoining plates, and in asthenosphere represent possible compensating flow in response to the downward movement of segments of lithosphere. An arc to arc transform fault appears at the left between oppositely facing island arcs, two ridge to ridge transform faults along the ocean ridge in the centre, and a simple arc structure at the right. From Isacks *et al.* (1968), *J. Geophys. Res.* **73**, 5857.

to a depth of 700 km also occur (fig. 83). These two seismic provinces, the one characterized by shallow focus earthquakes and the other by deeper focus earthquakes reflect different processes according to the sea-floor spreading hypothesis. New oceanic crust is created along the crests of the mid-ocean ridges and is then partly absorbed in the trench systems.

To a first approximation it is possible to divide the earth's surface into a number of essentially aseismic plates or blocks bounded by the seismicity associated with active ridge crests, faults, trenches and mountain systems (Le Pichon, 1968; Morgan, 1968). These plates can be composed entirely of continental crust, or of oceanic crust, or they might be made up of a combination of both. The sea-floor spreading hypothesis maintains that these plates are in constant relative motion and that seismic boundaries between them delineate zones where oceanic crust is created or destroyed, continental crust extended or compressed, and crustal plates translated laterally along faults without change in their surface area (fig. 83). In this 'new global tectonics' (Isacks *et al.*, 1968), now generally referred to as *Plate Tectonics*, it is supposed that a mobile, near-surface layer of strength (the lithosphere) plays a key role. Some of the principle points of this mobile lithosphere hypothesis are illustrated in fig. 84. Three flat-lying layers are distinguished. The *lithosphere*, which generally includes the crust and uppermost mantle, has significant strength, and is of the order of 100 km thickness. The *asthenosphere*, which is a layer of effectively no strength on the appropriate time-scale, extends from the base of the lithosphere to a depth of several hundred kilometres. The *mesosphere*, which may have strength, makes up the lower remaining portion of the mantle and is relatively passive, perhaps inert, in tectonic processes.

The boundaries between the layers may be gradational within the earth. The asthenosphere corresponds more or less to the low-velocity layer of seismology. It strongly attenuates seismic waves, particularly high-frequency shear waves. The lithosphere and mesosphere have relatively high seismic velocities and propagate seismic waves without great attentuation. At the principal zones of tectonic activity within the earth (ocean ridges, island arcs and major strike–slip faults) the lithosphere is discontinuous, whilst elsewhere it is continuous. The lithosphere is therefore composed of relatively thin blocks, which may be considered infinitely rigid laterally. The major tectonic features are the result of relative movement and interaction of these blocks, which spread apart at ocean ridges, slide past one another at large strike–slip faults, and are underthrust at island arcs and similar structures. A generalized version of the ideas of plate tectonics is shown in fig. 84. McKenzie & Parker (1967) used this concept to explain focal mechanisms of earthquakes, vulcanism, and other tectonic features in the northern Pacific, whilst a comprehensive study of the observations of seismology has provided widely based strong support for it (Isacks *et al.*, 1968), as have also observations of heat flow from the mid-ocean ridges (Le Pichon & Langseth, 1969).

5.2 MARINE MAGNETIC ANOMALIES

5.2.1 *Vine–Matthews crustal model*

Total magnetic field anomalies observed on crossing the mid-ocean ridges show three essential features. There is always a pronounced central anomaly associated with the median graben, whilst over the rugged flanks short wavelength anomalies are observed. On the other hand, over the exposed or buried foothills of the ridge long wavelength anomalies are observed. The short wavelength local anomalies can often be correlated with bathymetry but the long wavelength anomalies were originally regarded as a puzzling feature and difficult to explain.

The central anomaly can be reproduced if it is assumed that a block of material, very strongly magnetized in the present direction of the earth's magnetic field, underlies the median graben producing a positive susceptibility contrast with the adjacent crust. In analysing these magnetic anomalies, it is usual to speak of the effective susceptibility of the material which is given by the total intensity of magnetization (remanent plus induced) divided by the present magnetic field strength. Measurements of the magnetic properties of dredged rock samples (Ade-Hall, 1964*b*; Opdyke &

Hekinian, 1967) demonstrate the predominance of remanent magnetization, since the susceptibility is comparatively low. The induced magnetization is typically an order of magnitude less than the remanent magnetization. For basalts the mean value of effective susceptibility is of the order of 0.01, corresponding to an intensity of magnetization of about 0.005 emu cm^{-3}. The central anomaly will be positive or negative depending on the latitude of the region being studied. At the equator a block horizontally magnetized northwards will produce a field above it which opposes the geomagnetic field, so that a negative anomaly is observed. At high latitudes a block vertically magnetized in the same sense as the geomagnetic field will reinforce the field above it and produce a positive anomaly.

In order to account for the long period anomalies and to explain the high susceptibility contrast between the central block and the adjacent crust, Vine & Matthews (1963) combined the ideas of sea-floor spreading with reversals of the earth's magnetic field. According to their crustal model, as new crust is formed at the crest of a mid-ocean ridge and cools through the Curie temperature, it will acquire a remanent magnetization parallel to the ambient direction of the geomagnetic field. If the earth's magnetic field reverses polarity intermittently then stripes of crust of alternate polarity will be produced parallel to and, in the simplest case, distributed symmetrically about, the axis of the ridge. The model is illustrated in fig. 85.

Vine & Wilson (1965) first attempted to analyse ocean magnetic anomalies in this way following the publication of the first geomagnetic polarity time-scale (Cox *et al.* 1963a, 1964). The original model of Vine & Matthews (1963) was applied to the anomalies over the Juan de Fuca ridge off Vancouver Island. The normal and reversed blocks were assumed to extend from 3 to 11 km below sea level and thus to include both layers 2 and 3 of oceanic crust. The effective susceptibility assumed was ±0.0025 except for the central block where a value of +0.005 was used. Vine & Wilson (1965) showed that a better fit to the anomalies was obtained if it was assumed that all the magnetization was confined to layer 2 between 3.3 and 5 km depth, and if the effective susceptibility was ±0.01 except for the central block where +0.02 was assumed. These values, although high, are quite compatible with the magnetic properties of dredged rock samples (Ade-Hall, 1964b; Opdyke & Hekinian, 1967). The assumption that the bulk of the magnetization of the oceanic crust resides in layer 2 is probably compatible with the likely composition of layer 3 (Vine, 1968a). The possible compositions are serpentine or gabbro, both of which are likely to have remanent magnetization not as significant as the basaltic layer 2.

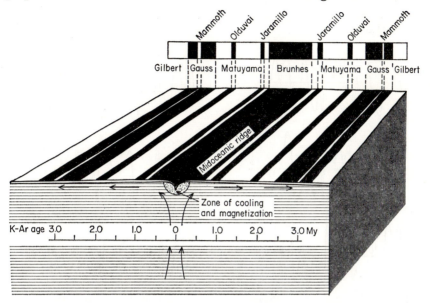

FIGURE 85   Schematic representation of the principle of sea-floor spreading and reversals of the earth's magnetic field as proposed by Vine & Matthews (1963). Normal polarity zones are shown shaded. After Allan (1969), *Earth Sci. Rev.* **5**, 230, fig. 7.

In order to account for the axial anomaly being about twice as large as those on either side, Vine & Wilson (1965) supposed that this was because the central block was the only one composed exclusively of young material. Vulcanism probably occurs over a wider zone than the central block, and all the other blocks will therefore be contaminated with younger material, often of the opposing polarity, and hence lowering or modifying its magnetic effect. However, samples collected from the mid-Atlantic ridge at 45N and up to 150 km from the axis of the ridge confirm that the median graben intensities of magnetization are much higher than elsewhere (Irving *et al.*, 1970). This decrease of intensity with distance from the axis occurs within about 5 km, which in terms of time means a period of the order of 10^5 years. Irving *et al.* (1970) propose that this decrease of intensity is due to chemical demagnetization processes including the oxidation of titano-magnetite by its partial replacement by titanomaghaemite, and the incipient submicroscopic unmixing of magnetite and ilmenite rich phases. In both processes parts of the original TRM are replaced with a much weaker CRM.

Improvements to the geomagnetic polarity time-scale, especially the discovery of the Jaramillo event (Doell & Dalrymple, 1966) led to improved analyses of the ridge anomalies. Four of these are shown in fig. 86 from

Sea-floor spreading and plate tectonics

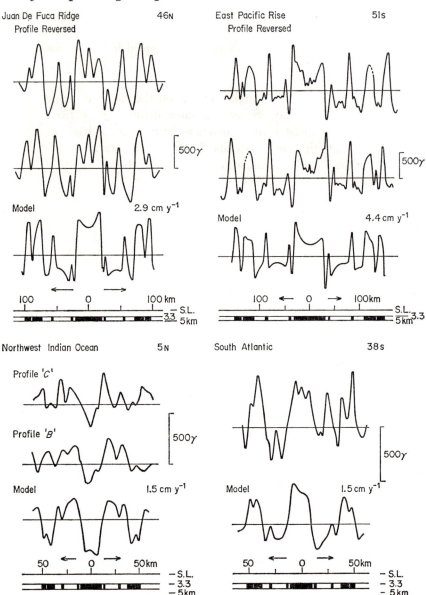

FIGURE 86 Observed magnetic profiles at various points on the mid-ocean ridge system are compared with simulated profiles based on the geomagnetic polarity time-scale. From Vine (1966), *Science*, **154**, 1405–15. © 1966 by the American Association for the Advancement of Science.

the analysis of Vine (1966) of four widely separated areas on the mid-ocean ridge system. In the first two profiles from the Juan de Fuca ridge and the East Pacific rise, the profile has also been reversed for comparison. The remarkable symmetry displayed is noteworthy. Note that the three profiles from mid-latitudes have central positive anomalies, whilst the equatorial profile from the northwest Indian Ocean on the Carlsberg ridge has a central negative anomaly. Spreading rates deduced vary from 1.5 to 4.4 cm y^{-1} over the past 4 My, which is as far back as the land-based geo-magnetic polarity time-scale extends. These fits, obtained for areas widely spaced over the world's oceans, were so convincing that the Vine–Matthews model was henceforth applied whenever magnetic profiles over a mid-ocean ridge were made. Vine (1966) showed that the model could be used beyond the limits of the known geomagnetic polarity time-scale. By con-structing a configuration of blocks to match the observed profile and assuming a constant rate of spreading, the reversal time-scale may be extrapolated to the limits of the area being surveyed.

Pitman & Heirtzler (1966) applied this criterion to four observed profiles across the Pacific–Antarctic ridge. They found a spreading rate of 4.5 cm y^{-1} for the past 3.5 My and, by assuming this rate to be constant, they traced the reversals of the geomagnetic field over the past 10 My. The model is shown in fig. 87. A comparison of the profile and its reverse shows the quite remarkable symmetry observed over a distance of 1000 km, whilst the agreement between the profile computed from the model and that observed convincingly demonstrates the validity of the extrapolation. When this model was applied to a profile across the Reykjanes ridge, at the other end of the globe, a good agreement was obtained if a spreading rate of 1 cm y^{-1} was assumed. The detailed linear pattern revealed by an aero-magnetic survey of the Reykjanes ridge made by Heirtzler *et al.* (1966) is a particularly good example of the pattern predicted by the Vine–Matthews crustal model (fig. 85).

5.2.2 *Application to the oceans*

Heirtzler and his colleagues at the Lamont–Doherty Geological Observa-tory have made extensive analyses of the magnetic anomalies observed in the various oceans. They have shown that a magnetic anomaly pattern, parallel to and bilaterally symmetric about the mid-ocean ridge system, exists over extensive regions of the North and South Pacific (Pitman *et al.*, 1968), the South Atlantic (Dickson *et al.*, 1968) and the Indian (Le Pichon & Heirtzler, 1968) Oceans. They demonstrated that the pattern is the same in each of these oceanic areas and may be simulated in each region by the

Sea-floor spreading and plate tectonics

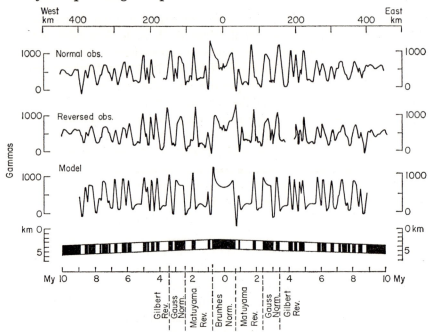

FIGURE 87 The observed and reversed magnetic profiles across the Pacific–Antarctic ridge over a distance of 1000 km. The reversal sequence beyond 4 My is inferred from the magnetic anomalies by assuming a constant spreading rate calculated from the first 4 My. Figure as redrawn by Allan (1969), after Pitman & Heirtzler (1966), *Science*, **154**, 1164–71. © 1966 by the American Association for the Advancement of Science.

same sequence of alternating stripes of normally and reversely magnetized material. Examples of some of these anomaly profiles are shown in fig. 88. The anomalies are on some of the longest profiles recorded and cover distances of up to 3000 km from the ridge crest. The magnetic anomaly pattern associated with the axial zone can always be related to the geomagnetic time-scale over the past 3.5 My by assuming a constant spreading rate. For the various oceans the axial spreading rates vary from 1.0 to 6.0 cm y^{-1} as summarized in table 10. When the axial spreading rates are extrapolated to the end of the long profiles in fig. 88, they suggest the dates indicated, which approach 100 My in some cases. Similarly shaped anomalies are identified by numbers at the top of the dashed vertical lines which connect them. Each of these is presumed to be related to the same absolute age, so that the time-scales deduced for each of the profiles obviously differ and are not linearly related. The relative spreading rate curves, though non-linear, are however continuous as shown in fig. 89, where the distance from the ridge crest of a given anomaly in the South Atlantic is compared

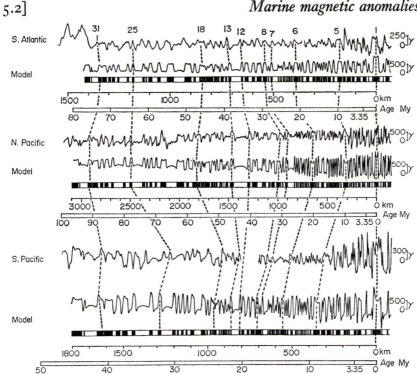

FIGURE 88 Long magnetic profiles from three widely separated regions are compared with those simulated from the pattern of crustal blocks found by assuming constant spreading rate over the first 3.5 My and extrapolating to the end of the profile. Dashed lines connect anomalies numbered according to the Lamont system. Modified after Heirtzler *et al.* (1968), *J. Geophys. Res.* **73**, 2120.

with its distance in the South Indian, North and South Pacific Oceans. The system of numbering the most readily identifiable anomalies in sequence adopted by the Lamont group has now come into general use. Until an absolute time-scale can be confidently attached to the anomaly sequence, this system is the simplest means of identifying sections of the ocean floor.

It is convenient to select one of the time-scales extrapolated from the first few million years as a standard. There is some suggestion (Vine, 1966) that the spreading rate at the Juan de Fuca ridge in the North Pacific has not been constant over the past 10 My, but has decreased from 4.4 to 2.9 cm y^{-1}. Relative spreading in the South Pacific also appears to have varied considerably with time, and palaeontological evidence suggests its time-scale is too young by a factor of two. The Indian Ocean anomalies have only recently been extended back to the Late Cretaceous (McKenzie & Sclater, 1971). They show that there were considerable variations in spreading rate during the Tertiary. There is every reason to believe,

TABLE 10   *Spreading rates at various points on the mid-ocean ridge system during the past 3.5 My*

Ocean	Lat.	Long.	Rate (cm y^{-1})	Reference
Atlantic	60N	28W	1.0	Vine (1966)
	27N	44W	1.25	Phillips (1967); Heirtzler *et al.* (1968)
	22N	45W	1.40	Phillips (1967); Heirtzler *et al.* (1968)
	25S	13W	2.25	Heirtzler *et al.* (1968)
	28S	13W	1.95	Heirtzler *et al.* (1968)
	30S	14W	2.0	Heirtzler *et al.* (1968)
	38S	17W	1.5	Vine (1966)
	38S	17W	2.0	Heirtzler *et al.* (1968)
	41S	18W	1.65	Heirtzler *et al.* (1968)
	47S	14W	1.60	Heirtzler *et al.* (1968)
	50S	8W	1.53	Heirtzler *et al.* (1968)
Indian	19N	40E	1.0	Vine (1966); Heirtzler *et al.* (1968)
	16N	41E	1.0	Vine (1966)
	13N	50E	1.0	Heirtzler *et al.* (1968)
	7N	60E	1.5	Heirtzler *et al.* (1968)
	5N	62E	1.5	Vine (1966); Heirtzler *et al.* (1968)
	22S	69E	2.2	Heirtzler *et al.* (1968)
	30S	76E	2.4	Heirtzler *et al.* (1968)
	45S	95E	3.0	Le Pichon & Heirtzler (1968)
Pacific	46N	130W	2.9	Vine (1966)
	42N	127W	1.0–3.0	Menard (1967)
	17S	113W	6.0	Heirtzler *et al.* (1968)
	40S	112W	5.1	Heirtzler *et al.* (1968)
	45S	112W	5.1	Heirtzler *et al.* (1968)
	48S	113W	4.7	Heirtzler *et al.* (1968)
	51S	117W	4.9	Heirtzler *et al.* (1968)
	51S	117W	4.4	Vine (1966); Pitman & Heirtzler (1966)
	58S	149W	3.8	Heirtzler *et al.* (1968)
	60S	150W	4.0	Heirtzler *et al.* (1968)
	63S	167W	2.3	Heirtzler *et al.* (1968)
	65S	170W	2.0	Heirtzler *et al.* (1968)
	65S	174W	2.8	Heirtzler *et al.* (1968)

however, that the South Atlantic spreading rate has been relatively constant, so that Heirtzler *et al.* (1968) selected this time-scale as a standard. Application of the Heirtzler time-scale to all the profiles from the oceans then yields the isochron map of the ocean floor as in fig. 90.

The isochron map has implications for continental drift. The theory of plate tectonics supposes that the continents do not move through the rigid crust of the earth, but along with it. In the South Pacific (Pitman *et al.*, 1968) the last anomaly observed running from the Pacific–Antarctic ridge to New Zealand is anomaly 32, estimated to be 80 My in age. It appears that New Zealand broke away from Antarctica during the Late Cretaceous and subsequently drifted to its present position. Anomaly 17 (estimated to

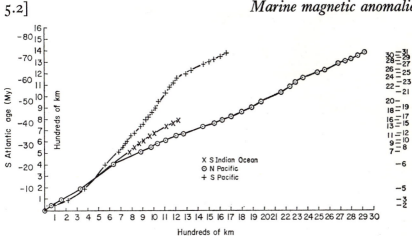

FIGURE 89   The distance to a given anomaly in the South Atlantic versus distance
to the same anomaly in the South Indian, North Pacific and South Pacific Oceans.
Numbers on the right refer to anomaly numbers, as in fig. 88. From Heirtzler
et al. (1968), *J. Geophys. Res.* **73**, 2121.

be 43 My old) is the last anomaly observed at the Australian continental
shelf (Le Pichon & Heirtzler, 1968). It might therefore be presumed that
the Great Australian Bight was adjacent to the Australia–Antarctic ridge
at that time. During the past 43 My Australia and Antarctica drifted apart
to their present positions. Although the anomalies in the South Atlantic
have not been measured up to the continental shelves of either Africa or
South America, extrapolation of the data suggests an age of 180–200 My
for the initial separation of these two continents (Dickson *et al.*, 1968).
Vine (1966) notes that there is no discernible anomaly pattern over the
southwest branch of the South Indian Ocean ridge, and suggests that it
may be a dead ridge and has not been the site of measurable spreading
since the Mesozoic era.

In the North Atlantic a distinct line, parallel to the continental slopes
on either side, separates two contrasting magnetic zones, referred to as the
disturbed and undisturbed magnetic regions (Heirtzler & Hayes, 1967).
The boundaries lie 2000 to 2500 km from the axis of the mid-Atlantic
ridge and roughly equidistant from it. The inner zone is the disturbed
region and its central portion, for example over the Reykjanes ridge, has
been interpreted in terms of the sea-floor spreading hypothesis and reversals
of the earth's magnetic field. The outer undisturbed or magnetic quiet zones
lie in a 400 km wide belt on either side of the North Atlantic and are characterized by fairly featureless low amplitude anomalies.

A number of explanations for these magnetic smooth zones have been
put forward, including lack of magnetic material in the crust (Drake &

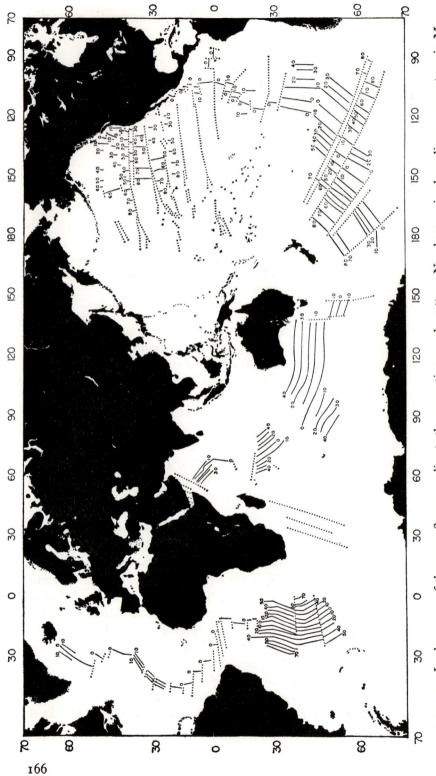

FIGURE 90 Isochron map of the ocean floor according to the magnetic anomaly pattern. Numbers on isochron lines represent age in My. Dotted lines represent fracture zones. From Heirtzler *et al.* (1968), *J. Geophy. Res.* **73**, 2123.

Nafe, 1969), stages of continental rifting producing plateau basalts with smaller specific intensities than submarine dykes (Irving, 1970), magnetization at low geomagnetic latitudes (Vogt *et al.*, 1970) and spreading during periods of few or no geomagnetic reversals (Heirtzler & Hayes, 1967; Emery *et al.*, 1970; Burek, 1970). Heirtzler & Hayes (1967) suggested the quiet zones reflected the Kiaman Magnetic Interval. Burek (1970) on the other hand suggested a Late Triassic to Early Jurassic age for initial rifting of the Atlantic Ocean, correlating the quiet zones with what was later to be called the Graham Magnetic Interval (McElhinny & Burek, 1971; see §4.2.4). More detailed anomaly profiles (Rona *et al.*, 1970) over the quiet zones correlate well with this interval. Emery *et al.* (1970) in attempting to correlate the quiet zones with the Kiaman Interval required that the spreading rate decrease from 1.43 cm y^{-1} to 0.80 cm y^{-1} about half way between the ridge and continental shelf. McElhinny & Burek (1971) point out however that a constant spreading rate places the continental slope at about 200 My and the quiet zone lies between 160 and 200 My within the Graham Magnetic Interval. Three prominent anomaly peaks in an otherwise featureless profile correlate very well with three reversed zones proposed to lie within the Graham Interval. There seems no reason therefore to invoke more elaborate explanations for the quiet zones when a relatively simple one exists.

These are some of the important features of the world-wide anomaly patterns as interpreted in terms of the sea-floor spreading hypothesis and geomagnetic reversals. The relationship of these features to palaeomagnetic and other data in terms of continental drift is discussed more fully in chapter 7.

5.2.3 *Cainozoic polarity time-scale*

The standard polarity time-scale derived from the South Atlantic by Heirtzler *et al.* (1968) is illustrated in fig. 91. The time-scale is simply deduced by dividing the distance to the South Atlantic model magnetized bodies by the calculated axial zone spreading rate of 1.9 cm y^{-1} (Dickson *et al.*, 1968). In fig. 91 the anomaly numbers are shown together with the corresponding geological periods following the time-scale of the Geological Society of London (Harland *et al.*, 1964). The ages of the normally magnetized intervals are shown in table 11. There are a total of 171 polarity intervals deduced for the past 76.3 My. These include some events as short as 30000 years and which are distinguishable from the anomaly pattern, whilst the longest intervals are about 3 My in length. The average normal interval is 0.42 My in length, and the average reversed interval is

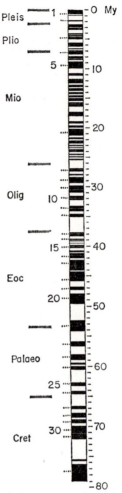

FIGURE 91 Geomagnetic polarity time-scale for the Cainozoic and Late Cretaceous deduced from the magnetic anomalies in the South Atlantic. Normal polarity zones are black, reversed are white. Geological periods are indicated on the left, numbers assigned to prominent magnetic anomalies are given in the centre, and the age given on the right. From Heirtzler *et al.* (1968), *J. Geophys. Res.* **73**, 2124.

0.48 My long. This is about twice as long as that deduced for the past 4.5 My from the geomagnetic polarity time-scale (§4.2.1). An analysis of the frequency of reversals over the past 75 My deduced from the time-scale is shown in fig. 92. There appears to have been an average of about three polarity changes per My during the past 40 My reducing to one per My during the Early Tertiary and latest Cretaceous. It is not clear to what

168

TABLE 11   *Intervals of normal polarity deduced from marine magnetic anomalies for the Tertiary and Late Cretaceous*

0.00– 0.69	18.91–19.26	40.03–40.25
0.89– 0.93	19.62–19.96	40.71–40.97
1.78– 1.93	20.19–21.31	41.15–41.46
2.48– 2.93	21.65–21.91	41.52–41.96
3.06– 3.37	22.17–22.64	42.28–43.26
4.04– 4.22	22.90–23.08	43.34–43.56
4.35– 4.53	23.29–23.40	43.64–44.01
4.66– 4.77	23.63–24.07	44.21–44.69
4.81– 5.01	24.41–24.59	44.77–45.24
5.61– 5.88	24.82–24.97	45.32–45.79
5.96– 6.24	25.25–25.43	46.76–47.26
6.57– 6.70	26.86–26.98	47.91–49.58
6.91– 7.00	27.05–27.37	52.41–54.16
7.07– 7.46	27.83–28.03	55.92–56.66
7.51– 7.55	28.35–28.44	58.04–58.94
7.91– 8.28	28.52–29.33	59.43–59.69
8.37– 8.51	29.78–30.42	60.01–60.53
8.79– 9.94	30.48–30.93	62.75–63.28
10.77–11.14	31.50–31.84	64.14–64.62
11.72–11.85	31.90–32.17	66.65–67.10
11.93–12.43	33.16–33.55	67.77–68.51
12.72–13.09	33.61–34.07	68.84–69.44
13.29–13.71	34.52–35.00	69.93–71.12
13.96–14.28	37.61–37.82	71.22–72.11
14.51–14.82	37.89–38.26	74.17–74.30
14.98–15.45	38.68–38.77	76.64–76.33
15.71–16.00	38.83–38.92	
16.03–16.41	39.03–39.11	
17.33–17.80	39.42–39.47	
17.83–18.02	39.77–40.00	

extent short events have been lost, especially since the spreading rate is fairly low and thus likely to smooth these out. The geomagnetic polarity time-scale for the past 4.5 My (Cox, 1969; fig. 63) suggests a rate of five polarity changes per My. Certainly there are some events now established for the past 4.5 My from the geomagnetic polarity time-scale that are not always seen on the anomaly patterns. Also Wellman *et al.* (1969) have shown from a 42 My old lava sequence in southeast Australia that the number of polarity changes per My was somewhat greater at that time than predicted by the time-scale. However, it certainly seems reasonable to suppose that there has been a decrease in reversal frequency back to the beginning of the Cretaceous. The rate was high in the Upper Tertiary, reduced during the Early Tertiary and uppermost Cretaceous, and very low during most of the Cretaceous (see §4.2.4, fig. 72).

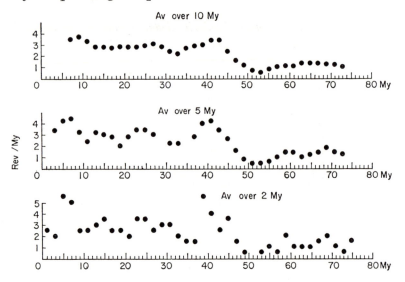

FIGURE 92 Frequency of geomagnetic reversals as a function of time derived from table 10 by averaging over intervals of 2, 5 and 10 My. From Heirtzler *et al.* (1968), *J. Geophys. Res.* **73**, 2125.

5.2.4 *Magnetization of seamounts*

Conventional palaeomagnetic measurements can be carried out on volcanic and sedimentary rocks from the various continental blocks. The results of these investigations are analysed in the following chapters. In the new concept of global tectonics, where the earth's surface is paved with crustal plates, it is important that palaeomagnetic results be obtained for the various geological epochs from each of these plates. The present distribution of plates is shown in fig. 83, from which it can be seen that the largest one existing at present, the Pacific plate of Le Pichon (1968), is entirely oceanic. Palaeomagnetic sampling is limited to the top of volcanic islands, which tend to be young and thus of limited value in evaluating possible movements of this oceanic plate. Vacquier (1962) has shown that, if a combined magnetic and bathymetric survey of a seamount is made, it is possible to calculate the direction of uniform magnetization of the seamount. These directions of magnetization are then equivalent to those determined palaeomagnetically by conventional methods, and it becomes possible to attempt to study the palaeomagnetism of ocean basins.

The growth of a small seamount takes place over 10^4 to 10^5 years. Secular variation will tend to be averaged out in a computation of its average direction of magnetization. Results from each seamount will not therefore

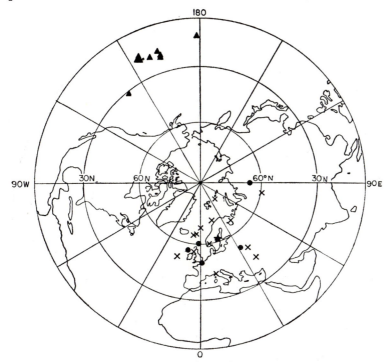

FIGURE 93　Palaeomagnetic pole positions for the Hawaiian seamounts. Seamount locations are marked by triangles. Crosses represent north poles, and solid circles represent south poles. The star is the mean pole position. After Francheteau *et al.* (1970), *J. Geophys. Res.* **75**, 2044.

TABLE 12　*Palaeomagnetic poles deduced for the Pacific plate from the study of seamounts. After Francheteau et al.* (1970)

Age (My)	Seamounts	Pole position (A_{95})
~ 10	7 Tripod seamounts (T)	87N, 90E (14)
~ 30	Midway atoll (Vine, 1968*b*)*	(75N, 12E)
~ 40	6 California seamounts (C)	71N, 354E (13)
85–90	17 Hawaiian seamounts (H)	61N, 16E (8)
Cretaceous	7 Japanese seamounts	56N, 326E (13)

* Results from two reef cores.

be analogous to sampling a number of individual lava flows, but rather equivalent to sampling a site in a sedimentary formation over, say, 1 m thickness. Larger seamounts may remain volcanically active over a period of 10^4 to 10^7 years (Menard, 1969), and the problem that arises is that several polarity changes may have taken place during this time. The sea-

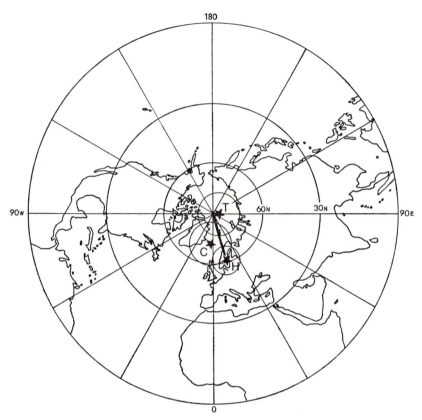

FIGURE 94   Polar-wander curve for the northeastern Pacific deduced from the magnetization of seamounts (table 13). T, Tripod seamounts; C, Californian seamounts; H, Hawaiian seamounts.

mount will not then be uniformly magnetized as will be found when the computed anomaly fails to match the observed anomaly. Francheteau *et al.* (1970) have summarized all the available computations of the magnetizations of seamounts in the Pacific ocean. Two criteria were used to decide on the suitability of each result. From their computations of the uniform magnetization direction, they calculated a standard deviation (circle of confidence). Results were rejected when this was greater than 10°. An index r of the goodness of fit was computed as the ratio of the mean observed anomaly to the mean residual anomaly. When $r < 1.3$, the results were rejected. The pole positions obtained from an analysis of seventeen seamounts in the vicinity of the Hawaiian islands are illustrated in fig. 93. These seamounts are Upper Cretaceous in age. K–Ar ages of 85–90 My being reported (Dymond & Windom, 1968). The mean pole position is

6IN, 16E with A_{95} of 8°, and is thus significantly different from the present north geographic pole.

All data pertaining to the Pacific ocean are summarized in table 12 after Francheteau *et al.* (1970). The results of Vine (1968*b*) from Midway atoll relate to palaeomagnetic results from two reef cores, estimated to be 30 My old. Although normally results from only two cores would be regarded as inedaquate in spite of demagnetization studies, these are of special interest because they confirm the results obtained from the Californian seamounts of similar age, by the use of conventional techniques. The pole positions are plotted in fig. 94. Except for the Japanese seamounts, the poles lie on a smooth curve back to the Upper Cretaceous. The difference between the poles for the Hawaiian and Japanese seamounts, which are of similar age, suggests that these seamounts have experienced relative displacement since the Upper Cretaceous. If this is so it means that the Pacific plate as at present constituted may not have remained entirely rigid for the past 100 My. The existence of the so-called 'Great Magnetic Bight' in the northeast Pacific lends some support to this (see §5.3.2), with the possibility that this represents an ancient triple junction of three plates (Vine & Hess, 1970).

5.3 PLATE MOTIONS

5.3.1 *Tectonics on a sphere*

On the theory of plate tectonics (McKenzie & Parker, 1967; Morgan, 1968; Le Pichon, 1968), it is supposed that the earth's surface is divided into a number of rigid crustal blocks (fig. 83). It is assumed that each block is bounded by ridges (where new surface is being formed), trenches or young fold mountains (where surface is being destroyed) and transform faults, and that there is no stretching, folding, or distortion of any kind within a given block. These individual aseismic areas move as rigid plates on the surface of the earth. The movement of these plates on the surface of a sphere is best understood in terms of rotations by applying Euler's theorem. If one of two plates is taken to be fixed, the movement of the other corresponds to a rotation about some pole (fig. 95). If the angular velocity of this rotation is $\boldsymbol{\omega}$ (reckoned as positive when the rotation is clockwise looking from the centre of the sphere, and taken to be a positive vector pointing outwards along the rotation axis), the velocity of one plate relative to the other will vary along their common boundary. If the angular distance from a point on one plate to the pole is θ, then the relative velocity $\boldsymbol{v}_\theta$ at that point is given by (fig. 96)

$$\boldsymbol{v}_\theta = a\boldsymbol{\omega}\sin\theta, \tag{5.1}$$

where a is the radius of the earth. The relative velocity thus has a maximum

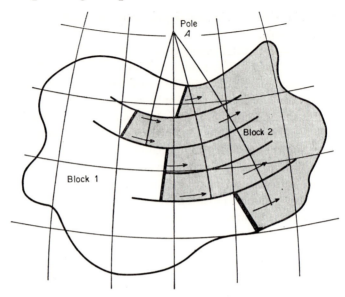

FIGURE 95   On a sphere the motion of block 2 relative to block 1 must be a rotation about some pole. Transform faults on the boundary between 1 and 2 must be small circles (lines of latitude) about the pole A. From Morgan (1968), *J. Geophys. Res.* **73**, 1962.

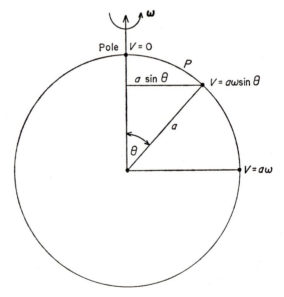

FIGURE 96   Cross-section through the earth showing the variation of spreading rate V with angular distance θ from the pole of rotation.

at the 'equator' and vanishes at the pole of rotation. The relative velocity vectors must lie along small circles or 'latitudes' with respect to the pole. The pole position itself has no significance other than being a construction point.

If these small circles, along which relative velocity vectors lie, cross the line of contact between two plates, the line must be either a ridge or a trench depending upon the sense of the rotation. Neither of these structures conserves crust. If, however, the line of contact is itself a small circle, then it is a transform fault (fig. 95). This property of transform faults is very useful in finding pole positions and is a consequence of the conservation of crust across them. This is essentially an extension of Wilson's (1965 a) concept of transform faults to rotations on a sphere. Morgan (1968) and Le Pichon (1968) have analysed the transform faults which offset the mid-ocean ridges in terms of this property and have deduced instantaneous centres of rotation describing the relative rotation between pairs of plates (table 13). When several plates are in relative motion, as is occurring at the present time on the surface of the earth, then it is possible to use the property of angular velocities that they behave like vectors. Around any closed circuit crossing a number of plates, A, B, C, D etc. the sum of the angular velocities must be zero, that is

$$_A\boldsymbol{\omega}_B + {_B\boldsymbol{\omega}_C} + {_C\boldsymbol{\omega}_D} + {_D\boldsymbol{\omega}_A} = \mathbf{0}, \tag{5.2}$$

where $_A\boldsymbol{\omega}_B$ is the angular velocity of the rotation which describes the magnitude and direction of relative motion between plates A and B. The sense of the rotation is that found by moving from plate A to plate B and so on. Both Morgan (1968) and Le Pichon (1968) deduced the relative motion between the Antarctic and African plates by closure around the Africa–North America–Pacific–Antarctica–Africa circuit. This result is also shown in table 13. The analysis suggests Africa and Antarctica are separating with a maximum rate of around 3 cm y^{-1} (half-rate 1.5 cm y^{-1}).

5.3.2   *Plate evolution*

The principal features of ridges, trenches and transform faults are the direct consequence of the relative motion of rigid plates. In order to discuss the geometry or plate evolution it is convenient to define trenches, ridges and transform faults in terms of the destruction and creation of plates, rather than in terms of topographic features. Trenches are therefore defined as structures which consume the lithosphere from only one side, and ridges as structures which both produce lithosphere symmetrically and generally lie at right angles to the relative velocity vector between two separating

TABLE 13   *Instantaneous poles of rotation as given by Morgan* (1968) (*M*),
Le Pichon (1968) (*LeP*), *and McKenzie & Parker*(1967) (*MP*)

Parameters used	Lat.	Long.	Equatorial opening rate (cm y^{-1})
1. *South Atlantic* (*America–Africa*)			
Strike of 18 fracture zones (M)	62N	36W	3.6
Strike of 18 fracture zones (LeP)	58N	37W	4.1
9 spreading rates (LeP)	69N	32W	4.1
2. *North Pacific* (*America–Pacific*)			
Strike of 33 fracture zones (M)	53N	53W	−8.0
Strike of 32 fracture zones (LeP)	53N	47W	−6.7
2 slip vectors (MP)	50N	85W	—
3. *South Pacific* (*Antarctica–Pacific*)			
Strike of 6 fracture zones (M)	71S	118E	11.4
Strike of 6 fracture zones (LeP)	70S	118E	12.0
11 spreading rates (LeP)	68S	123E	12.0
4. *Arctic Ocean* (*America–Eurasia*)			
Strike of 4 fracture zones (LeP)	78N	102E	3.1
5. *NW Indian Ocean* (*Africa–India*)			
Strike of 5 fracture zones (LeP)	26N	21E	4.4
6. *SW Indian Ocean* (*Antarctica–Africa*)			
Closure via Africa–America–⎫ (M)	25S	35W	3.2
Pacific–Antarctica–Africa ⎭ (LeP)	42S	14W	3.6

plates. Transform faults are defined as active faults parallel to the relative slip vector.

The simplest form of plate evolution is that of a trench as illustrated in fig. 97*a*. Evolution occurs because a trench consumes lithosphere on only one side; the upper part of the trench *ab* consumes the plate *Y*, whereas the lower part *bc* consumes *X*. The arrows show the relative motion between *X* and *Y*, and are on the plate which is being consumed. As the motions continue in the direction of the arrows, *Y* is consumed between *a* and *b*, but not between *b* and *c*, so that *bc* must therefore be steadily offset from *ab* to form two trenches joined by a transform fault (fig. 97*b*). The Alpine fault in New Zealand is an example of such a transform fault joining two trenches which consume different plates (fig. 97*c*). To the north of North Island the Kermadec trench consumes the Pacific plate, whereas the Tasman Sea is being consumed between South Island and the Macquarie ridge (McKenzie & Morgan, 1969).

There are several points on the surface of the earth where three plates meet, called *triple junctions*. It is convenient to analyse the relative velocities *v* at a triple junction (as given by (5.1)) rather than the angular

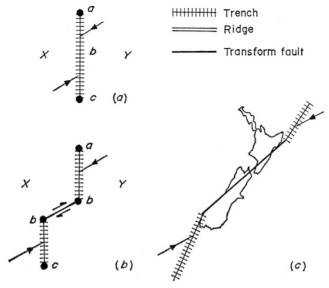

FIGURE 97   The evolution of a trench. The arrows show the relative motion vector and are on the plates being consumed, Y is consumed between a and b, X between b and c, so that the trench evolves to form two trenches joined by a transform fault. The Alpine fault in New Zealand shown in (c) is a trench–trench transform fault of the type in (b). From McKenzie & Morgan (1969), *Nature*, **224**, 125–33.

velocity $\boldsymbol{\omega}$. At triple junctions the relative motions of the plates is not completely arbitrary, because given any two velocity vectors the third can be determined. Fig. 98 shows how a ridge and a trench can meet to form a transform fault, the triple junction separating three plates A, B and C. Starting at point x on A and moving clockwise the relative velocity of B, $_A\boldsymbol{v}_B$, is in the direction AB in the vector diagram. Similarly the relative velocities $_B\boldsymbol{v}_C$ (at right angles to the ridge) and $_C\boldsymbol{v}_A$ (along the transform fault) are represented by BC and CA. The vector diagram must close because the circuit returns to x, so that

$$_A\boldsymbol{v}_B + {}_B\boldsymbol{v}_C + {}_C\boldsymbol{v}_A = \mathbf{0}. \tag{5.3}$$

If the triple junction is required to look the same at some later time, there are important restrictions on the possible orientations of the three plate boundaries. Unless these conditions are satisfied the junction can exist for an instant of time only, and is therefore defined as unstable. If evolution is possible without a change in geometry, then the vertex is defined as a stable junction. Only the movement of stable junctions permits continuous plate evolution.

McKenzie & Morgan (1969) have discussed the stability conditions of

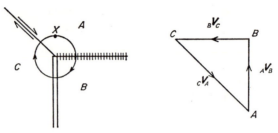

FIGURE 98   A ridge and a trench can meet to form a transform fault. A closed circuit around the triple junction produces the vector velocity diagram shown (symbols as in fig. 97). From McKenzie & Parker (1967), *Nature*, **216**, 1276–80.

triple junctions, restricting themselves to those cases in which ridges spread symmetrically at right angles to their strike. Although the analysis can readily be extended to the general case, the discussion below will be restricted to the seven most common types of the sixteen analysed by McKenzie & Morgan (1969). An example of each of these seven types can be found at the present day in the Pacific (figs. 99 and 100). The first case is the triple junction between three ridges, called an RRR junction (fig. 99 *a*). The lengths AB, BC and CA in the velocity triangle are proportional to and parallel to the velocities $_A\boldsymbol{v}_B$, $_B\boldsymbol{v}_C$ and $_C\boldsymbol{v}_A$ respectively. The triangle is therefore in velocity space, and represents the condition imposed by (5.3). In the cases being considered ridges spread symmetrically at right angles to their strike, so that a point on the axis of the ridge AB will move with velocity $\frac{1}{2}\,_A\boldsymbol{v}_B$ relative to A. The velocity corresponds to the mid-point of AB in fig. 99 *a*. Consider a reference frame moving with a velocity corresponding to a point on the perpendicular bisector ab of AB. ab is parallel to the ridge AB, so in the frame the ridge will move along itself and will have no velocity at right angles to AB. The same is true of the plate boundaries BC and CA when observed from reference frames whose velocities lie on bc and ca respectively. The perpendicular bisectors of the sides of *any* triangle meet at the centroid, and this point (J) in velocity space gives the velocity with which the triple junction moves. It is therefore always possible to choose a reference frame in which the triple junction does not change with time. Such a junction between three ridges is therefore stable for all ridge orientations and spreading rates.

A more complicated junction is that of three trenches, TTT in fig. 99 *b*. The arrows are on the plates being consumed and show the relative vector velocities between plates. The velocity triangle is formed as before, but points in velocity space corresponding to reference frames in which the position of the plate boundaries is fixed no longer lie on the perpendicular bisectors of the sides of the triangle. Consider the trench between plates A

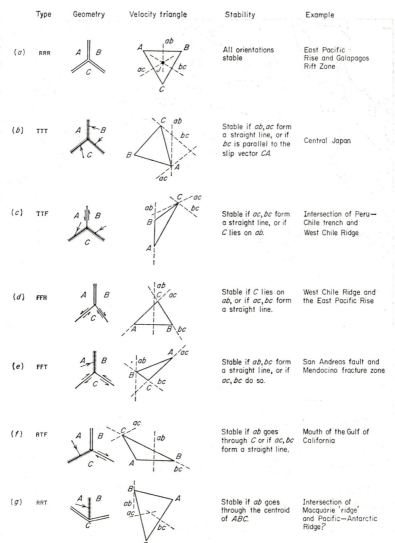

Type	Geometry	Velocity triangle	Stability	Example
(a) RRR			All orientations stable	East Pacific Rise and Galapagos Rift Zone
(b) TTT			Stable if *ab*, *ac* form a straight line, or if *bc* is parallel to the slip vector *CA*.	Central Japan
(c) TTF			Stable if *ac*, *bc* form a straight line, or if *C* lies on *ab*.	Intersection of Peru—Chile trench and West Chile Ridge
(d) FFR			Stable if *C* lies on *ab*, or if *ac*, *bc* form a straight line.	West Chile Ridge and the East Pacific Rise
(e) FFT			Stable if *ab*, *bc* form a straight line, or if *ac*, *bc* do so.	San Andreas fault and Mendocino fracture zone
(f) RTF			Stable if *ab* goes through *C* or if *ac*, *bc* form a straight line.	Mouth of the Gulf of California
(g) RRT			Stable if *ab* goes through the centroid of *ABC*.	Intersection of Macquarie 'ridge' and Pacific—Antarctic Ridge?

FIGURE 99 The geometry (symbols as in fig. 97) and stability of seven triple junctions, each of which have a possible example in the Pacific Ocean today as indicated in fig. 100. Dashed lines *ab*, *bc* and *ac* in the velocity triangles join points the vector velocities of which leave the geometry of *AB*, *BC* and *AC*, respectively, unchanged. After McKenzie & Morgan (1969), *Nature*, **224**, 125–33.

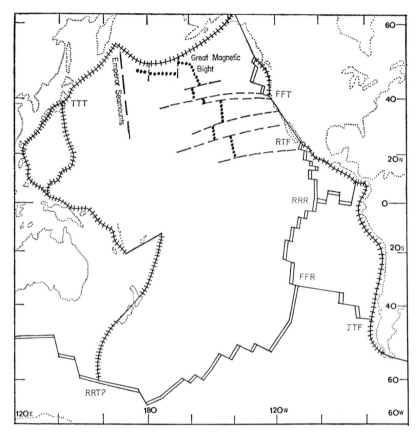

FIGURE 100   The ridge and trench systems (symbols as in fig. 97) in the Pacific region with possible examples of the triple junctions analysed in fig. 99 indicated. Some fracture zones in the North Pacific are indicated by dashed lines and the magnetic anomalies corresponding to the Cretaceous–Tertiary boundary (65 My) are shown by the dotted lines.

and B. Because A is not consumed, the trench does not move relative to A. This condition is clearly also satisfied by any reference frame with a velocity parallel to the plate boundary AB. Such velocities correspond to points on ab, a line through A parallel to the trench AB. The lines bc and ca are constructed in the same way. Unlike the triple junction of three ridges, ab, bc and ca do not intersect at a point unless certain conditions are satisfied. One of these is that if bc goes through A, and therefore the plate boundary BC is parallel to $_A\boldsymbol{v}_C$, the junction is stable and fixed to plate A. Another possible stable arrangement occurs if ab and ac are the same line, requiring that boundaries AB and AC form one straight line. A triple junction between three trenches can thus be stable. The principles outlined for the preceding two cases are readily applied to the remaining five cases as illustrated in fig. 99. Each of the seven cases analysed in fig. 99 has an example at the present time in the Pacific and they are indicated in fig. 100.

The analysis of plate evolution through the stability of triple junctions has some bearing on the interpretation of magnetic anomalies in the North Pacific. Here the guidelines of separating continents are absent. The configuration of linear magnetic anomalies in this region is also shown in fig. 100 as summarized by Vine & Hess (1970). Two points of particular interest are the origin of the so-called 'Great Magnetic Bight' at 50N, 160W (Elvers *et al.*, 1967). Vine & Hess (1970) have shown that the bight can be explained by spreading from a triple junction of ridge crests (fig. 99*a*), and this implies that a great deal of the spreading geometry of the North Pacific has been lost. Vast quantities of oceanic crust must have been destroyed between the Alaskan peninsula and the Great Magnetic Bight during the Cainozoic. Note that this view is consistent with the movement of the Pacific plate deduced from the magnetization of seamounts (§5.2.4). As a result, the Aleutian trench is consuming magnetic anomalies in the reverse order to what one might expect (that is younger anomalies first), although in some respects this is no different from the situation off California where a trench system appears to have overridden a ridge crest. The anomalies east of Japan are not readily correlated with those of the northwest Pacific, and it seems probable that they are older than those south of the Aleutians. The boundary between these two is probably the Emperor seamount chain, a former transform fault that accommodated a younger phase of spreading to the east.

5.3.3 *Mountain belts and geosynclines*

The concepts and consequences of sea-floor spreading and the theory of plate tectonics have been discussed without reference to the continents. Within the framework of this theory the role of the continents is a passive one, but the distribution of continental crust is important in determining the geometry of plates and the nature of boundaries at the leading edges of continents.

Rifting of the lithosphere may be initiated beneath a continent or simply beneath older oceanic crust. In the latter case new oceanic crust is formed and the lithosphere and older oceanic crust is consumed in the marginal trenches with the formation of an island arc (e.g. Japan) behind the trench. The ridge need not be median within the oceanic area as is the case in the Pacific Ocean at the present time. When rifting takes place beneath

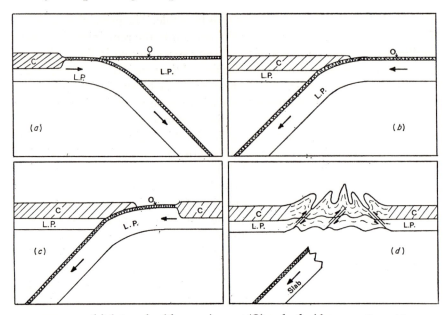

FIGURE 101   (*a*) A trench with oceanic crust (O) on both sides may attempt to consume a continent. Continental crust (C) cannot sink, so that the direction of overthrusting changes (*b*) to consume the oceanic crust originally behind the island arc. (*c*) If the trench in (*b*) attempts to consume a continent, then (*d*) mountains are built over a wide zone by overthrusting etc., and regeneration of the sinking slab of lithospheric plate (L.P.) ceases. After McKenzie (1969).

a continent a new ocean basin begins to be formed by spreading (e.g. the Red Sea) with the continental fragments riding passively on lithospheric plates. A continent drifting on a lithospheric plate may eventually arrive at a trench (fig. 101*a*), which will then attempt to consume it. Because continental crust is made of much lighter material, it is too buoyant and cannot sink. If the trench and island arc originally had oceanic crust on both sides, then the island arc will 'flip' so as to consume the oceanic crust behind the arc (fig. 101*b*). A continental arc is then formed (e.g. the western side of South America) consuming oceanic crust along the edge of the continent from the oceanic side. Another continent may then arrive at the trench (fig. 101*c*). The continental arc then collides with the approaching continent and mountains are built over a wide range by overthrusting (e.g. the Himalayas). Regeneration of the sinking slab of lithosphere ceases (fig. 101*d*) and the plate boundaries subsequently change to permit consumption of oceanic lithosphere at another region. Essentially this produces an ocean basin with a median ridge and surrounded by recently undeformed trailing edges of drifting continents, typical of the Atlantic and Indian Oceans today.

The consequences of the various forms of plate evolution just outlined are that oceanic crust is continually being created and destroyed, and that continental crust is continually being rifted apart and rewelded together. The distribution of continental crust is thus the only record that remains of the plates of the past. Comparisons of the palaeomagnetic poles for the various continental blocks provide a means of detecting these past plate motions and of reconstructing the surface of the globe at various times. Before the palaeomagnetist can tackle the problem, it is necessary at least to have some idea whereby the past plate boundaries can be recognized within the present continental crust. All the classical ideas of mountain belts and geosynclines need to be re-examined in the light of the theory of plate tectonics as has been fully discussed by Dewey & Bird (1970a, b). Drake *et al.* (1959) summarized the problems of mountain belt evolution in the following way. If the continental margin of eastern North America is accepted as a future mountain system, it is necessary to outline the major requirements needed to convert it to such a system. The single most important process in an orogenic cycle is the one which increases crustal thickness in this area to continental proportions. Once the sources of heat, which are the key to orogenesis, are determined and the method by which it is focused on small areas is established, the nature of orogenic movements should be revealed.

On plate tectonics mountain belts develop by the deformation and metamorphism of the sedimentary and volcanic assemblages of Atlantic-type continental margins. These assemblages result from the events associated with the rupture of continents and the expansion of oceans by lithospheric plate generation at oceanic ridges. The earliest assemblages thus developed are volcanic rocks and coarse clastic sediments deposited in fault-bounded troughs on separating continental crust. As the continental margin moves away from the ridge, marine sediments (sandstones and limestones) accumulate on the continental shelf, whilst turbidites accumulate on the continental rise. This type of continental margin is transformed into an orogenic belt in one of two ways (Dewey & Bird, 1970a). If a trench develops near, or at, the continental margin to consume lithosphere from the oceanic side, a mountain belt (cordilleran type) grows by dominantly thermal mechanisms related to the rise of calc-alkaline and basaltic magmas. Cordilleran-type mountain belts are characterized by paired metamorphic belts (blueschist on the ocean side and high temperature on the continental side) and divergent thrusting and synorogenic sediment transport from the high temperature volcanic axis. If the continental margin collides with an island arc, or with another continent, a collision-type mountain belt develops by dominantly mechanical processes. Where a continent–island arc collision occurs, the resulting mountains will be small, such as the Tertiary fold belt of northern New Guinea, and a new trench will develop on the oceanic side of the arc. Where a continent–continent collision occurs, the mountains will be large, such as the Himalayas, the single trench zone of plate consumption being replaced by a wide zone of deformation. Collision-type mountain belts do not have paired metamorphic belts; they are characterized by a single dominant direction of thrusting and synorogenic sediment transport, away from the site of the trench over the underthrust plate (Dewey & Bird, 1970*a*).

It is possible to accommodate the classification of geosynclines by Kay (1951) within the framework of plate tectonics, with the notable exception of entirely continental eugeosynclines (Dewey & Bird, 1970*b*). Plate tectonics does not necessitate the concept of forelands with geosynclinal zones between them, such as has been commonly accepted for the Appalachians, Caledonides and Urals. In any case continuity of continental crust under such geosynclinal zones from foreland to foreland throughout its history has never been demonstrated.

Although the cordilleran–island arc and collision mechanisms are probably the fundamental ways by which mountain building occurs, mountain belts and their associated geosynclines are generally the result of complex combinations of these mechanisms. According to Bird & Dewey (1970), the evolution of the Appalachians involved Ordovician cordilleran and island arc mechanisms and Devonian collision. The Alpine–Himalayan system has been developing since early Mesozoic times by multiple collision resulting from the motion of subcontinents and island arcs across the Tethyan–Indian Ocean. Mountain belts such as the Urals and the Verkhoyansk, lying at present within a single continent, may be the sites of successive collision zones between subcontinents, the former occurring in the Palaeozoic (Hamilton, 1970) and the latter in the Mesozoic. In either case the collision may have resulted from the contraction of a major ocean basin. It is these aspects of plate tectonics that can be analysed through palaeomagnetic investigations. Palaeomagnetism provides the basic method of estimating the past motions of plates and the past distribution of continental crust as is fully discussed in chapter 7.

6 Apparent polar-wandering

6.1 AXIAL GEOCENTRIC DIPOLE

This chapter is concerned largely with a review of the palaeomagnetic data which are summarized in the appendix. The interpretation of all results in terms of the past motions of plates is given in chapter 7. The important assumption in palaeomagnetism is that of an axial geocentric dipole, so that the first stage in the analysis of the results is to investigate whether or not this model describes the palaeomagnetic data for the past few million years. Attempts to show that the model holds over geological time rely on palaeo-climatic comparisons with palaeomagnetic data. A description of these comparisons follows. This ultimate justification of the model leads to the concept of apparent polar-wander paths, the most useful method of describing the palaeomagnetic results from various parts of the world.

6.1.1 *Upper Tertiary to Recent*

Before considering the longer term aspects of the geomagnetic field over the Upper Tertiary, it is instructive first to consider the relatively short-term behaviour over the past 1000 years or so as deduced from archaeomagnetic information. In fig. 14 (§ 1.2.2) the field variations at London deduced from archaeomagnetic and historical measurements were plotted (Aitken, 1970). If a series of lava flows were extruded at London at equal intervals of 100 years, say, the palaeomagnetists would measure a series of spot readings of the geomagnetic field, average them, and then deduce a palaeomagnetic pole. As an exercise the field values at 100 year intervals (ten values from A.D. 1000 to A.D. 1900 and three values from A.D. 100 to 300) have been converted to virtual geomagnetic poles. These are illustrated in fig. 102, from which it is apparent that they cluster about the geographic pole rather than the geomagnetic pole. The statistics of these field directions and VGPs are given in table 14. Within the limits of error the time average over the past 10^3 years for ten roughly equally spaced measurements would have provided a quite reasonable estimate of the present geographic pole position. In general however lava flows very rarely are extruded at such convenient equally spaced intervals and one problem is to decide just how many lavas might be required to provide sufficient time to average to the axial geocentric dipole field. In some cases it has been shown that even a large number of flows might in palaeomagnetic terms represent only

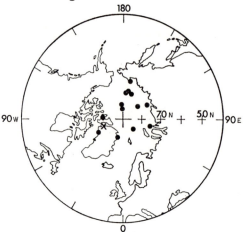

FIGURE 102 Virtual geomagnetic poles corresponding to the field variations at London deduced from archaeomagnetic measurements (Aitken, 1970). Values are taken at 100 year intervals (A.D. 100–300 and A.D. 1000–1900) from the data given in fig. 14 (§1.2.2). Polar stereographic projection north of 40N; diagonal cross is the present geomagnetic pole.

TABLE 14 *Mean of* VGPs *calculated from British archaeomagnetic measurements found by averaging values at* 100 *year intervals from fig.* 103. (N, *number of* VGPs; *K, precision parameter;* A_{95}, *circle of* 95 *per cent confidence*)

Period	N	K	A_{95}	Pole position
A.D. 1000–1900	10	37	8	86N, 148E
A.D. 100–300 and A.D. 1000–1900	13	46	6	86N, 162E

a few spot readings of the geomagnetic field in the past (Wilson, 1970a).

There are seventy results listed in the appendix which are classified as Plio-Pleistocene and Quaternary. It is estimated that they cover the past 5 My. The palaeomagnetic poles are illustrated in fig. 103 and they quite clearly cluster about the geographic pole rather than the geomagnetic pole. The world average of palaeomagnetic results for this period thus clearly conforms with the axial geocentric dipole hypothesis. Three poles (JA13.2, NA13.2 and OC13.29) are widely divergent with latitudes of less than 60°. The mean of the sixty-seven poles lies at 88.8N, 131.9E (table 15a) with A_{95} of 1.9° and therefore is not significantly different from the geographic pole. Opdyke & Henry (1969) have analysed the mean inclinations observed in fifty-two deep-sea sediment cores from all the oceans. They argue that

186

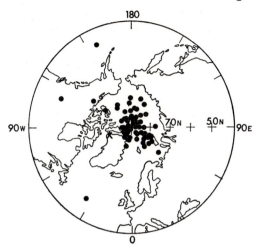

FIGURE 103   Palaeomagnetic poles for the Plio-Pleistocene to Recent (5–0 My). Polar stereographic projection north of 40N; diagonal cross is the present geomagnetic pole.

the mean inclination in each core is a time average over a period of the order of 10^5 to 10^6 years, since each core selected included the Brunhes–Matuyama boundary known to be at 0.7 My. A plot of mean inclination versus latitude of the cores should follow the curve of fig. 17 (§1.2.4) for an axial geocentric dipole. The best fitting *geocentric* dipole had north polar co-ordinates of 89N, 211E. The 1° offset from the rotational pole was within the limits of error and no significant departure from the *axial* geocentric dipole was observed. One very important conclusion from this result is that there can be no significant 'inclination error' in sediments from depositional effects (§2.3).

Palaeomagnetic studies are not in general made on a world-wide basis, but rather from limited regions of the earth's surface. The comparison of results between regions is the basis of the method of deciding what relative movement has taken place between the regions. It is thus of some importance to establish to what extent the *regional* Upper Tertiary and Plio-Pleistocene to Quaternary results conform with the axial geocentric dipole hypothesis. In the analyses that follow the earth's surface has been divided into thirteen regions of roughly equal size, but controlled to a certain extent by the availability of palaeomagnetic results. When considering all the 194 poles for the past 20 My (Upper Tertiary to Quaternary) there are sufficient data to analyse for each region. However when the data are split into two parts, Plio-Pleistocene to Quaternary and Upper Tertiary, then in each case there are only sufficient data to average from ten and nine

187

TABLE 15 *Analysis of world-wide palaeomagnetic poles for the Upper Tertiary and Quaternary. N, number of poles (number omitted); K, precision parameter; A_{95}, circle of 95 per cent confidence. The site mean longitude is given after each region*

(a) *Plio-Pleistocene and Quaternary (5–0 My)*

Region	N	Mean pole	K	A_{95}
1. Western North America (126w)	4 (1)	82.1N, 172.5E	113.0	8.7
2+3. North Atlantic (20w)	5	87.3N, 126.1E	89.5	8.1
4. Western and southern Europe (14E)	8	84.6N, 153.8E	142.6	4.6
5. Southern U.S.S.R. (47E)	13	85.9N, 69.6E	97.9	4.2
6. South Atlantic (30w)	4	83.1N, 6.0E	263.6	5.7
7. Indian Ocean (59E)	5	83.5N, 200.1E	199.6	5.4
8. South Pacific (170E)	6 (1)	86.7N, 346.6E	258.9	4.2
9. The Far East (131E)	9 (1)	87.1N, 244.6E	35.8	8.7
10. Aleutians and Kamchatka (176E)	7	87.7N, 124.0E	196.0	4.3
11. Hawaiian islands (156w)	4	86.3N, 10.4E	166.9	7.1
10 regions combined	10	89.0N, 137.4E	254.1	3.0
10 common site longitude poles	10	88.9N, 116.9E	255.0	3.0
All poles combined	67 (3)	88.8N, 131.9E	86.4	1.9

(b) *Upper Tertiary (Miocene and Pliocene; 20–5 My)*

Region	N	Mean pole	K	A_{95}
1. Western North America (126w)	9 (2)	86.8N, 140.1E	55.9	7.0
2+3. North Atlantic (20w)	14	83.8N, 119.1E	78.3	4.5
4. Western and southern Europe (14E)	19 (1)	83.7,N 141.6E	58.5	4.4
5. Southern U.S.S.R. (47E)	32	78.4N, 195.0E	43.7	3.9
6. South Atlantic (60w)	6	86.4N, 104.0E	172.1	5.1
8. South Pacific (170E)	5	82.3N, 307.6E	23.2	16.2
9. The Far East (131E)	17 (2)	86.7N, 283.1E	54.3	4.9
10. Aleutians and Kamchatka (176E)	6	87.9N, 16.5E	33.7	11.7
12. East Africa and S. Arabia (37E)	7	88.3N, 187.2E	105.2	5.9
9 regions combined	9	88.1N, 169.8E	187.2	3.8
9 common site longitude poles	9	85.6N, 149.9E	375.4	2.7
All poles combined	119 (5)	85.8N, 182.4E	41.6	2.0

(c) *Upper Tertiary to Quaternary (20–0 My)*

Region	N	Mean pole	K	A_{95}
1. Western North America (126w)	13 (3)	85.5N, 157.1E	65.8	5.2
2. Iceland and Jan Mayen Land (18w)	5	85.4N, 11.5E	103.3	7.6
3. Cape Verde Islands etc. (21w)	14	82.2N, 131.4E	114.6	3.7
4. Western and southern Europe (14E)	27 (1)	84.0N, 144.8E	72.4	3.3
5. Southern U.S.S.R. (47E)	45	82.4N, 187.6E	39.4	3.4
6. South Atlantic (50w)	10	86.7N, 46.9E	138.4	4.1
7. Indian Ocean (59E)	7	86.4N, 197.2E	100.7	6.0
8. South Pacific (170E)	11 (1)	85.0N, 321.0E	49.0	6.6
9. The Far East (131E)	26 (3)	87.0N, 271.1E	47.9	4.1
10. Aleutians and Kamchatka (176E)	13	88.7N, 80.8E	64.8	5.2
11. Hawaiian islands (156w)	4	86.3N, 10.4E	166.9	7.1
12. East Africa and S. Arabia (37E)	8	88.3N, 181.3E	122.1	5.0
13. Siberia (91E)	3	(72.3N, 234.0E)	51.5	17.4
12 regions combined	12 (1)	88.8N, 146.7E	285.4	2.6
12 common site longitude poles	12 (1)	87.2N, 141.0E	413.7	2.1
All poles combined	186 (8)	87.0N, 175.9E	50.2	1.5

regions respectively. Eight poles with latitudes of less than 60° are excluded from the analyses. The thirteen regions are:

1. Western North America, including Alaska;
2. Iceland and Jan Mayen Land;
3. Cape Verde Islands, Canary Islands, Madeira and the Azores;
4. Western and southern Europe, including Poland, the Balkans and Greece;
5. Southern u.s.s.r. and Turkey;
6. The South Atlantic, including the Argentine, Tristan da Cunha, South Shetlands and western Antarctica;
7. Indian Ocean (Réunion, Mauritius and Heard Island);
8. The South Pacific, including eastern Australia, New Zealand, Cook Islands, Tonga, Samoa and the New Hebrides;
9. The Far East, including Taiwan, Korea, Japan, Sakhalin and the Kharbarovsk region of the far eastern u.s.s.r.;
10. Kamchatka and the Aleutian islands;
11. The Hawaiian islands;
12. East Africa and South Arabia;
13. Siberia.

A statistical analysis of the poles for each region is given in table 15. Table 15a gives the data for the Plio-Pleistocene and Quaternary (5–0 My) whilst table 15b gives the Upper Tertiary results (Miocene and Pliocene; 20–5 My). All results are combined in table 15c for the Upper Tertiary and Quaternary (20–0 My). For the whole period there are only three results from Siberia, so that region 13 is not in fact used. Wilson & Ade-Hall (1970) first pointed out that there is a tendency for the mean poles for different parts of the earth's surface, for the Upper Tertiary and Quaternary, to lie on the far side of the geographic pole from the region being sampled. This property has been analysed in more detail by Wilson (1970b, 1971), who restricted his analysis to data derived from at least twenty oriented samples and whose polar errors were less than or equal to 12°. About twice as much data is involved in the present analysis, but the sampling requirements are less stringent. To analyse the effect Wilson (1971) introduced the concept of the *common site longitude*, in which all observers (collection sites) are displaced about the rotation axis so that they have a common longitude taken as zero. Each observer remains at his original latitude and his observed palaeomagnetic poles are rotated along with him. The method makes it possible to see all the poles as they are seen by their own observers. Table 15 gives the site mean longitude for each of the regions, so that the common site longitude for each pole is simply obtained by subtracting the eastwards site mean longitude from the longitude of the mean pole position.

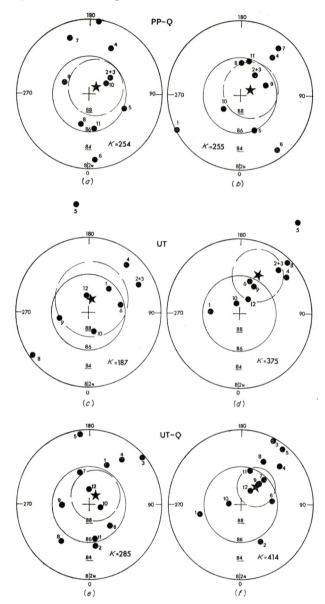

FIGURE 104. For legend see facing page.

The analyses of table 15 are illustrated in fig. 104. For the Plio-Pleistocene and Quaternary the original site longitude regional poles are randomly scattered about a polar projection (fig. 104*a*). No improvement in grouping (change in precision parameter K) is observed when referred to their common site longitudes (fig. 104*b*). The means lie 1.0° and 1.1° respectively from the geographic pole, well within the 3.0° error attached to each (table 15*a*). The Upper Tertiary poles are likewise scattered about the projection when referred to their original site longitudes (fig. 104*c*), with the mean lying 1.9° away from the geographic pole – not significant with an error of 3.8°. When referred to their common site longitudes (fig. 104*d*), *all* nine regional poles lie on the far side of the observer and their grouping has improved significantly. Their mean now lies 4.4° away from the geographic pole, significant when the error is 2.7° (table 15*b*). When all 186 Upper Tertiary and Quaternary poles are considered together, then ten of the twelve regional means lie on the far side of the observer when referred to their common site longitudes (fig. 104*f*). There is an improvement in grouping and the common site longitude mean lies 2.8° away from the geographic pole, just outside the 2.1° error limit. The data thus suggest that an important change in the configuration of the geomagnetic field took place over the past few million years.

Both Wilson & Ade-Hall (1970) and Wilson (1970*b*, 1971) point out that it is difficult to imagine that the far-sided effect observed in the Upper Tertiary data could be due to sea-floor spreading over the past 20 My. This would require that all motions have been essentially northwards, the effects of which would surely be observed in the Arctic today. Setting aside the possibility that the observation is a coincidence, Wilson (1970*b*, 1971) suggests that a possible explanation is that the time average of the geomagnetic field is not that of an axial *geocentric* dipole, but that of an axial dipole shifted north of the equator (fig. 105). The essential feature of this model is that the inclination is shallower than the axial geocentric dipole inclination in the northern hemisphere, but steeper in the southern hemi-

---

FIGURE 104 Regional mean pole positions (numbered as in table 15) for the Upper Tertiary and Quaternary plotted on a polar stereographic projection north of 82N. Left hand figures (*a*), (*c*), (*e*) are the original site longitude representation of the poles, and the right hand figures (*b*), (*d*), (*f*) are the common site longitude representation.
(*a*), (*b*) Plio-Pleistocene and Quaternary (5–0 My)
(*c*), (*d*) Upper Tertiary (Miocene and Pliocene; 20–5 My)
(*e*), (*f*) All Upper Tertiary and Quaternary (20–0 My)
The star is the regional mean pole position and the dashed circle the ninety-five per cent confidence limit giving unit weight to each region.

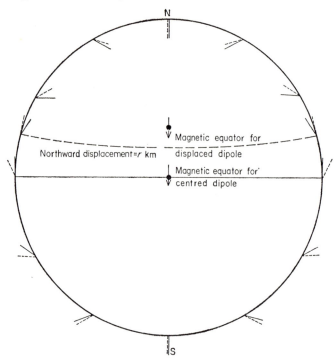

FIGURE 105   Offset dipole model of Wilson (1970*b*, 1971). Inclinations at the earth's surface are everywhere more negative (dashed lines) than the inclination of the centred dipole (solid lines). From Wilson (1971).

sphere. The observations require that this model has applied back to 20 My ago during the Upper Tertiary, but that over the past few million years the field has been much closer to that of an axial geocentric dipole. If the explanation is correct, it would indicate that non-dipole components may persist for a few tens of millions of years.

The axial geocentric dipole relation between magnetic inclination I and colatitude θ (see §1.2.4) is given by:

$$\tan I = 2\cot\theta. \qquad (6.1)$$

The equivalent relation between I and θ for a dipole displaced northwards r km along the rotational axis (Wilson, 1971) is:

$$I = \tan^{-1}\left[2\cot\left\{\theta + \sin^{-1}\left(\frac{r\sin\theta}{\rho}\right)\right\}\right] - \sin^{-1}\left(\frac{r\sin\theta}{\rho}\right), \qquad (6.2)$$

where $\rho = (r^2 + R^2 - 2Rr\cos\theta)^{\frac{1}{2}}$ and R = radius of the earth.

On this model the departure from the geocentric axial dipole inclinations are least at the poles and greatest in equatorial and temperate latitudes.

Wilson's (1971) analysis of ninety-six Upper Tertiary and Quaternary poles gave a value of $r = 285 \pm 74$ km to produce minimum scatter in the poles. The data analysed here essentially confirm this result. Note that the first approximation to the earth's magnetic field over Upper Tertiary to Recent times is still an axial geocentric dipole, the model above provides the second approximation. If this type of second approximation to the average geomagnetic field has persisted throughout geological time, it suggests that, no matter how accurately one might determine the palaeomagnetic pole position for a particular region at a particular epoch, the palaeomagnetic pole might still be displaced from the rotational pole by up to about 5°.

6.1.2 *Palaeoclimates and palaeolatitudes*

Irving (1956a) first suggested that comparisons between palaeomagnetic results and the geological evidence of past conditions could provide a test for the hypothesis of an axial geocentric dipole over geological time. The essential point is that both palaeomagnetic and palaeoclimatic data provide evidence for past latitudes and the factors contolling climate are quite independent of the earth's magnetic field. For the purpose of the comparison it is necessary to recognize some feature called a *palaeoclimatic indicator* which can reasonably be assumed to indicate some element of climate. At the present time the mean annual equatorial temperature is about $+25$ °C, and the polar value is about -25 °C. Whilst the range in temperature from the equator to the pole may have varied in the past, the simplest palaeoclimatic model is derived from the fact that the net solar flux reaching the earth's surface has a maximum at the equator and a minimum at the poles. Temperature therefore follows the same pattern. The density distribution of very many climatic indicators at the present time shows a maximum at the equator and either a polar minimum or a whole high latitude zone from which the indicator is absent (such as reefs, evaporites and carbonates). A less common distribution, seen today in the distribution of glacial phenomena and some deciduous trees for example, has a maximum in polar or intermediate latitudes.

The determination of the palaeomagnetic pole for any region on the assumption of an axial geocentric dipole enables palaeolatitude lines to be drawn across the region. These are then compared with the occurrences of various palaeoclimatic indicators. These comparisons can be made in four distinct ways.

(1) The time variation of palaeolatitude for any single place may be compared with palaeoclimatic evidence from that place (Irving, 1956a; Blackett, 1961).

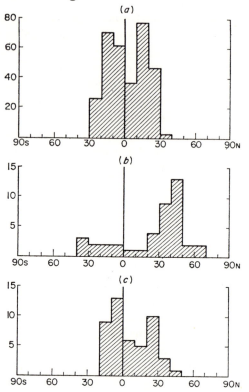

FIGURE 106   Equal angle latitude histogram for organic reefs. (*a*) Present latitude of modern reefs, compiled from Termier & Termier (1952); (*b*) present latitude of fossil reefs; (*c*) palaeolatitude of fossil reefs. After Briden & Irving, *Problems in palaeoclimatology*. © Interscience Publishers, Inc. 1964.

(2) The space variations of palaeolatitude over a given region for a given time may be represented as maps with the palaeolatitudes and the occurrences of palaeoclimatic indicators compared (Runcorn, 1961; Opdyke, 1962; Briden & Irving, 1964).

(3) The latitude (or palaeolatitude) range of the main occurrences of palaeoclimatic indicators may be plotted against geological time (Briden & Irving, 1964).

(4) The palaeolatitude values for a particular occurrence may be compiled in the form of equal angle or equal area histograms to give the palaeolatitude spectrum of the particular indicator (Irving & Gaskell, 1962; Irving & Briden, 1962; Briden & Irving, 1964).

The fourth method listed above is the most useful approach to adopt. The examples which follow are from the compilations of Briden & Irving (1964). These examples are based upon palaeomagnetic data of about seven or

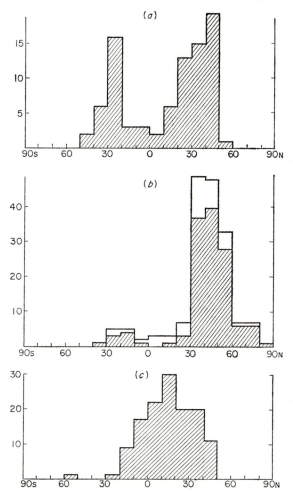

FIGURE 107   Equal angle latitude histogram for evaporites. (*a*) Present latitude of modern terrestrial evaporites, compiled from Lotze (1957); (*b*) present latitude of fossil evaporites, shaded portions being those for which palaeomagnetic control is available; (*c*) palaeolatitude of fossil evaporites. From Briden & Irving, *Problems in palaeoclimatology*. © Interscience Publishers, Inc. 1964.

more years ago, but for the purposes of the comparisons, these are still reasonably valid.

The distribution of modern coral reefs is symmetrical about the equator (fig. 106*a*), the maximum frequency being between 10° and 20° latitude and most occurrences lying within 30° of the equator. The present latitude of fossil reefs (fig. 106*b*) does not show this distribution, but when referred to their palaeolatitudes (fig. 106*c*) the spectrum is very similar to that of modern reefs, over ninety-five per cent of the occurrences falling within 30° of the equator. The deposition of evaporites is thought to require high temperatures, the most notable modern deposits being in the tropical and temperate deserts or semi-deserts, associated with the dry trade wind belts and with the arid centres of large continents. No occurrences are recorded north of 53°N or south of 43°S (fig. 107a) and there are only two regions, one in east Africa and the other in Peru, in which evaporite deposition occurs within 10° of the equator. The present latitudes of fossil evaporates through the geological column show a spread from over 30°S to over 80°N (fig. 107b), but when referred to their palaeolatitudes these occurrences condense into a distribution with seventy-five per cent lying within 30° of the palaeoequator (fig. 107c).

The only example so far analysed which relates to plant distribution is that of coalfields. Coal deposits indicate the existence of moist conditions, which could arise either from heavy precipitation in warm environments or modest rainfall in cold climates. Coal forms where the accumulation of vegetation exceeds its removal or decay. This occurs either in hot rain forests where, although decay is rapid, growth rates are high, or in cold places where, although growth may be less rapid, decay is inhibited by cold winters. The palaeolatitude spectrum of coalfields shows this effect (fig. 108). In the figure both hemispheres are plotted together using an equal area histogram in which the class intervals are as the sine of the latitude and therefore contain equal areas of the earth's surface. The low palaeolatitude group is predominantly the Carboniferous coals of western Europe and North America, whilst the high palaeolatitude group is mainly the Permian and younger coals from Canada, Siberia and the southern continents. The two groups are also distinct in the fossil floras they contain.

An example of the palaeolatitude distribution of a faunal group has been given by Irving & Brown (1964) for labyrinthodont reptiles. This was an attempt to show latitude variation in taxonomic diversity. The present latitudes show a group in each hemisphere, the northern group being much the more numerous. The palaeolatitude distribution shows fair symmetry about the palaeoequator (fig. 109). This result was the cause of much discussion (Stelhi, 1966; Irving & Brown, 1966), which served to illustrate the point that one of the most important aspects of this type of investigation is the problem of sampling and definition. Stehli (1968, 1970) has proposed that the study of taxonomic diversity is the most reliable palaeoclimatic parameter to use to test the axial dipole hypothesis in the past. For large groups of widely distributed organisms, diversity reaches a maximum at or near the equator showing a smooth decrease into higher latitudes. Applying this model to the distribution of Permian brachiopods, Stehli (1970) claims

196

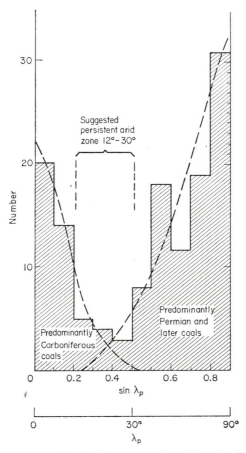

FIGURE 108   Equal area latitude histogram of coal deposits. Palaeolatitude values are plotted irrespective of sign. The dashed line shows the tentative division into two groups centred near the equator and poles respectively. Using the data of Briden & Irving (1964). From Irving, *Paleomagnetism and its application to geological and geophysical problems.* © John Wiley & Sons, Inc. 1964.

that the present latitude framework is a far better description of the diversity data than one derived from Permian palaeomagnetic data. He concludes that, unless the palaeontological or palaeomagnetic data are incorrect, the earth's magnetic field in Permian time was not axial. However, during the Permian land–sea distribution was longitudinally asymmetrical due to the existence of Pangaea (§7.3). Because of the earth's rotation one might expect climatic zones across the single supercontinent to be oblique to geographic latitude lines. In this case 'diversity gradient poles' might be expected to differ systematically from the geographic poles.

197

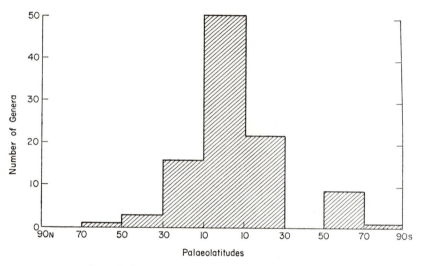

FIGURE 109   Generic frequency versus palaeolatitude for Upper Carboniferous and Permian labyrinthodonts. Using the data of Irving & Brown (1966). From Briden (1968), *History of the earth's crust*. Princeton Univ. Press, 178–94.

The study of palaeowind directions determined from cross stratification in aeolian sandstones is another palaeoclimatic indicator which can be compared with palaeolatitudes deduced from palaeomagnetism. Opdyke & Runcorn (1960) have shown that in the upper Palaeozoic aeolian sandstones of Europe and North America, the directions observed correspond to a trade wind belt if compared with the palaeomagnetic equator. This type of comparison is limited to the occurrences of aeolian sandstones which are not very abundant in the geological column, so that its usefulness is somewhat restricted. At the other extreme, evidence of glacial conditions provides an indication of a polar or near polar position of the region concerned. This can provide especially useful confirmation of a palaeomagnetic pole deduced from rocks associated with the glacial deposits. The presence of the pole in a region does not necessarily mean that evidence of glaciation will be found, but the converse is certainly true. In certain instances this evidence can suggest a possible position of the pole in the absence of any palaeomagnetic information. Information such as this will be used in interpolating the position of the palaeomagnetic pole between two periods later on in this chapter.

The basic assumption made by palaeoclimatologists is that ancient climatic belts run essentially parallel to ancient lines of latitude. There are wide departures from such a pattern today caused by perturbations in planetary atmospheric circulation introduced by the distribution of land

198

and sea, the so-called monsoon effect. If the distribution of land and sea was markedly different in the past, as must surely follow from the theory of plate tectonics, there could be quite considerable departures from the present day climatic belts and from simple parallelism. This rather interesting new approach has been pursued by Robinson (1971), who shows that if some commonly accepted Triassic and Permian reconstructions are assumed, then the low pressure equatorial belt responsible for the monsoon would be substantially displaced. This more realistic way of tackling past climatic problems could well provide better explanations for the past distributions of faunas and floras.

6.1.3 *Apparent polar-wander paths*

There are two principle ways of presenting palaeomagnetic information for a given region over a number of epochs. For each epoch a palaeogeographic map of the region can be drawn showing the directions of the palaeo-meridians and lines of latitude drawn across the region. These maps are generally found to be most useful by climatologists for comparison with relevant information, but a simple view of the overall variation from time to time is not easy because a whole series of maps, one for each epoch, are required. A much simpler approach, and one which is inherently more useful, is to plot the successive positions of the palaeomagnetic pole from epoch to epoch on the present latitude–longitude grid. The path traced out by the palaeomagnetic poles is called the *apparent polar-wander path* for that region. It should be noted that the procedure adopted is to *assume* the region fixed and plot the movement of the pole. This polar movement could represent polar-wandering or continental drift and the data from a single region cannot alone distinguish between the two. The first estimates of apparent polar-wandering were made by Creer *et al.* (1954) relative to the British Isles. The bulk of this chapter deals with an analysis of palaeomagnetic poles listed in the appendix in terms of apparent polar-wander paths.

To draw apparent polar-wander paths, it is simplest to consider a continent presently lying in the northern hemisphere. For the Upper Tertiary a palaeomagnetic pole will be determined lying close to either the north or south geographic pole. Obviously the pole lying closest to the present north pole is regarded as the north palaeomagnetic pole, even though this is strictly speaking indeterminate due to field-reversals. During the Lower Tertiary the palaeomagnetic pole might now lie 20 degrees away from the north geographic pole, and its antipole will lie 160 degrees away. The principle of minimum movement is applied. It is always assumed that the path is the one found by joining successive poles through the shortest distance, going back in time, from the present geographic pole. There are thus two paths, one starting at the north geographic pole and is presumed to represent the path of the north pole, and the other starting at the south geographic pole representing the path of the south pole. If the north (south) pole wanders through an angle of greater than 90° it could cross the present equator and lie in the southern (northern) hemisphere. In these circumstances, if the path were not known it might be then quite wrongly assumed that the pole which lies in the northern (southern) hemisphere was, by definition, the north (south) pole. This would result in the polarity of the rocks being wrongly defined. The usual definition of polarity is that if the north-seeking pole falls in the northern (southern) hemisphere, the corresponding direction of magnetization is normal (reversed). This definition is valid for all Devonian and younger rocks, but for pre-Devonian rocks this breaks down in certain instances. Strictly speaking polarity should be defined with respect to the apparent polar-wander paths.

For the purposes of the analyses that follow, the earth's surface has been divided into a number of continental blocks or regions. These blocks have been chosen either because they are simply continents surrounded by oceans, such as Africa or Australia, or because they are regions within present continents separated from adjacent regions by an important geological feature. Examples of the latter are the Russian platform and the Siberian platform separated by the Urals, and the Kolyma block separated from the Siberian platform by the Verkhoyansk Mountains. These are essentially geological features which could, on the theory of plate tectonics, represent collision zones between two continents and therefore mark the boundaries of some past plate. This is an important point to bear in mind. It is unlikely that all points on two continents will meet at the same time or to the same degree on collision. A large strike–slip component of motion between two blocks might produce rotation of small intervening blocks. This could be a feature of collision zones. Some of the anomalous results from near these boundaries could be explained in this way as is discussed in each case.

One final point is to decide on the most appropriate time groupings of the paleomagnetic poles. There is no reason to suppose that these could follow geological periods, although of course because ages are defined by these periods this is the simplest approach to adopt. If it is supposed that typical polar groupings might cover a period of 10–20 My, then the regional results of table 15 c can be used as a guide line for analysing past regional results. From this it appears that polar groupings with K of around 30–100 might typically be expected. Values lower than this might suggest that a finer subdivision is necessary because the poles are being averaged over too great a time interval. These are the essential principles that have been used in determining the polar groupings and averages for each block.

Palaeogeographers, palaeoclimatologists and others wishing to compare palaeomagnetic results from various regions with other information will of course find it most useful to draw a palaeogeographic map of the region. There is insufficient space in this book to contemplate drawing such maps, but these are very simply constructed by using the palaeomagnetic poles for each region. The palaeomagnetic pole places the region in latitude and azimuthal orientation, the longitude is indeterminate although the directions of the meridian are clearly towards the pole. With the pole as centre it is a simple matter to draw a series of latitude lines across each region, and this essentially is the map required.

6.2 THE NORTHERN HEMISPHERE

In each of the regions analysed, the north palaeomagnetic pole path has been plotted. The mean poles are those which lie on this path irrespective of whether they fall in the northern or southern hemisphere.

6.2.1 *North America*

Much of the early work on the palaeomagnetism of North American rocks was undertaken by Graham (1949, 1954, 1955, 1956) and later by Runcorn (1955, 1956a), Howell & Martinez (1957), Howell *et al.* (1958) and Collinson & Runcorn (1960). Many of these early results have now been superseded and apart from the lower Palaeozoic, the coverage of the geological time scale is fairly good.

There are 155 poles listed in the appendix for North America and these constitute the largest body of palaeomagnetic data for any of the regions. The discussion of the apparent polar-wander path will be given in two parts, the first for the Phanerozoic (eighty-four poles) and the second for the Precambrian (seventy-one poles). These two groups are treated in different ways. For the Phanerozoic the ages of the beds studied are known fairly precisely, so that it is possible to group poles according to age and calculate group mean poles. Although the Precambrian results have been selected on the basis of isotopic age information, these are of variable quality and often represent only the minimum ages (for example from K–Ar determinations). The approach in this case is to define a trend in the pole positions grouped according to five major time intervals.

The results for the Phanerozoic are summarized as group mean poles in table 16. Two poles in the Upper and one in the Lower Tertiary have been

TABLE 16 *Summary of North American Phanerozoic poles. N is the number of poles averaged, with the number excluded given in brackets. K is the precision of the N poles and A_{95} the circle of 95 per cent confidence about the mean*

Period	N	K	A_{95}	North pole	Poles used (not used)
Upper Tertiary	9 (2)	56	7	87N, 140E	NA12.1–9 (12.10–11)
Lower Tertiary	6 (1)	23	14	85N, 197E	NA11.2–7 (11.1)
Cretaceous	9	73	6	64N, 187E	NA10.1–9
Jurassic	2	—	—	76N, 142E	NA9.1–2 (9.3)
Upper Triassic	11	61	6	68N, 97E	NA8.9–19
Lower Triassic	8	89	6	56N, 104E	NA8.1–8
Permian	6	86	7	46N, 117E	NA7.1–6
Carboniferous	13	89	4	37N, 126E	NA6.1–13
Silurian–Devonian	7	32	11	29N, 123E	NA4.1–2, 5.1–5
Ordovician	1	—	—	28N, 192E	NA3.1
Cambrian	3 (1)	31	22	7N, 140E	NA2.1, 2, 4 (2.3)

omitted, because they have rather low latitudes and deviate markedly from the others. The Upper and Lower Triassic appear to form distinct groups 12° apart. It is unlikely that this is due to inclination error (Creer, 1970a), since it has already been argued that such effects are not observed in practice (§2.3.6 and §6.1.1). There is a 33° difference between the Upper Triassic and Cretaceous group means, and although the Jurassic is not well defined, the mean of the two Jurassic poles falls midway between these groups. Eleven of the thirteen Carboniferous poles can be classified as Middle to Upper Carboniferous, as they cover the time range Late Mississippian to Late Pennsylvanian. There appears to be no obvious way in which the Carboniferous results could be subdivided into two groups as is found in other regions. However, Roy (1969), in discussing the very careful and thorough investigations of sedimentary rocks from eastern Canada made at the Dominion Observatory, provides good evidence for a polar shift of 16° between the Late Mississippian and the earliest Permian, or 27° between the Late Silurian and earliest Permian. The problem becomes one of comparing a larger body of data of variable quality with a smaller body of data of very good quality. The Silurian and Devonian poles form an indistinguishable group and have therefore all been averaged. Of the four Cambrian poles, that for the Lodore formation is widely divergent from the other three, and as it is the only result which was not derived from cleaning, it has been omitted. The single Lower Ordovician pole is a well substantiated result from Newfoundland, and it does not lie on the smooth curve between the Cambrian and later Palaeozoic poles. It has previously been suggested from palaeomagnetic results (Nairn *et al.*, 1959) that Newfoundland has rotated 20° since the Carboniferous. Du Bois (1959) however concluded no

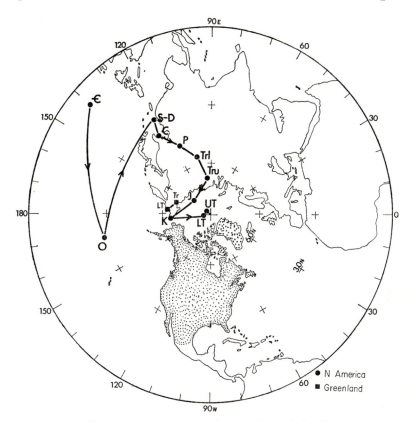

FIGURE 110   Phanerozoic apparent polar-wander path for North America.
Two results from Greenland are also shown on a polar stereographic projection.

such rotation had occurred. Later Black (1964) also confirmed there had
been no post-Carboniferous rotation, but suggested a 30° rotation in Middle
to Late Devonian time. It should be noted that one of the four Cambrian
poles is from Newfoundland, and this is in general agreement with the
other results from the mainland. This would certainly suggest there has
been no post-Cambrian rotation, and in reviewing the evidence for such
a rotation Hospers & Van Andel (1969) also doubt that any such conclusion
can now be drawn from the available data. It should be noted however that
the Appalachian orogen lies along the eastern continental margin of North
America from Newfoundland to Florida. Bird & Dewey (1970) have re-
interpreted the tectonic evolution of the Appalachians in the light of plate
tectonics. If their interpretation is correct, then the lower Palaeozoic results
from the Appalachians need to be viewed with some caution because of the
possibility that more than one plate might have been involved. With this in

mind, the Ordovician result for the moment suggests an easterly excursion between the Cambrian and Silurian as is illustrated in fig. 110.

A pole from the Fransiscan formation (NA9.3) diverges widely from those of similar age for the rest of North America. This is due to tectonic movement (largely rotation) within the western cordillera region. The diverging Lower Tertiary pole from the Siletz Volcanics of Oregon (NA11.1) is also explained in this way. Two results from Greenland conform with the minimum criteria, and are shown in fig. 110 for comparison.

Spall (1971 a) has reviewed the palaeomagnetic information for the Precambrian of North America. Taking the view that palaeomagnetic poles from Precambrian rocks suffer from a great number of unknowns, such as the effects of tectonic and orogenic events, and the difficulty of precise age determinations, Spall looks at the overall trends in apparent polar-wander rather than attempting to determine group means for various ages. This approach has been followed here so that the seventy-one Precambrian poles have been divided into five major time groupings, which, with minor modifications, follow Spall (1971 a). These are

(1) Keweenawan trend (1000–1150 My),
(2) Mackenzie trend (1150–1300 My),
(3) Elsonian trend (1300–1475 My),
(4) Animikie–Sudbury trend (1600–2060 My),
(5) Matachewan–Stillwater trend (2080–2500 My).

All the poles are plotted in fig. 111, a different symbol being used to distinguish each of the five major trends. The poles plotted are those which provide continuity with the Phanerozoic curve of fig. 110 and minimum apparent wander between successive age trends. The overall Precambrian polar-wander path suggested in fig. 111 is therefore considered to be that of the north pole.

Du Bois (1962) originally summarized the poles obtained from Keweenawan rocks of the Lake Superior region, and his determinations have been broadly confirmed by subsequent more detailed studies (e.g. Beck & Lindsley, 1969; Palmer, 1970). The Keweenawan igneous activity is dated at around 1100 My in this region with a range from 1000 to 1200 My (Goldich, 1968), so that the spread in pole positions is taken to indicate apparent polar-wandering of about 40° in a southerly direction during this time. This interpretation is favoured by the only two results from rocks outside the Lake Superior region (NA1.43 and NA1.61) which show the same trend of polar movement with age. Good evidence is also provided therefore that the earth's magnetic field was at least that of a geocentric dipole at that time. Robertson (1969) and Fahrig & Jones (1969) conclude that there is good evidence for basic igneous activity in the Canadian shield

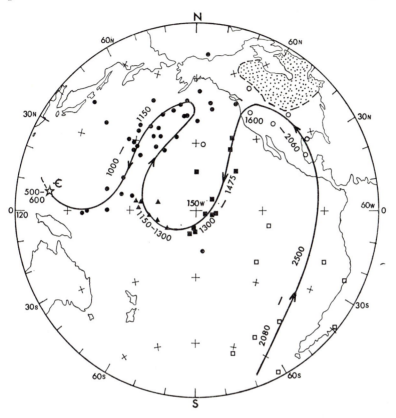

FIGURE 111 Precambrian apparent polar-wander path for North America. Poles
are plotted on an equatorial equal area projection using different symbols to indicate
the five major trends as defined by Spall (1971*a*). Open squares, Matachewan–
Stillwater trend (2080–2500 My); open circles, Animikie–Sudbury trend (1600–
2060 My); solid squares, Elsonian trend (1300–1475 My); solid triangles, Mac-
Kenzie trend (1150–1300 My); solid circles, Keweenawan trend (1000–1150 My).
The path is chosen so as to join up with the early part of the Phanerozoic path of
fig. 110 indicated by the star.

at about 1200 My and that rocks associated with this event give poles in the
vicinity of 4N, 189E. This is the Mackenzie trend. Much of the work on rocks
older than 1150 My relates to studies of various diabase dyke swarms of
the Canadian shield (e.g. Fahrig *et al.*, 1965; Larochelle, 1966, 1967).
For rocks older than about 2100 My the poles are more scattered (Matachewan–Stillwater trend), but they all lie in the region of the southeast
Pacific consistent with the extension of the curve back from the Animikie–
Sudbury trend. The oldest result is for about 2500 My ago from the Stillwater Complex of Montana (Bergh, 1970). There appears to be a rapid

Apparent polar-wandering

polar shift through about 40° in the vicinity of 2100 My ago between the Animikie–Sudbury and Matachewan–Stillwater groups. The significance of all these features in the apparent polar-wander curves is discussed in chapter 7.

6.2.2 Western and southern Europe

The distinction between western Europe and the Russian platform has been made partly because the techniques used by Russian workers have tended to be somewhat different from those used by western workers and because there is a remarkable lineament which divides these two regions. The regions south of the Pyrenees, the Alpine front and the Carpathians are considered separately under the heading of southern Europe. The early studies of various rocks through the geological column in western Europe were undertaken by groups in Britain (Clegg *et al.*, 1954, 1957; Creer *et al.*, 1954, 1957) and on the continent (Nairn, 1957*a*; Roche, 1957; As & Zijderveld, 1958). In spite of two decades of investigation there are still important gaps in the Phanerozoic, where information for the lower Palaeozoic and the Jurassic and Cretaceous needs improvement.

Phanerozoic results for western Europe are summarized in table 17 and are illustrated in fig. 112. There are eighty-two results for the Phanerozoic and eleven for the Precambrian. Going backwards in time, the Upper and Lower Tertiary poles lie progressively southwards along the 150–160E meridians, but during the Cretaceous the pole seems to have backtracked to the other side of the present north pole. There are only three Cretaceous

TABLE 17 *Summary of western Europe Phanerozoic poles.
Symbols as in table 16*

Period	N	K	A_{95}	North pole	Poles used (not used)
Upper Tertiary	9 (1)	153	4	80N, 157E	WE12.2–10 (12.1)
Lower Tertiary	13	84	4	75N, 151E	WE11.1–13
Cretaceous	3	1660	3	86N, 000E	WE10.1–3
Jurassic	1	—	—	(36N, 50E)	WE9.1
Triassic	4	88	10	45N, 143E	WE8.1–4
Permian	13	95	4	45N, 160E	WE7.1–13
Upper Carboniferous	14	39	6	38N, 161E	WE6.5–18
Siluro-Devonian to Lower Carboniferous	7	87	6	17N, 161E	WE4.1, 5.1,6, 6.1–4
Upper Silurian (to Lower Devonian)	6	36	11	00N, 136E	WE4.2–3, 5.2–5
Ordovician (1)	4	71	11	10N, 176E	WE3.1–3,5
Ordovician (2)*	1	—	—	2N, 212E	WE3.4
Cambrian	2	—	—	22N, 167E	WE2.1–2

* From Ireland, see text.

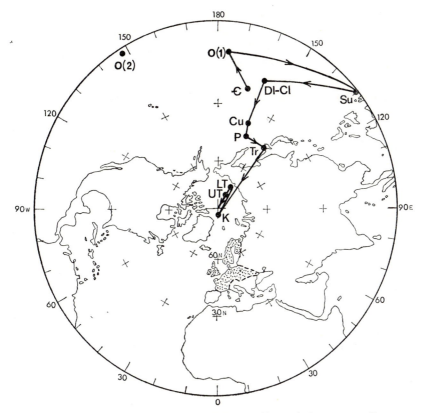

FIGURE 112   Phanerozoic apparent polar-wander path for western Europe.
Polar stereographic projection.

poles, but the results from England and Czechoslovakia are in excellent
agreement and they lie 45° away from the Triassic pole. The single Jurassic
pole was not derived from cleaning and appears to be anomalous. The
Carboniferous poles were noted originally by Irving (1964) to form two
groups. Although there was no clear stratigraphic distinction between the
groups, Irving suspected they reflected two 'magnetic' ages and inferred an
apparent polar shift of about 20° during the Carboniferous. Concentrating
on the few data supported by laboratory stability evidence, Storetvedt
(1967) came to a similar conclusion, but it was then clear that the two groups
also had a stratigraphic basis. A similar conclusion can also be drawn from
the results listed in the appendix. The first four poles (WE6.1–6.4), essentially Lower to Middle Carboniferous, have latitudes around 15N, whilst
the remaining fourteen (WE6.5–6.18), essentially Middle to Upper Carboniferous, have latitudes around 34–50N.

Apparent polar-wandering

Analysis of lower Palaeozoic results from western Europe is complicated by the presence of the Caledonian orogen. Dewey (1969) has inferred three distinct structural zones for the Caledonides of Great Britain, and the palaeomagnetic data all refer to one or more of these zones. There is perhaps no *a priori* reason why the palaeomagnetic poles from these three zones should give one set of pole positions. For example the Siluro-Devonian poles from Norway (WE4.1, 5.1, 5.6) are quite distinct from the Siluro-Devonian poles from Scotland and Wales (WE4.2–4.3, 5.2–5.5). The difficulty is one of deciding whether the poles reflect different ages or whether the differences arise from movements between the regions. Since the youngest Devonian pole is from Norway (WE5.6) and the Norwegian poles are similar to the Lower Carboniferous poles, it is simplest as a first step to suppose the differences are due to differences in age. The age relationship between the Old Red Sandstone of the Anglo-Welsh Cuvette and the Ringerike Sandstone of Norway is very doubtful because of the difficulties of precise palaeontological correlation of terrestrial deposits. The problem is made more difficult by the rather uncertain position of the Siluro-Devonian boundary on the absolute time-scale (Harland *et al.*, 1964). Following McElhinny & Briden (1971), the Siluro-Devonian sediments from Norway are regarded as being younger than the Old Red Sandstone of Wales and the lavas and intrusives of 'O.R.S.' age in Scotland. K–Ar ages suggest a Late Silurian age for these igneous rocks. The Siluro-Devonian results from Norway and the Lower Carboniferous results have been combined to form a single group in table 17. A polar shift is therefore recorded close to the Siluro-Devonian boundary. It should be emphasized however that this interpretation is by no means unique, it is the simplest to adopt at present.

The four results from Ordovician igneous rocks in Great Britain are in good agreement and are combined in table 17. Note that there is a polar shift of $45°$ between this group and the Late Silurian group, a shift which takes place between the Late Ordovician and Late Silurian. This general interpretation of results from Europe in terms of two polar shifts due to an excursion during the Silurian has been made previously by McElhinny & Briden (1971). The Ordovician result from Ireland (WE3.4) is quite distinct, so that the possibility that the northern part of Britain and Ireland formed a distinct crustal plate in the Ordovician (Dewey, 1969) must be countenanced. Cambrian and Ordovician results from Czechoslovakia (Bohemian Massif) are so inconsistent in the various publications in which they have been reported (Bucha, 1961, 1965; Andreeva *et al.*, 1965), that they have not been included in the appendix. The important Cambrian result from the Caerfai Series, Wales (Briden *et al.*, 1971a) includes a fold

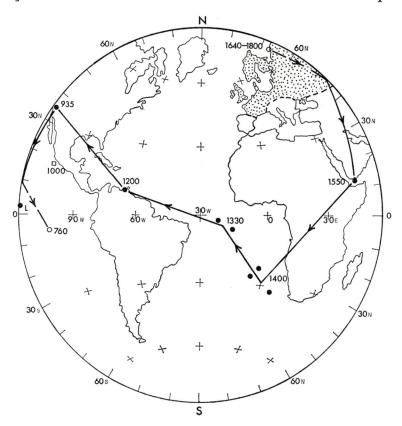

FIGURE 113   Precambrian apparent polar-wander path for Scotland and the Baltic
shield. Equatorial equal area projection, solid lines and symbols refer to the projection
shown, open symbols and dashed lines refer to points on the other half of the globe.
The square is a single pole from the Russian platform. Ages in My are indicated
at various points.

test which shows that the magnetization is at least pre mid-Devonian, and
indicates that the position of the Cambrian pole for western Europe was
very similar to its Devonian–Lower Carboniferous position (table 17). Two
results from Jurassic igneous rocks in Spitsbergen do not agree and these
are not considered further.

There are only eleven Precambrian results from western Europe, restricted to Great Britain and Scandinavia. The poles are illustrated in fig.
113. The large polar shift of 40° found by Irving & Runcorn (1957) between
the Lower and Upper Torridonian Sandstone of Scotland has now been
demonstrated to be due to a time interval of 170 My represented by the
unconformity between the Lower and Upper sediments (Moorbath, 1969).

TABLE 18 *Summary of palaeomagnetic poles for various blocks south of the Alpine front, southern Europe. Symbols are as in table 16*

Period	N	K	A_{95}	North pole	Poles used (not used)
(a) Spain/Portugal					
Lower Tertiary	3	137	11	72N, 191E	SP11.1–3
Cretaceous (Upr)	1	—	—	76N, 174E	SP10.1
Triassic	2 (1)	—	—	59N, 188E	SP8.1–2 (8.3)
Permo-	6	74	8	46N, 213E	SP7.1–6
Carboniferous					
Silurian (Upr)	2	—	—	29N, 216E	SP4.1–2
(b) Alps/Italy					
Lower Tertiary	2	—	—	82N, 201E	IT11.1–2
Triassic	4 (1)	183	7	56N, 243E	IT8.1–4 (8.5)
Permian	11 (3)	85	5	47N, 237E	IT7.1–11 (7.12–14)
(c) Corsica/Sardinia					
Corsica Permo-Carb.	2	—	—	46N, 221E	CS7.1–2
Sardinia Permian	2	—	—	30N, 256E	CS7.3–4
Sardina Lr Tertiary	1	—	—	54N, 263E	CS11.1
(d) Pannonia					
Upper Tertiary	10	41	8	86N, 110E	PN12.1–10
Lower Tertiary	1	—	—	76N, 105E	PN11.1
Cretaceous (Lr)	2	—	—	76N, 237E	PN10.1–2
Triassic	1	—	—	51N, 190E	PN8.1
Permo-Triassic	2 (1)	—	—	41N, 158E	PN7.1–2 (7.3)

The Longmyndian pole falls fairly close to that for the Upper Torridonian. A very tentative Precambrian curve has been drawn from the Torridonian poles back through those from Sweden and Finland on the basis of the age information, but this assumes that Scotland and the Baltic shield were all part of the same crustal plate during this time. Apart from the Torridonian result, the only polar shift that might be considered seriously at this stage is the one between a group of three Swedish poles dated at 1400 My and two Finnish poles at 1330 My. The only Precambrian result from the Russian platform (RP1.1) from the Katav River Suite gives a pole lying half-way between the Torridonian poles. In view of the rather approximate date assigned by the Russians (1000 My), its position is not unreasonable.

The results from the various blocks in southern Europe south of the Alpine front and the Carpathians are given in table 18 and illustrated in fig. 114. These areas have been the subject of much investigation by Dutch workers over the past decade. They have discovered the major discrepancies between the palaeomagnetic data for these regions and those of

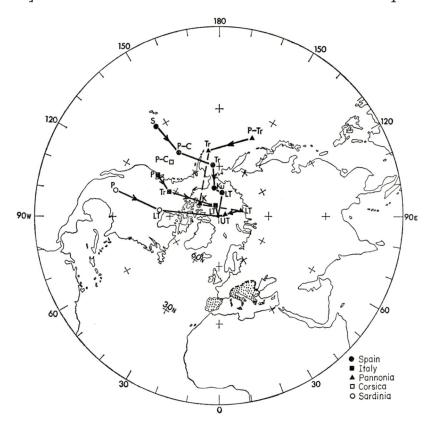

FIGURE 114   Phanerozoic apparent polar-wander paths for various blocks of southern Europe. A dashed line is used where the path is uncertain. Polar stereographic projection.

'stable' Europe (Van Hilten, 1962a; de Boer, 1963; Van Hilten & Zijderveld, 1966; Van Dongen, 1967; Van der Voo, 1967, 1969; Zijderveld & de Jong, 1969; Zijderveld *et al.*, 1970a, b). One problem that arises in interpreting these results is that in some cases the sampling sites are situated within and very close to the centre of the Alpine front or the Pyrenees. Several results from these regions refer neither to stable Europe to the north nor to the block to the south. It seems that some rotation of small blocks has occurred within these highly disturbed regions, and anomalous results from them are marked with a dagger in the appendix and not considered in the analysis. The data are limited largely to Permian and Triassic rocks, but a comparison of figs. 112 and 114 for 'stable' western Europe and the various regions of southern Europe shows the clear discrepancies between the pole for both regions. The significance of these differences is discussed in §7.2.

6.2.3   *Russian and Siberian platforms*

An important summary of palaeomagnetic work in the U.S.S.R. was given by Kalashnikov (1961) covering the period up to a decade ago. More recently a complete summary of all palaeomagnetic poles for the U.S.S.R has been given by Khramov & Sholpo (1967). Many of the results listed in this compilation have been reported in the summary lists given in the Geophysical Journal of the Royal Astronomical Society. The Russian summary list is regarded as superseding all these and is regarded as being the most complete reference to Russian palaeomagnetic work, apart from work which has obviously been published subsequent to the compilation made by Khramov and Sholpo. All the Russian data are those given by this compilation, and where the result has also been given in the Pole Lists of the Geophysical Journal, both reference numbers are given.

It is a difficult problem comparing Russian work with that from the western world because the techniques employed are different. In particular, considerable use is made of stability tests from steady field demagnetization studies and the determination of coercivity of NRM (§3.4.5). Whilst this is regarded as an acceptable stability criterion in the present context, it is certainly very minimal. The data are divided into results from the Russian platform and Siberian platform, the dividing line being taken at the Urals. Problems of course arise from results from rocks within the Urals, but these are discussed as they arise period by period. A few results from the

TABLE 19   *Summary of Phanerozoic poles for the Russian platform. Symbols are as in table 16*

Period	N	K	A_{95}	North pole	Poles used (not used)
Upper Tertiary	25	38	5	78N, 191E	RP12.1–25
Lower Tertiary	4	44	14	68N, 192E	RP11.1–4
Cretaceous	2	—	—	66N, 166E	RP10.1–2
Jurassic (?)	4	25	19	65N, 138E	RP9.1–4
Triassic	10	52	7	51N, 154E	RP8.1–10
Permian	17	118	3	44N, 162E	RP7.1–17
Upper Carboniferous	12	106	4	43N, 168E	RP6.6–9, 6.30–37
Middle Carboniferous	13 (3)	69	5	31N, 172E	RP6.4, 13, 14, 16, 18, 19, 22, 24–29 (6.17, 20, 23)
Lower Carboniferous	7 (2)	255	4	22N, 168E	RP6.3, 5, 10–12, 15, 21 (6.1–2)
Devonian	10 (3)	162	4	36N, 162E	RP5.1–6, 9–12 (5.7–8, 13)
Ordovician–Silurian	5 (1)	25	16	28N, 149E	RP3.1–2, 4.1–3 (3.3)
Cambrian	4	38	15	8N, 189E	RP2.1–4

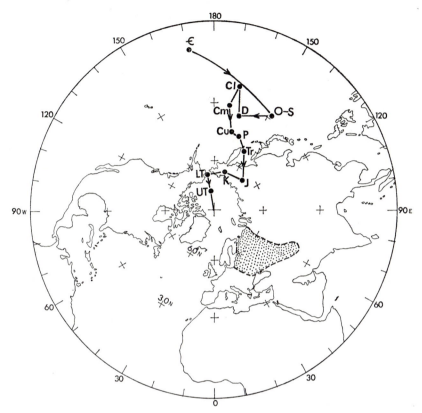

FIGURE 115   Phanerozoic apparent polar-wander path for the Russian
platform. Polar stereographic projection.

region east of the Verkhoyansk and Sikhote Alin Mountains are discussed
separately in the following section (§6.2.4).

There are 134 results available from the Phanerozoic of the Russian
platform. An important paper by Khramov (1967) summarizes the extensive investigations of Carboniferous and Permian rocks. The results are
summarized in table 19, and the resulting polar-wander path is shown in
fig. 115. Tertiary and Cretaceous results are almost exclusively derived
from rocks in the southern U.S.S.R., in Southern Kazakhstan, Turkmenia and
the region of the Sea of Azov, whilst the Jurassic(?) results come exclusively
from volcanic rocks of the Crimea. Extensive studies have been made
of Carboniferous rocks from Donbass and the Moscow basin. Excluding
five of the thirty-seven Carboniferous results, which deviate from the main
group, the remainder can be subdivided into three time groupings. The
pole shows a gradual movement from Lower through Middle to Upper

Carboniferous through an angle of 20°. Note that this is similar to the 17° difference observed between the Lower and Upper Carboniferous of western Europe. The Devonian pole lies north of the Lower Carboniferous pole. The Ordovician to Lower Silurian results become more scattered, one deviating result from the Urals has been omitted. For the Cambrian there are only four results from the Asha River Suite in the southern Urals. Although reasonably well grouped their location in the Urals would indicate that they should be treated with caution until confirmation is forthcoming from other parts of the Russian platform.

There are 119 results from the Phanerozoic of Siberia. Extensive studies of lower Palaeozoic formations have been made by Rodionov (1966) and Gurariy (1969), whilst Aparin & Vlassov (1965) in particular have studied middle and upper Palaeozoic formations. Apart from the lower Palaeozoic, analysis of palaeomagnetic results from Siberia is complicated by two factors. First, it is not yet clear how one should define the southwest boundary of a Siberian block. One might presume that the Urals bend in a southeasterly direction to join up with the Tien Shan Mountains. Further, there are many regions bordering the Siberian platform which have been subjected to severe tectonism. In analysing the results samples collected from the region of Central and Northern Kazakhstan are excluded, because

TABLE 20   *Summary of Phanerozoic poles for the Siberian platform and adjacent regions. Symbols are as in table 16*

Period	N	K	A_{95}	North pole	Poles used (not used)
Upper Tertiary	2	—	—	66N, 234E	SB12.1–2
Lower Tertiary	1	—	—	(57N, 152E)	SB11.1
Cretaceous	6	15	18	77N, 176E	SB10.1–6
Jurassic	0 (4)	—	—	—	(SB9.1–4)
Triassic	16 (4)	31	7	47N, 151E	SB8.1–5, 8.9–19 (8.6–8, 20)
Permian and Carboniferous	11 (8)	25	9	34N, 144E	SB6.2–4, 6, 13–4, 17, 7.1–2, 5–6 (6.1, 7–9, 11–12, 18, 7.7)
Tuva–Minusinsk–Kuznets					
Devonian to Permian	7 (2)	50	9	04N, 146E	SB5.2, 13, 6.5, 10, 15–16 7.3 (6.19, 7.4)
Devonian	7 (6)	65	8	28N, 151E	SB5.1, 3–4, 8–9, 11, 14 (5.5–7, 10, 12, 15)
Silurian	3 (3)	30	23	24N, 139E	SB4.3–4, 6 (4.2, 5, 7)
Lower Silurian	1	—	—	(2N, 98E)	SB4.1
Ordovician	15	49	6	25S, 131E	SB3.1–15
Upper Cambrian	14	141	3	36S, 127E	SB2.9–22
Middle Cambrian	4	61	12	44S, 157E	SB2.5–8
Lower Cambrian	4	16	24	35S, 188E	SB2.1–4

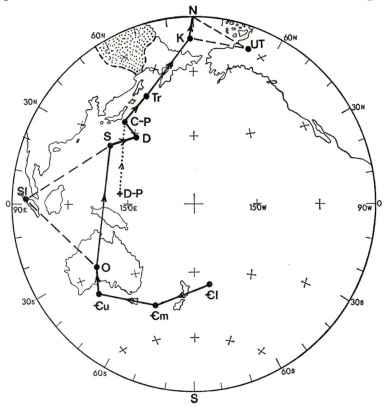

FIGURE 116   Phanerozoic apparent polar-wander path for Siberia. The solid line
joins those points regarded as the most reliable, and dashed lines indicate possible
alternative paths through less reliable information. Equatorial equal area projection.
The cross refers to results from the Tuvan massif, Minusinsk depression and the
Kuznets basin to the south of the Siberian platform.

it is not clear to which block they refer. Hope (1967) in giving a translator's
summary of Russian views on the Baikal Rift System demonstrates the
extreme complications of these regions in Central Asia. Devonian to
Permian results fall into two distinct groups. A group of results having poles
with almost equatorial latitudes and longitudes around 145E is associated
with the region of the Tuvan massif, the Minusinsk depression and the
Kuznets basin to the south of the Siberian platform. Permian and Carboniferous results from just north of these regions however have poles
around 35N and 145E, and it is not possible to distinguish separate groups
for these two periods, so they have been combined. Inconsistent results are
obtained from the River Kureika sections north of the Krasnoyarsk region
and these have therefore been omitted. Results from the Urals are not

considered nor are those from the Taimyr Peninsula, which is probably its northern extension. When analysing data from east of the Urals, it is essential to refer to the tectonic map of Eurasia. One needs to be wary of averaging out all results from regions east of 60E and suppose they refer to Siberia as has often been done in the past.

Table 20 summarizes all the results analysed in the manner outlined above. The apparent polar-wander path which results is illustrated in fig. 116. The deviating upper Palaeozoic poles from Tuva and the Kuznets basin suggest these regions were part of a separate plate at that time. The Jurassic data are inconsistent, so that between the Triassic and the present there is only the mean Cretaceous pole that has any real significance. There are too few Tertiary results at the present time. The poles for the lower Palaeozoic all lie across the equator in the southern hemisphere, about 45° separating the Silurian and Ordovician poles. A single pole from the Llandovery Stage of the River Lena (SB4.1) could represent a position intermediate between them. Extensive studies of Cambrian rocks suggest considerable polar movement with respect to Siberia during this period.

The Precambrian data are illustrated in fig. 117 in which the Phanerozoic polar-wander curve has been extended back to the Lower Sinian group of poles. The interpretation is based upon the rather approximate ages assigned by the Russians, but suggests an extension back to about 1500 My or so. A single Lower Sinian pole from China (CHI.1) is also plotted in fig. 117.

6.2.4 *China and the Far East*

Very little palaeomagnetic work has been carried out in China. Gurariy *et al.* (1966) have investigated Cambrian and Cretaceous sediments from North Korea, whilst Lee *et al.* (1963) and Chen Zhiqiang *et al.* (1965) have studied Mesozoic and Tertiary redbeds from China. Although none of the Chinese work involved any cleaning procedures, there is remarkably good agreement between results from widely separated localities in China. The consistency suggests these results are reasonable estimates of the palaeomagnetic field in China during the Jurassic, Cretaceous and Tertiary. The information for China is summarized in table 21 and a preliminary polar-wander path is shown in fig. 118. The Cambrian and Jurassic poles have been joined along the great circle path between them, because of the lack of any further information. Future work on rocks for the intervening periods will undoubtedly show the path to be more complicated than this.

Vlassov & Popova (1963) have produced some measurements from the

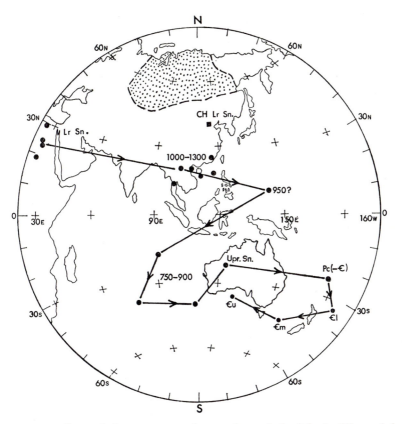

FIGURE 117 Precambrian apparent polar-wander path for Siberia. The path has been chosen so that the youngest poles join up with the three Cambrian poles as in fig. 116. A single pole for China is indicated by the solid square. Equatorial equal area projection with approximate ages in My indicated at various points.

Soviet Far Eastern Maritime Province to the east of the Sikhote Alin Mountains in the Vladivostok region, whilst Pecherskiy (1970a) has investigated the Kolyma block to the east of the Verkhoyansk Mountains. In this latter region there are both tectonic and metamorphic complications. Four results listed in the appendix from the West Verkhoyan region within the Mountain belt are obviously anomalous because of their locality. In fact the two Triassic results from the western side of the mountains might more appropriately refer to the Siberian platform. The late Palaeozoic results from the Umkuveyen depression (vs6.1) need to be treated with some caution, because according to Pecherskiy (1970a) this depression was formed as a result of block movements within the Yablonskiy massif, and he suggests some rotation has occurred. Pecherskiy attempts to allow

217

TABLE 21   *Summary of palaeomagnetic poles for China and the Far East. Symbols are as in table 16*

Period	N	K	A_{95}	North pole	Poles used (not used)
(a) China					
Lower Tertiary	I	—	—	75N, 38E	CH11.1
Cretaceous	3	44	19	69N, 182E	CH10.1–3
Jurassic	I	—	—	55N, 149E	CH9.1
Cambrian (N. Korea)	I	—	—	11S, 137E	CH2.1
(b) Sikhote Alin (Vladivostok region)					
Cretaceous (Lr)	I	—	—	58N, 146E	VS10.1
Triassic	I	—	—	54N, 186E	VS8.7
Permian (Lr)	2	—	—	18N, 198E	VS7.1–2
(c) Kolyma block					
Cretaceous	6	43	10	60N, 166E	VS10.2–7
Jurassic	7 (1)	15	16	72N, 144E	VS9.2–8 (9.1)
Triassic	5 (2)	35	13	63N, 237E	VS8.2–6 (8.1, 8)
Permian (Upr)	I (1)	—	—	52N, 279E	VS7.3 (7.4)
Upper Devonian–Lower Permian	0 (1)	—	—	—	(VS6.1)

for any metamorphic effects by making corrections to his results on the basis of observations of the preferred directions of the long axes of grains within his samples. The results given in the appendix are those which conform with the palaeomagnetic procedures outlined in chapter 3. They are the cleaned values without any 'corrections' applied to them. They are summarized in table 21 and fig. 118 together with the Sikhote Alin results.

One outstanding feature can be noted of the three apparent polar-wander paths plotted in fig. 118. They are widely divergent in the late Palaeozoic, approach one another in the Jurassic and converge in the Cretaceous. This of course suggests that each of these regions were separated from one another but subsequently became welded onto Asia towards the end of the Mesozoic. This aspect of the results will be discussed more fully in §7.2.4. The information for Japan is not summarized at this point because the results suggest that some bending of the Japanese islands has occurred. This discussion is left until §7.2.4.

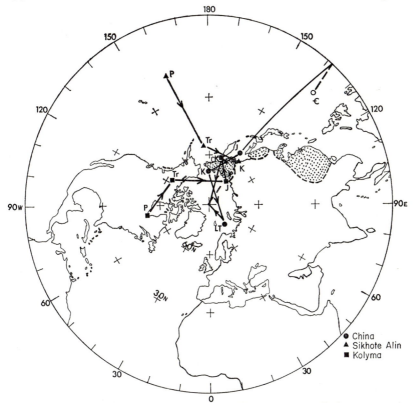

FIGURE 118   Phanerozoic apparent polar-wander paths for three regions of the far east. Polar stereographic projection; solid symbols lie on the projection shown, open symbols lie on the other half of the globe in the southern hemisphere.

6.3   THE SOUTHERN HEMISPHERE

For the purposes of discussing the results, India and Arabia (including Turkey) are included in the southern hemisphere, because they formed part of the southern supercontinent of Gondwanaland. In these analyses it is easiest to consider the path of the south pole, because this is the pole nearest to the continents concerned. The summary tables for each of the Gondwana regions therefore list the south pole positions.

6.3.1   *Africa and Arabia*

Some of the first investigations of the palaeomagnetism of African rocks were undertaken by Gough (1956), Graham & Hales (1957), Gough & van Niekerk (1959) and Nairn (1956, 1957*b*, 1960). Nairn's approach of

219

TABLE 22   *Summary of Phanerozoic poles for Africa.*
Symbols are as in table 16

Period	N	K	A_{95}	South pole	Poles used (not used)
Upper Tertiary	4	1890	2	87S, 332E	AF12.1–4
Lower Tertiary	2	—	—	85S, 6E	AF11.1–2
(Cretaceous)	2	—	—	(61S, 80E)	AF10.1–2
Mesozoic (Tr–J–K)	10	123	4	65S, 82E	AF8.1–4, 9.1–4, 10.1–2
Lower Permian	1	—	—	27S, 89E	AF7.1
Upper Carboniferous	1	—	—	46S, 40E	AF6.2
Permo-Carb. combined	2	—	—	40S, 64E	AF6.2, 7.1
(Lr) Carboniferous	1	—	—	26S, 26E	AF6.1
(Pc–) Cambrian and Ordovician	4	14	25	24N, 345E	AF2.1–2, 3.1–2

surveying the geological column with the minimum amount of sampling was soon found to be ineffective in African conditions (Nairn, 1957*b*). Palaeomagnetic work in Africa was however only undertaken in a systematic way at the time when cleaning procedures were first being developed, so that much of the work has been able to stand the test of time. Reviews have been given by Gough *et al.* (1964) and McElhinny *et al.* (1968). The Phanerozoic results have changed little over the past few years and are summarized in table 22 and the polar-wander path is illustrated in fig. 119. Four lower Palaeozoic poles form a loose group around northwest Africa. Two of the poles could represent latest Precambrian, whilst the other two are not especially well defined. Whether the scatter represents the difference in ages or poorly defined poles or a combination of both is not clear. At present it is simplest to average them to form a lower Palaeozoic group. From northwest Africa the pole moves about 60° southwards to southern Africa in the Carboniferous, represented by results from the Dwyka varves of Central Africa. The age of these varves is not known except by correlation with the Dwyka glacial beds in South Africa. Du Toit (1954) regards the Dwyka Series as being Upper Carboniferous, but Haughton (1963) refers to it as mid-Carboniferous. Plumstead (1969) has, however, discovered fossil plants lying upon Dwyka tillite dated as Lower Carboniferous or even Upper Devonian. The Dwyka varves are referred to a Lower Carboniferous in table 22.

Both the lower Palaeozoic pole off northwest Africa and the Dwyka pole in southern Africa are consistent with the occurrence of the spectacular Late Ordovician glacial horizons in the Sahara (Beuf *et al.*, 1966; Biju-Duval & Gariel, 1969) and the Dwyka glaciation in southern Africa (du Toit,

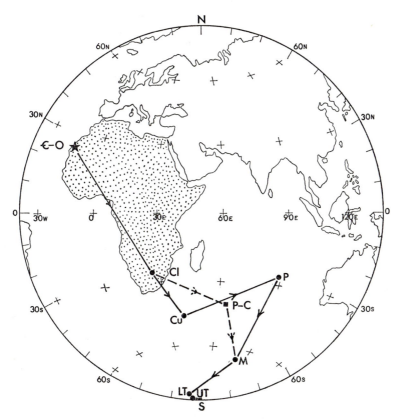

FIGURE 119   Phanerozoic apparent polar-wander path for Africa. The dashed lines show an alternative interpretation of late Palaeozoic results in which the path passes through the mean Upper Carboniferous (Cu) and Permian (P) poles. M is the mean Mesozoic pole, and the star represents the mean of a loose group of latest Precambrian and lower Palaeozoic poles. Equatorial equal area projection.

1954). The timing of the movement of the pole from north to south across Africa is of some interest and can be inferred from palaeoclimatic evidence. The earliest evidence of glaciation in southern Africa during the Palaeozoic occurs in the Table Mountain Group of the Cape System in the Pakhuis formation (170 m of tillite containing boulders with striated and flattened surfaces) which immediately overlies the Lower Shales studied to give pole AF3.2 falling off northwest Africa. The overlying Cedarberg formation includes an 11 m band containing fossils dated as lying close to the Ordovician–Silurian boundary (Cocks *et al.*, 1970). The tillite is thus of Late Ordovician age and this stratigraphic evidence suggests the pole made its transition from north to south across Africa during the Late Ordovician.

Apparent polar-wandering

The interpretation of the results from the 'Ecca' redbeds of southern Tanzania is in dispute. Originally Opdyke (1964) combined all results from both the Galula, Songwe–Kiwira and Ketewaka–Mchuchuma coalfields together into a single Lower Permian result, although he noted there were two distinct groups, one for the Galula beds and one for the Songwe and Ketewaka beds. It is the Ketewaka–Mchuchuma K_3 beds that are well dated on fossil evidence as not younger than Middle Artinskian (Hart, 1960), but the Galula K_3 beds are merely correlated with them. The Galula beds are younger than the Dwyka varves (K_1 beds at Galula) but have no upper limit in age. McElhinny & Opdyke (1968) suggested the sequence of ages followed poles AF6.1 – AF6.2 – AF7.1, so that the Galula beds might be Upper Carboniferous. This is disputed by Creer (1970a) and Embleton (1970), who prefer to accept the original mean pole given by Opdyke (1964). Both possibilities are illustrated in fig. 119, because they lead to different interpretations of the comparisons of results between continents (§7.3). Much more Palaeozoic information is needed for Africa before this problem can be resolved satisfactorily.

There are ten results for the Mesozoic covering the period from Lower Triassic to mid-Cretaceous (about 110 My). These results suggest that the pole with respect to Africa remained essentially in the same position during this period. These poles have been averaged to form the Mean Mesozoic Pole for Africa following McElhinny et al. (1968). The Ethiopian traps (AF11.1, 11.2) were originally quoted by McElhinny et al. as being Upper Cretaceous to Eocene on the basis of the greatest K–Ar age of 69 My cited for them. Brock et al. (1970) point out that they are now regarded almost certainly as being not older than Eocene–Oligocene, and more recent information suggests they might even be as young as Miocene. There still therefore appears to be no information for the period from about 110 My to at least Oligocene, say 30 My ago. At some time(s) during this interval the pole moved from its Mesozoic position through about 25° to near the present pole.

For the Precambrian, there is information covering two time intervals, the first from 2700 My to about 2200 My and the second between 1950 My and 1300 My. Because there is no information for the period between 1300 My and about 600 My, it is not obvious which of the two possibilities for the pole position at 1300 My might be the north or south pole. In the absence of any other information the youngest pole (AF1.14, 1300 My) is plotted in the position which lies closest to the oldest points on the Phanerozoic curve. The resulting apparent polar-wander curve is shown in fig. 120. All the information is derived from the Transvaal and Rhodesian shields, and since the close of major orogenic events in the intervening

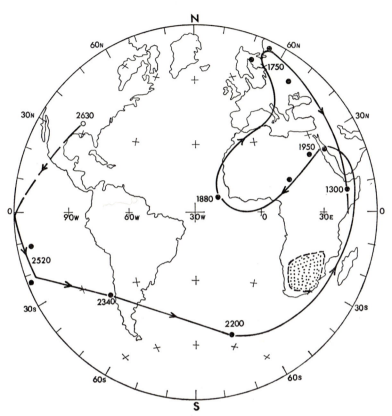

FIGURE 120 Precambrian apparent polar-wander path for southern Africa. Equatorial equal area projection, solid lines and symbols refer to the projection shown, open symbols and dashed lines refer to points on the other half of the globe. Ages in My are indicated at various points.

Limpopo belt was 2000 My ago (Van Breemen *et al.*, 1966), they can be considered as a unit at least during the interval 1950 My to 1300 My. Two series of results are summarized by this part of the curve, those from a wide variety of dated igneous rocks and those from the sediments of the Waterberg System (AFI.11). The sediments do not give a single palaeomagnetic pole covering the time of deposition of the redbeds. Instead a polar-wander path covering their time-span can be drawn. Jones & McElhinny (1967) showed that the polar-wander paths for the sediments and the igneous rocks can be superimposed, and the path drawn in fig. 120 follows the one they derived. The path for the older part of the curve is also shown in fig. 120. A great circle distance of 80° separates the youngest pole of this group (about 2200 My) from the oldest of the other group (1950 My).

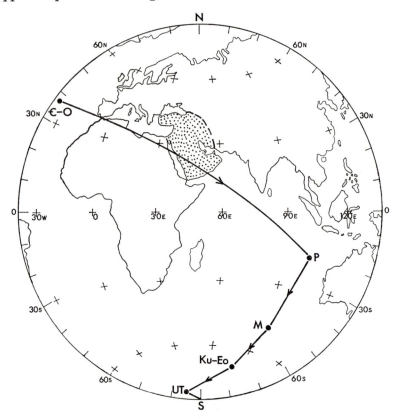

FIGURE 121 Phanerozoic apparent polar-wander path for Arabia (including Turkey). M is the mean Mesozoic pole. Equatorial equal area projection.

This gap becomes 100° if the antipoles of the older part of the curve are taken. The difference between the two parts of the curve is sufficiently close to 90° that the manner in which the two parts should be joined remains ambiguous (McElhinny *et al.*, 1968).

Since the formation of the Red Sea has been a fairly recent geological event, it seems clear that Arabia was originally part of Africa and it is therefore appropriate to discuss the results from Arabia with those from Africa. The problem that arises is where to draw the northern boundary of Arabia from a geological–tectonic point of view. Although the northern boundary at the present time might be taken at the Taurus Mountains in southern Turkey, palaeomagnetic results from the Cretaceous of Turkey and Arabia seem to justify including Turkey within Arabia and taking the northern boundary at the Caucasus Mountains. The Cretaceous poles from Arabia proper (ARI0.1, 10.2, 10.5, south of the Taurus) and those from Turkey

TABLE 23 *Summary of Phanerozoic poles for Arabia (including Turkey). Symbols are as in table 16*

Period	N	K	A_{95}	South pole	Poles used (not used)
Upper Tertiary	10	71	6	82S, 35E	AR12.1–10
Upper Cretaceous–Eocene	3	47	18	69S, 84E	AR11.1–3
(Cretaceous)	5	32	14	(51S, 95E)	AR10.1–5
(Jurassic)	1 (1)	—	—	(39S, 85E)	AR9.1 (9.2)
Mesozoic (J–K)	6 (1)	33	12	49S, 93E	AR9.1, 10–1–5 (9.2)
Permian	1	—	—	18S, 102E	AR7.1
Cambro-Ordovician	1	—	—	37N, 323E	AR2.1

(AR10.3, 10.4, north of the Taurus) do not differ significantly. Also the single Jurassic pole from Turkey falls within the spread of the Cretaceous poles, as might be expected for a region that was part of Africa (constant Mesozoic pole position).

The investigations in Turkey and the Lebanon have been undertaken by Dutch workers as part of a general study of the Alpine tectonics of Eurasia (Gregor & Zijderveld, 1964; Van Dongen *et al.*, 1967; Van der Voo, 1968). The results are all summarized in fig. 121 and table 23. One rather obviously anomalous result from the Lebanese Mountains has been omitted. Some of this region has almost certainly been rotated as a result of movement along the Dead Sea fault system (Quennell, 1958). The similarity within the African curve (cf. fig. 119) is striking, even to the extent that the Permian result lies well out into the Indian Ocean like the Songwe pole. The point is discussed again in §7.2.1.

6.3.2 *South America*

Palaeomagnetic work in South America over the past decade or more has been undertaken almost exclusively by Creer (1958, 1962, 1965, 1967*b*, 1970*b*), although in recent years other workers have also contributed (Valencio & Vilas, 1970; Creer *et al.*, 1970; Embleton, 1970). A report of his work over many years has been given by Creer (1970*b*). In analysing the data it has been necessary to combine the results in such a way that they conform with the minimum critera set out in §3.5. A summary of the resulting mean palaeomagnetic poles for various periods is given in table 24 and the resulting apparent polar-wander path illustrated in fig. 122.

The Cambrian and Ordovician results produce a loose group of poles off northeast South America. McElhinny & Briden (1971) argue that the

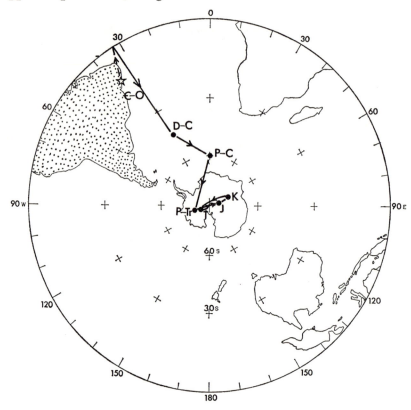

FIGURE 122 Phanerozoic apparent polar-wander path for South America. The star, lying in the northern hemisphere, is the mean of a loose group of lower Palaeozoic poles. Polar stereographic projection.

TABLE 24 *Summary of Phanerozoic poles for South America. Symbols are as in table* 16

Period	N	K	A_{95}	South pole	Poles used (not used)
Tertiary	3	37	20	82S, 242E	SA11.1–2, 12.1
Cretaceous (Lr)	1 (1)	—	—	78S, 56E	SA10.1 (10.2)
Jurassic	1	—	—	84S, 76E	SA9.1
Permo-Triassic	7	76	7	80S, 251E	SA7.4–5, 8.1–5
Permo-Carboniferous	4	141	8	60S, 000E	SA6.4, 7.1–3
Devonian–Carboniferous	3 (1)	24	26	43S, 331E	SA5.2, 6.1, 3 (6.2)
Cambrian–Ordovician	5	18	18	11N, 323E	SA2.1, 3.1–3, 5.1

Bolivian sediments cited as Lower Devonian (SA5.1) and the Urucum formation cited as Silurian (SA3.3) are likely to be older than stated so that the ages are confined to the Cambrian and Ordovician. Creer (1970a) suggests there is a Cambrian group to the east and an Ordovician group to the west, but this must be regarded as speculative at this stage. Each of the poles is not very well defined because of the difficulty in isolating a primary component from lower Palaeozoic rocks in South America. Both the African and South American lower Palaeozoic groups have similar scatter ($K = 14$ and 18 respectively), reflecting the uncertainty involved with the data for this period.

The pole from the Picos and Passagem Series of Brazil (SA5.2) is not very well defined, but it has been combined with those from the Taiguati and Piaui formations to form a middle Palaeozoic group of poles. As observed in Africa, there is a large polar shift between the lower and middle Palaeozoic, but the timing is not clear because of the uncertainty in some of the ages of the formations studied. The pole moved essentially along the east of South America from north to south. There is evidence of glaciation from the Silurian and Devonian of Brazil, whilst glaciers covered regions in the Andean belt and Parana basin during the Early Carboniferous (Frakes & Crowell, 1969; Crowell & Frakes, 1970). The glacial conditions find considerable support in the palaeomagnetic data.

The work of Embleton (1970) on Permo-Carboniferous rocks from the Paganzo formation appears to indicate no easterly excursion of the pole as suggested for Africa via the Songwe pole (cf. fig. 119). One problem is that the equivalent pole for South America would fall very close to that for the Cretaceous Serra Geral formation, near the present south pole. Permo-Triassic poles form a good group close to the present south pole, but an excursion to the other side of the pole is indicated by the Jurassic and Cretaceous results. The Cretaceous pole for the Serra Geral formation (SA10.1) is the pole preferred in table 24. At present no conclusions can be drawn from the two results for the Precambrian.

6.3.3 *Australia*

Palaeomagnetic work in Australia was started by Irving (1956b, 1957b) and Irving & Green (1957, 1958). An important review of work carried out on Mesozoic rocks in eastern Australia is given by Irving et al. (1963), whilst reports on late Palaeozoic investigations are given by Irving & Parry (1963) and Irving (1966). Very little work was carried out on lower Palaeozoic rocks apart from some unsuccessful attempts by Briden (1965), who found that many lower Palaeozoic rocks in southeast Australia had been remagnetized.

TABLE 25 *Summary of Phanerozoic poles for Australia. Symbols are as in table 16. K–Ar ages are given for parts of the Tertiary*

Period	N	K	A_{95}	South pole	Poles used (not used)
Miocene (17–24 My)	3	413	6	77S, 95E	AU12.1–12.3
Oligocene (34 My)	1	—	—	71S, 96E	AU11.3
Eocene (52 My)	1	—	—	70S, 126E	AU11.2
Palaeocene (–Eocene)	1	—	—	63S, 140E	AU11.1
(Cretaceous)	2	—	—	(53S, 149E)	AU10.1–2
Mesozoic (Tr–J–K)	10	47	7	48S, 151E	AU7.3, 8.1–2, 9.1–5, 10.1–2
Permo-Carboniferous	5	96	8	46S, 135E	AU6.3–5, 7.1–2
(Upr Carboniferous)	1	—	—	(73S, 147E)	AU6.2
Lower Carboniferous	1	—	—	73S, 214E	AU6.1
Upper Silurian–Upper Devonian	4	23	20	72S, 354E	AU4.2, 5.1–3
Middle Silurian	1	—	—	54S, 271E	AU4.1
Cambrian–Ordovician	3	7	40	2S, 8E	AU2.1–2.3, 1

McElhinny & Luck (1970a) describe an extensive study of Lower Cambrian volcanics and give a review of the state of palaeomagnetic work in Australia, but the most recent information for the Palaeozoic is given by Luck (1971), many of whose new results are given in the appendix. The results are summarized in table 25 and the polar-wander path is illustrated in fig. 123.

The Cambrian and Ordovician poles have possibly slightly greater scatter on them ($K = 7$) than their African and South American counterparts. In this case however the ages of the formations are quite firmly based, and the three formations studied are related stratigraphically. All three results are from northern Australia, whereas the remainder are from southeast Australia. Whether or not results for the lower Palaeozoic of Northern Australia can be related to southeast Australia becomes debatable on plate tectonic considerations. No interpretation has yet been carried out of South Australian tectonic features in terms of plate theory, but recent work in Central Australia (Wells *et al.*, 1970) shows that a Carboniferous event took place in that region. Whether or not this region marks the site of a collision zone is a matter of some conjecture.

A 90° arc separates the lower Palaeozoic group from the oldest of the Palaeozoic results from southeast Australia (Middle Silurian). The Middle Silurian to Lower Devonian data are from the sequences in the Canberra–Yass region whereas the Lower Carboniferous (Visean) results are from the

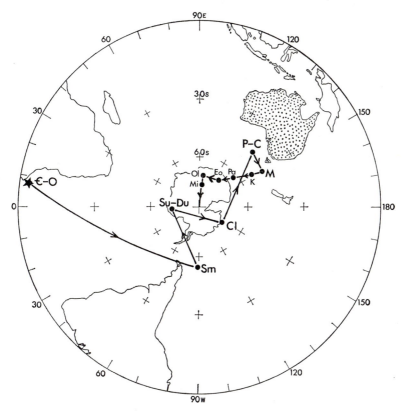

FIGURE 123 Phanerozoic apparent polar-wander path for Australia. The star is
the mean of a loose group of lower Palaeozoic poles and M is the mean Mesozoic
pole. Polar stereographic projection.

Hunter Valley to the northeast. The interpretation of the geology of eastern
Australia in terms of plate tectonic models (Oversby, 1971) could well
lead to a quite different interpretation of the most recent palaeomagnetic
results for the period from the Silurian to the Carboniferous. For example
three serpentine belts cut across the eastern part of Australia becoming
progressively younger from west to east. They are approximately Cambro-
Ordovician to the west, Siluro-Devonian centrally and Devonian–
Carboniferous represented by the Great Serpentine Belt in the east
(Brown *et al.*, 1968). It might be possible to interpret these as representing
successive continent–island arc collisions, in which further material was
accreted onto the eastern part of Australia. This speculation could explain
some of the discrepancies that are now arising in comparisons of results
from eastern Australia and the rest of Gondwanaland for these periods.

The problem is mentioned at this stage in anticipation of the discussion which follows in §7.3.2.

Irving (1966) identified a rapid polar shift with respect to southeast Australia during the Carboniferous, about 40° was represented and almost entirely a latitudinal change. Intermediate latitudes in the Lower Carboniferous gave way to high latitudes in the Upper Carboniferous and Permian. This was consistent with palaeoclimatic evidence, because coral reefs in the Lower Carboniferous gave way to glacial deposits near the Permo-Carboniferous boundary. Irving's polar shift is slightly modified by more recent results from the Hunter Valley, but its magnitude still remains. An intermediate pole (AU6.2) from the Paterson toscanite is bracketed because it is probably only a VGP and not a true palaeomagnetic pole. The late Palaeozoic glaciation observed in Australia and recently summarized and discussed by Crowell & Frakes (1971) is of course consistent with the near polar position deduced from the palaeomagnetism of Upper Carboniferous and Permian rocks (Irving & Parry, 1963; Irving, 1966).

Irving (1966) averaged all the Permo-Carboniferous (Kiaman) and Mesozoic poles together, claiming there was no distinguishable difference between them. However McElhinny & Luck (1970 a) demonstrated that they can be regarded as two distinct groups separated by about 12°. Their means are given in table 25. As was observed for Africa, there was no polar movement during the Mesozoic. From a position just south of Australia during the Mesozoic, the pole moved progressively towards the present south pole during the Tertiary. The details of this movement have been investigated by Wellman et al. (1969).

Only three Precambrian results have radioisotope dates associated with them, although there have been studies of a number of haematite ore bodies in South Australia (Chamalaun & Porath, 1968) and Western Australia (Porath & Chamalaun, 1968). The absolute ages of these deposits are not known at present, except that they are of Precambrian age. At the present time it is not possible to interpret the results in terms of a Precambrian apparent polar-wander path.

6.3.4   *India*

Clegg et al. (1956, 1958) and Deutsch et al. (1958, 1959) first investigated the palaeomagnetism of Indian rocks. Later more detailed work was undertaken on the Deccan traps (Sahasrabudhe, 1963; McElhinny, 1968c), and the Rajmahal traps (Radhakrishnamurty, 1963). More recently some work has been carried out on sediments of the Gondwana System (Verma & Pullaiah, 1967; Verma & Bhalla, 1968; Bhalla & Verma, 1969; Wensink, 1968; Wensink & Klootwijk, 1968). The data are summarized in table 26

TABLE 26   *Summary of Phanerozoic poles for India.*
Symbols are as in table 16

Period	N	K	A_{95}	South pole	Poles used (not used)
Upper Cretaceous–Palaeocene	1 (1)	—	—	37S, 102E	IN11.2 (11.3)
Upper Cretaceous	4	120	8	30S, 110E	IN10.5–7, 11.1
Lower Cretaceous	4	86	10	13S, 119E	IN10.1–4
Triassic	3	62	16	20S, 128E	IN8.2–4
Permo-Triassic	1	—	—	7N, 124E	IN8.1
Permo-Carboniferous	2	—	—	26N, 132E	IN6.1, 7.1
Cambrian	1	—	—	28S, 32E	IN2.1

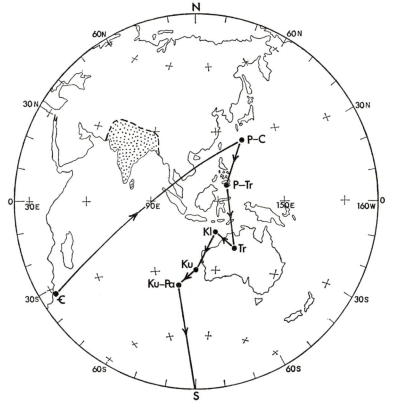

FIGURE 124   Phanerozoic apparent polar-wander path for India.
Equatorial equal area projection.

and the polar-wander path is illustrated in fig. 124. The only early Palaeozoic result so far published for the Indian subcontinent is reported by McElhinny (1970*b*) from Cambrian rocks of the Salt Range. Following the review of the Indian apparent polar-wander curve by McDougall & McElhinny (1970), the Kamthi beds sampled in the Wardha Valley and the Godavary Valley are regarded as having different ages, the former Upper Permian and the latter Lower Triassic. The dating of the Rajmahal traps at 105 My necessitates placing the Tirupati and Satyavedu Sandstones into the Upper Cretaceous.

On the apparent polar-wander path in fig. 124 the Cambrian pole has been joined to the Permo-Carboniferous (Kiaman) poles through a great circle distance of about 110°, whereas normally the procedure is to join successive poles through the shortest possible path. This procedure has been adopted (as was also the case for Arabia, see fig. 121), because comparisons between the Gondwanic continents suggest this to be the correct interpretation (§7.3.2). It is interesting to note that the results from the Talchir Series (Wensink & Klootwijk, 1968) indicate a palaeolatitude of much of India lying between 50° and 40°. The Talchir boulder beds are regarded as being of glacial origin, and represent part of the late Palaeozoic glaciation observed in all the southern continents (Crowell & Frakes, 1970). It seems that this widespread glaciation extended up into intermediate latitudes in the Permo-Carboniferous, a feature also observed of course during the Pleistocene glaciation.

On the apparent polar-wander path the Mangli beds (IN8.1) are regarded as being intermediate in age between the late Palaeozoic (Kiaman) group and the group of three Triassic poles. The Cretaceous results are split into two groups on the basis of their ages and distinct groupings. They appear to differ from the Triassic group. There are no true Tertiary results from India, but results from the Deccan traps dated at 60–65 My or more (Wellman & McElhinny, 1970) show that about 50° of polar shift is required between the Cretaceous–Tertiary boundary and the present day. The few Precambrian results (Athavale *et al.*, 1963) suffer from lack of isotopic age information, and at the present time they do not justify drawing any polar path.

6.3.5 *Antarctica and Madagascar*

There are only a limited number of results for Antarctica and these have been subdivided into East and West Antarctica. Whereas East Antarctica is clearly a large continental block, it appears that West Antarctica could be a number of isolated regions (Schopf, 1969). For East Antarctica most

TABLE 27   *Summary of Phanerozoic poles for Antarctica.*
Symbols are as in table 16

Period	N	K	A_{95}	South pole	Poles used (not used)
(a) East Antarctica					
Upper Tertiary	1	—	—	81s, 274E	AN(E) 12.1
Jurassic	5	76	9	53S, 215E	AN(E) 9.1–5
Cambro-Ordovician	2	—	—	15S, 20E	AN(E) 2.1, 3.1
(b) West Antarctica					
Antarctic Peninsula (Upr Cretaceous)	1	—	—	86s, 358E	AN(W) 10.3
Marie Byrd Land (Lr Cretaceous)	1	—	—	47S, 120E	AN(W) 10.2
Jones Mountains (Mesozoic, Tr–J–K)	2	—	—	84S, 137E	AN(W) 9.1, 10.1

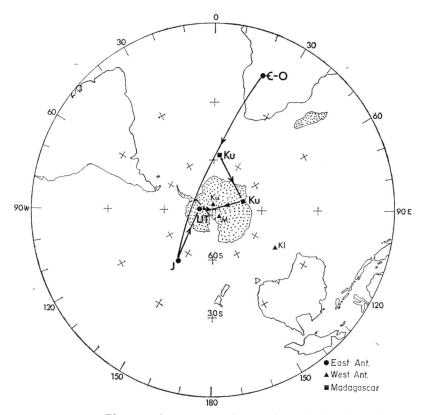

FIGURE 125   Phanerozoic apparent polar-wander paths for Antarctica
and Madagascar. Polar stereographic projection.

TABLE 28   *Summary of palaeomagnetic poles for Madagascar.*
Symbols are as in table 16

Period	N	K	A_{95}	South pole	Poles used
Upper Cretaceous (Santonian–Campanian)	2	—	—	71S, 71E	MD10.3–4
Upper Cretaceous (Turonian)	2	—	—	57S, 7E	MD10.1–2

of the work has revolved around studies of the Ferrar dolerites (Turnbull, 1959; Bull *et al.*, 1962). Attempts by these investigators to determine primary directions from Permian and Triassic sediments proved unsuccessful, both workers suggesting they were remagnetized at the time of intrusion of the dolerites. More recent work has been carried out in West Antarctica by Scharon *et al.* (1969, 1970). The data are summarized in table 27 and illustrated in fig. 125. The different results from various parts of West Antarctica suggest that several separate regions are involved, not necessarily related to East Antarctica. The East Antarctic data are therefore restricted to the Jurassic and the Cambro-Ordovician. Although the West Antarctic results are plotted in fig. 125, no path has been drawn through them, because of doubt about their relationship. Although no pole position is yet available for the late Palaeozoic of Antarctica, it is worth noting that there is considerable evidence for glacial conditions in Antarctica at that time (Frakes & Crowell, 1968). This would be consistent with near polar conditions for this period.

For Madagascar the only reliable results are from studies of Upper Cretaceous volcanics (Roche & Cattala, 1959; Andriamirado & Roche, 1969). They are separated into two age groups as summarized in table 28. The poles are illustrated in fig. 125 with the Antarctic results.

7 Palaeomagnetic poles and plate tectonics

The relative motion between two plates is simply described by means of a rotation about some pole, as was shown in §5.3.1. By definition, a pole of rotation for the relative motion between two plates is fixed to these two plates. As was stressed by McKenzie & Parker (1967), plate theory is only an instantaneous theory which applies to motions at present taking place. McKenzie & Morgan (1969) demonstrated the important point that it is impossible for three contiguous plates A, B, C to rotate simultaneously through finite angles about their instantaneous relative rotation axes. As Le Pichon (1968) pointed out, however, it is possible for two of the relative rotations, say A–B and A–C, to remain simple and be described by a single pole whereas the relative motion B–C is changing continuously in a complex manner. Bullard *et al.* (1965) have shown how to reconstruct the original relative positions of two continents by finite rotations about suitably chosen poles, but no physical significance can be attached to these rotations as they are essentially mathematical entities. They do however provide an important constraint because they represent an integral of the motion. If there is no sea floor or magnetic record between two plates A and B, the understanding of the relative motion between them would require that the timing of the motion between say, A and C, and B and C be accurately known as the order of the rotations is important. This whole problem has been discussed by Francheteau (1970) with respect to the opening of the Atlantic Ocean.

All the methods above describe the relative motion between plates, but in palaeomagnetism measurements are made which describe the relative motion between a plate and the earth's spin axis. A palaeomagnetic pole places the plate in latitude and azimuthal orientation. The longitude remains indeterminate and can only be defined by reference to some arbitrarily chosen point on the plate. The component of the angular velocity vector describing the motion of the plate relative to the spin axis (palaeomagnetic pole) and which lies along this axis therefore makes no difference to this motion. This axial component can thus have any value and can conveniently be set to zero (McKenzie, 1972). This means that the motion of plate A relative to the geographic north pole G (taken to be fixed) can always be described by an angular velocity vector $_G\Omega_A$ passing through the equator. If the angular velocity of another plate B relative to the plate A ($_A\boldsymbol{\omega}_B$) is known, then the angular velocity of B relative to the pole, $_G\boldsymbol{\Omega}_B$ is obtained from the relation

$$_G\boldsymbol{\Omega}_B = {_G\boldsymbol{\Omega}_A} + {_A\boldsymbol{\omega}_B} - ({_A\boldsymbol{\omega}_B} \cdot \boldsymbol{a}_z)\boldsymbol{a}_z, \qquad (7.1)$$

where $\boldsymbol{a}_z$ is a unit vector along the rotational axis. The last term on the right hand side of (7.1) is merely the component of $_A\boldsymbol{\omega}_B$ along the spin axis, and therefore causes $_G\boldsymbol{\Omega}_B$ to pass through the equator. The above suggests a simple constructional method of placing a plate in its correct latitude and orientation with respect to the north pole for some past epoch. If the palaeomagnetic north pole for this epoch lies at latitude λ and east longitude ϕ, a rotation about a point on the equator with longitude $(\phi - \tfrac{1}{2}\pi)$ through an angle $(\tfrac{1}{2}\pi - \lambda)$ in the usual positive sense will send the palaeomagnetic pole to the north pole (Francheteau, 1970). The plate is then in its correct latitude and orientation.

In analysing the apparent polar-wander paths summarised in chapter 6, it is important to note that variations in the rate of apparent polar-wandering with respect to a plate do not by themselves indicate the *same* variations in the motion of the plate concerned. Indeed the very opposite might be the case. It is possible for a point on one plate to move rapidly over great distances with respect to a point on another plate, but for no apparent polar-wandering to be observed. Likewise it is possible for a large amount of polar-wandering to be observed, even over a short period of time, and yet the motions between the plates might be quite small. The application of (7.1) to a few simple cases serves to illustrate these points. Three examples are illustrated in fig. 126.

In the first case (fig. 126a), no apparent polar-wandering is observed for each of two plates A and B. Thus in (7.1) $_G\boldsymbol{\Omega}_B = {_G\boldsymbol{\Omega}_A} = \mathbf{0}$, and the angular velocity vector must therefore lie along the rotational axis in the direction of $\boldsymbol{a}_z$. This means that $_A\boldsymbol{\omega}_B$ can have any value, so that the motion between the two plates can continue at any rate without change in the position of the palaeomagnetic pole with respect to either of the plates being recorded. An example of this type of situation can be imagined with respect to South America. The Triassic palaeomagnetic pole for South America (§6.3.2) lies close to the present geographic pole and remains close to it up to the present time. Yet it is known that South America separated from Africa during that time but this separation has been largely along lines of present latitude. No change in palaeomagnetic pole position need necessarily be observed.

In the other cases (figs. 126b, c), the pole of rotation between plates A and B lies on the equator. There is therefore no component of $_A\boldsymbol{\omega}_B$ along

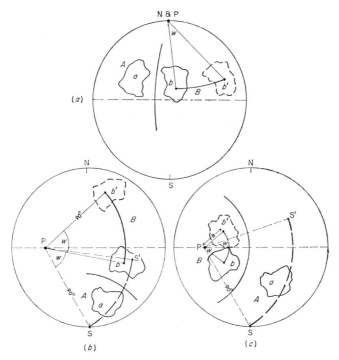

FIGURE 126   To illustrate the motions between plates and the resulting movement of the palaeomagnetic pole. Continent a on plate A is regarded as fixed to the north (N) or south (S) pole. The pole of rotation P describes the relative motion between plate B and plate A. Continent b on plate B moves through the angle w to position b'. In (a) P is at the north pole (N), so that no polar motion is observed, whilst in (b) and (c) P is on the equator and so coincides with the pole of rotation for the polar motion. In (b) continent b lies $90°$ away from P so that the polar motion SS' has the same length as the continental motion bb'. In (c) continent b lies near P, so that the polar motion SS' is greatly amplified compared with the continental motion bb'.

the spin axis, so that the last term in (7.1) is zero. If the geographic pole remains fixed to A, then $_G\Omega_A = 0$ and in this case (7.1) becomes

$$_G\Omega_B = {}_A\omega_B. \tag{7.2}$$

The angular motion of plate B with respect to the palaeomagnetic pole is the same as the angular motion between plate B and plate A. When dealing with the past motions of plates in terms of their motion with respect to the palaeomagnetic pole, only a part of the original plate (the portion of continental crust) is now discernible. This means that the angular distance θ between, say, the centre of this continent and the axis of the rotation vector

237

is not known. The velocity $_A\boldsymbol{v}_B$ between the centre of the continent on plate B and plate A is given by (5.1) in §5.3.1 as

$$_A\boldsymbol{v}_B = a_A\boldsymbol{\omega}_B \sin\theta, \tag{7.3}$$

where a is the radius of the earth. The velocity $_G\boldsymbol{v}_B$ between the centre of the continent and the palaeomagnetic pole is likewise

$$_G\boldsymbol{v}_B = a_G\boldsymbol{\Omega}_B. \tag{7.4}$$

Combining (7.2), (7.3) and (7.4)

$$_G\boldsymbol{v}_B = \operatorname{cosec}\theta \cdot {_A\boldsymbol{v}_B}. \tag{7.5}$$

Since cosec θ lies between 1 and ∞, the motion of the continent on plate B (b in fig. 126) with respect to the palaeomagnetic pole either equals its motion with respect to plate A or may greatly amplify it. The second and third cases shown in fig. 126 illustrate this point.

In the second case (fig. 126b), the pole of rotation lies 90° away from the continent (cosec$\theta = 1$), so that $_G\boldsymbol{v}_B = {_A\boldsymbol{v}_B}$ and the polar motion with respect to the continent on plate B (b) reflects its motion with respect to A. This is essentially the situation involving the separation of Australia and Antarctica during the Tertiary. The position of Antarctica changes very little with respect to the geographic pole during this time, whereas with respect to Australia the palaeomagnetic pole (the south pole) moves essentially north–south (§6.3.3 and §6.3.5). According to McKenzie & Sclater (1971) the pole of rotation between Australia and Antarctica lies at 6s, 41E, that is almost on the equator and roughly 90° from the centre of Australia.

In the third case (fig. 126c) the continent on plate B lies near the pole of rotation between A and B (θ small, cosecθ large). The continent need only move through a short distance, but the palaeomagnetic pole could move through a large distance. This is essentially the situation involved in the rotation of the Iberian peninsula with respect to Europe, together with the other regions of southern Europe (§7.2.3). Note that this apparent polar-wander does not even require that sea floor be created between A and B, the motion could be along a transform fault between them. In general of course the actual situations are more likely to lie between the three extremes outlined and illustrated in fig. 126, but they serve to illustrate the difficulty in drawing conclusions about the past motions between plates from apparent polar-wandering when the poles of rotation describing the motions are not known.

On the theory of plate tectonics, continental crust is continually being rifted apart and welded together. It is possible in favourable circumstances

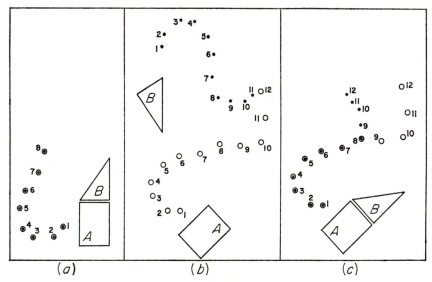

FIGURE 127   Reconstruction of the relative positions of two continents, A and B, which were previously part of the same plate. (a) Originally when A and B occupied the same plate, the apparent polar-wander paths for both A and B were the same and successive polar positions from times 1 through to 8 are shown. (b) After time 8, A and B split apart, each 'carrying' its apparent polar-wander path with it, until the present day situation at time 12 is reached as shown. (c) To reconstruct the previous relative positions, A and B are moved until the initial portion of the paths coincide. The time at which the paths diverge then also dates the time at which they separated. From Graham *et al.* (1964), *J. Geophys. Res.* **69**, 3896.

to determine the past relative positions of two continents, if they were previously part of the same plate. For the period of time in which they occupied the same plate, these blocks should have the same apparent polar-wander path. Matching the two paths enables the previous relative positions of the blocks to be determined uniquely. The method is illustrated in fig. 127 after Graham *et al.* (1964) and is based upon the proposal by Irving (1958*a*), although of course not originally stated in plate tectonic terms. In practice one rarely has a whole series of points on two paths that can be matched. Usually only two or possibly three points are available. It is important not to try to match the pole paths of two blocks, which were in relative motion because this will obviously lead to erroneous conclusions. Just because two pole paths appear to be similar it does not necessarily mean that a unique solution to the former relative positions of the blocks can be found. If at any time the palaeomagnetic pole lay near the pole of relative motion between the blocks, an undetected component of motion can result, and the similarity in paths is coincidental. The matching of two pole paths should therefore be undertaken with some additional information in mind, such as geological comparisons between the blocks or the matching of coast lines. Many of the analyses of the apparent polar-wander paths given in this chapter are based upon these considerations.

The method illustrated in fig. 127 applies to the case of an expanded ocean, such as the Atlantic and Indian Oceans of today. The reverse of this method can be applied to the case of a contracted ocean and is a means of demonstrating the existence of a collision zone between former subcontinents. The present day situation corresponds to that shown in fig. 127c, where time 1 now refers to the present day. Going back in time through to time 8, the paths then diverge showing that the two regions *A* and *B* did not occupy the same plate prior to time 8, which therefore dates the time of collision.

7.2 MAJOR BLOCKS AND COLLISION ZONES

In this section certain known situations will first be examined. The application of the methods outlined above will then show the extent to which comparisons of apparent polar-wander paths can be used to determine relative movements.

7.2.1 *Africa and Arabia*

The Red Sea and the East African Rift Valley were the first major features to be recognized as having been produced by extension of the earth's crust (Gregory, 1921). The plate tectonics of this region has been analysed by McKenzie *et al.* (1970). The seismicity of the Red Sea, Gulf of Aden and the East African Rift shows that three plates meet at the south end of the Red Sea. These have been called the Arabian, Nubian and Somalian plates by McKenzie *et al.* (1970) and are shown in fig. 128. In addition a small plate, the Sinai plate, is proposed separated from the Arabian plate by the Dead Sea fault system. There is considerable evidence that the entire Red Sea trough has been formed by the movement of Arabia away from Africa. The pole of rotation between the Arabian and Somalian plates has been calculated to lie at 26.5N, 21.5E, whilst the pole between the Arabian and Nubian plates lies at 36.5N, 18E (McKenzie *et al.*, 1970). Rifting in this part of the world is estimated to have taken place over the past 20 My, so that the Red Sea and Gulf of Aden represent the first stages in the expansion of an ocean.

The first attempt to apply a palaeomagnetic test to the structural problems

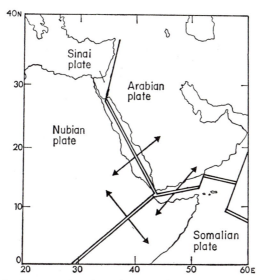

FIGURE 128   Present plate boundaries and spreading directions between
Africa and Arabia. From McKenzie *et al.* (1970), *Nature*, **226**, 1–6.

of the Red Sea and the Gulf of Aden was made by Irving & Tarling (1961)
in a study of the Aden Volcanics. Although these volcanics are certainly
younger than the initial rifting in this region, it is possible that they could
have recorded at least part of the tectonic movements. Irving & Tarling
(1961) observed that the mean inclination does not differ significantly from
that expected from an axial geocentric dipole but that the mean declination
was westerly (N 7W). This discrepancy in declination disappears if Arabia
is rotated 7° in a clockwise direction, by which movement the Red Sea is
closed. Further palaeomagnetic results from southern Arabia have been
reported by Tarling *et al.* (1967) but these contributed nothing further to
the problem. Hospers & Van Andel (1969) have argued that the result of
Irving & Tarling (1961) is inconclusive because comparative data are
necessary from northeast Africa. It is now possible, however, to make
a more substantial comparison of the African and Arabian palaeomagnetic
results as is described below.

Prior to 20 My ago, it can be assumed that Arabia was attached to
Africa along the coast lines of the Red Sea and the Gulf of Aden. The pre-
Miocene palaeomagnetic poles for Arabia should therefore show a systematic
departure from the corresponding African poles. Both apparent polar-
wander paths are compared in fig. 129. In discussing the results from
Arabia, it was noted in §6.3.1 that there was no difference between Mesozoic palaeomagnetic poles (largely Cretaceous) from Turkey and Arabia,

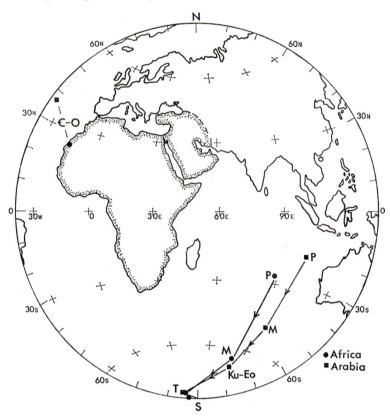

FIGURE 129 Comparison of the apparent polar-wander paths for Arabia (including Turkey) and Africa (cf. figs. 119 and 121, §6.3.1). Only the African Lower Permian Songwe–Ketewaka pole is shown. Equatorial equal area projection.

so that it was proposed that they together formed part of the same plate during that time (Van der Voo, 1968). What then was the former extent of the Arabian plate? Stocklin (1968) suggests that the Zagros thrust zone to the east is the surface expression of a deep split in the formerly coherent Arabian–Iranian platform. The presence of ophiolites and the recent volcanic activity beyond the Zagros thrust zone suggest this represents a subduction zone partly taking up the extension occurring in the Red Sea and Gulf of Aden. Activity along the Dead Sea fault system has occurred largely since the Miocene (Quennell, 1958), so it seems that the former Arabian plate, presumably part of a larger African plate, extended up to the southern foothills of the Caucasus Mountains. These might represent the collision zone resulting from the closing of the Tethys, relics of which are seen in the Black Sea and Caspian Sea. How far westwards in the Balkans or eastwards into Iran the plate extended is difficult to judge, since the palaeomagnetic information is limited to Turkey and Arabia.

In fig. 129 the Mesozoic pole for Arabia is separated from the African pole in the correct sense required from the opening of the Red Sea and Gulf of Aden. In fact the whole apparent polar-wander curve for Arabia from the Permian to the Lower Tertiary is consistently displaced to the east of the African as would be expected. The average displacement is 10–12° rather than the 7° required to close the Red Sea and this is consistent with the view expressed in §6.1 that palaeomagnetic poles are unlikely to be more accurate than 5°. The Cambro-Ordovician pole, although falling in the same region as the African equivalents, is not shifted in the correct sense. It might be somewhat optimistic to expect this comparison to hold as far back as the lower Palaeozoic (see §7.3.2). Note that the Permian pole from Turkey is displaced by the required amount from the African Songwe–Ketewaka Lower Permian pole. If one accepts the hypothesis that Turkey and Arabia were previously part of Africa, then support is provided for the easterly excursion of the pole with respect to Africa during the Permian. Gregor & Zijderveld (1964), in discussing their Permian pole from Turkey admit there are similarities with African Permian results. Because the inclination observed is similar to that expected for this region on the basis of the European Permian data, they prefer the view that the declination is anomalous and that this part of Turkey has rotated 90° with respect to 'stable' Europe. The interpretation given here would seem to represent a simpler and less drastic viewpoint, but further discussion revolves around the interpretation of the African Permian result in the light of comparisons with other continents (§7.3.2 and §7.3.3).

7.2.2   *Europe and Siberia*

Results from western Europe and the Russian platform were analysed separately in §6.2.2 (table 17) and §6.2.3 (table 19) respectively. There is every reason to suppose that these two regions have remained as a unit for most of the Phanerozoic. The similarity of the palaeomagnetic poles for each region since the Silurian confirms this view. However some evidence of movements within these regions is provided by the study of Birkenmajer *et al.* (1968), who, in a very thorough investigation of the Upper Carboniferous rocks of the Inner Sudetic Basin and the Bohemian Massif, concluded that the latter had rotated with respect to stable Europe by about 17°. Hospers & Van Andel (1969), whilst not disputing that there was evidence for relative rotation between the two regions, dispute the contention that it was necessarily the Bohemian Massif which had rotated with respect to

TABLE 29 *Combined palaeomagnetic poles for western Europe and the Russian platform. Symbols as in table 16; combined data of tables 17 and 19*

Period	N	K	A_{95}	North pole position
Upper Tertiary	34	44	4	79N, 183E
Lower Tertiary	17	53	5	75N, 164E
Cretaceous	5	19	18	80N, 162E
Jurassic(?)	4	25	19	65N, 138E
Triassic	14	52	6	50N, 150E
Permian	30	108	3	44N, 161E
Upper Carboniferous	39	46	3	37N, 167E
Silurian–Lower Carboniferous combined	32	39	4	27N, 162E
Cambrian	4	38	15	8N, 189E

Europe. Results from both these regions are used in establishing the overall data for western Europe. Although undoubtedly there will be such movements within large regions, thorough sampling should help to average out their effects. The Hercynian orogeny in Europe could well have been the cause of some of these movements.

Comparison of results from the Russian platform and western Europe during the lower Palaeozoic is difficult. Although the Cambrian and Ordovician poles for the two regions each differ, this is undoubtedly related to the sampling regions involved. The Cambrian and Ordovician of western Europe are all derived from the Caledonides in Britain and could therefore relate to a different plate at that time (§6.2.2). The same applies to the Late Silurian to Early Devonian poles from Scotland. In table 29 the results of tables 17 and 19 are combined together to produce an overall summary of results from Europe west of the Urals. Results from the British Caledonides have not been included. The procedure has been to combine the Silurian to Early Carboniferous results into a single pole. When the European data are compared with those from North America in §7.3.1, this pole will correspond to the Siluro-Devonian pole from that continent (table 16, §6.2.1). In Russian stratigraphy the Middle Carboniferous is essentially the lower part of the Upper Carboniferous of western Europe (see Harland *et al.*, 1964). Thus the Middle and Upper Carboniferous of the Russian platform have been combined with the Upper Carboniferous of Western Europe. This is convenient for later comparisons with North America, because the North American Carboniferous pole covers the span Late Mississipian to Late Pennsylvanian and therefore contains no real Lower Carboniferous equivalents. Note that the Jurassic and Cretaceous of Europe are the most poorly defined periods from a palaeomagnetic point of view.

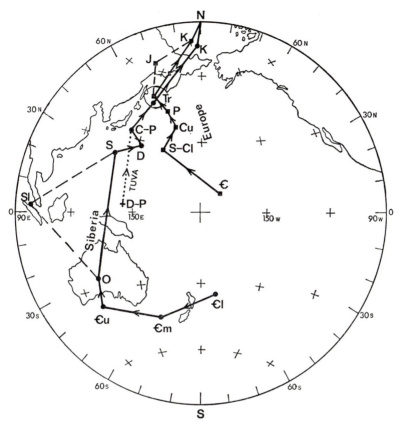

FIGURE 130   Comparison of the apparent polar-wander paths for Europe (data of table 29) and Siberia (table 20 and fig. 116, §6.2.3). The Triassic group of poles when the paths converge is circled. The cross refers to results from the Tuvan massif and regions to the south of the Siberian platform. Equatorial equal area projection.

Fig. 130 compares the apparent polar-wander paths for Europe and Siberia. The Siberian Cambrian and Ordovician poles differ widely from the few lower Palaeozoic results from west of the Urals. The two paths approach one another in the Silurian, but the Silurian to Permian Siberian poles lie consistently westwards of their European counterparts. The two paths only converge ultimately in the Triassic. These observations are in accord with the geology of the Uralides (Hamilton, 1970), and suggest that Europe and the Siberian platform were far apart during the early Palaeozoic, approached one another during the middle Palaeozoic and collided in the Permian or Triassic. The Urals therefore mark the site of a collision zone between two formerly separated continents, Angaraland and Euramerica. Note that the concept of a single supercontinent of Laurasia extending

245

from North America to the Siberian platform is untenable during the Palaeozoic. This supercontinent was restricted to North America and Europe only and will be referred to as Euramerica (§7.3.1). Late Palaeozoic results from the Tuvan massif and adjacent regions to the south of the Siberian platform suggest these regions only became welded to the Siberian platform during the Permian.

7.2.3   Southern Europe

It has long been suggested by both du Toit (1937) and Carey (1958) that the Iberian peninsula has undergone an anticlockwise rotation with respect to Europe, as a result of which the Bay of Biscay was formed. The first palaeomagnetic test of this concept was carried out by Clegg *et al.* (1957), who compared results from Triassic rocks in England and Spain. Their results lent some support to such a rotation in post-Triassic times, but the final interpretation of them remained inconclusive. Considerable work by the Dutch over the past decade (§6.2.2) now confirms that such a rotation has indeed occurred, the most recent information being summarized by Van der Voo (1969). Matthews & Williams (1968) report the presence of a fan-shaped pattern of linear magnetic anomalies in the Bay of Biscay, supporting the idea that new oceanic crust was formed during the anti-clockwise rotation of Spain. Bullard *et al.* (1965) found that a clockwise rotation of Spain through 32° is required to close the Bay of Biscay at the 500 fathom isobath.

The pole of rotation of Spain with respect to Europe must lie in the Pyrenees and close to the point where they meet the Bay of Biscay (say 43N, OE). The palaeomagnetic poles for the Permian and Triassic of Spain both show that the region of the Pyrenees lay very near the equator at that time. In other words the palaeomagnetic poles lie roughly 90° away from the pole of rotation. This is a situation analogous to that outlined in fig. 126c and represented by (7.2). The angular rotation of Spain will be equal to the angular rotation of its palaeomagnetic pole with respect to that for Europe, and is simply given by the great circle distance between the poles for the two regions and this is 36°.

The procedure outlined above can also be applied to other regions of southern Europe which are supposed to have been rotated with respect to Europe, namely Italy (southern Alps), Corsica and Sardinia. Within the limits of accuracy of the palaeomagnetic method, it can be assumed that each of these regions lie close enough to the palaeoequator to justify application of the simple analysis outlined above. The great circle distances between the Permian and Triassic poles for each region and those for

TABLE 30   *Rotation of various regions of southern Europe with respect to stable Europe. The angle is given by the great circle distance between the Triassic and/or Permian pole for each region (table 18) and the corresponding pole for Europe (table 29)*

Region	Age	Rotation
Spain	Permo-Carboniferous	36°
	Triassic	30°
Corsica	Permo-Carboniferous	42°
Italy	Permian	52°
(Southern Alps)	Triassic	60°
Sardinia	Permian	74°

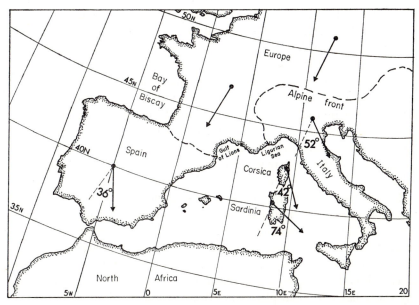

FIGURE 131   Rotations of various regions of southern Europe with respect to stable Europe from a comparison of their Permian poles. The average European declination direction is shown at two arbitrary points, whilst the declination observed for each region is indicated by the vector drawn at the sampling localities, except for Spain for which a central point is used.

Europe are summarized in table 30. The Permian figures are the more reliable in each case, but they suggest rotations between 35° and 70° have occurred since the Triassic as is illustrated in fig. 131. The rotation deduced from the Permian of Spain is 36°, very close to that expected from the opening of the Bay of Biscay. A 70° anticlockwise rotation of Sardinia is consistent with an original location in the Gulf of Lions and a pole of rotation situated in the Ligurian Sea in the region of 43N, 9E. Carey (1958) has previously suggested that Sardinia was once situated against the present Mediterranean coast of France and Spain. Attempts to detect linear magnetic anomalies in the Mediterranean (Vogt *et al.*, 1971) have so far been unsuccessful, so that the timing of these movements is not known with any certainty. An analysis of the evolution of the northeast Atlantic (Williams & McKenzie, 1971) from magnetic data suggests the Bay of Biscay was formed by the rotation of Spain in the Upper Jurassic or Lower Cretaceous. This aspect is discussed again in relation to the break-up of Pangaea in §7.3.3.

The method outlined above is not the usual one that has been used for comparing the results from southern Europe with those from stable Europe. The usual method used by Dutch workers has been to draw palaeoisoclines (lines of equal palaeoinclination) across the region (Van Hilten, 1962*b*) using data from stable Europe. De Boer (1963, 1965) in comparing data from the southern Alps (northern Italy) showed that this region had a mean Permian inclination of about $-30°$. This meant that the area of the Vicentinian Alps must, in Permian time, have been situated on the $-30°$ palaeoisocline for stable Europe. The orientation of this isocline is such that the area of the Vicentinian Alps would need to have been nearly 5000 km east of its present position. Hospers & Van Andel (1969) and Zijderveld *et al.* (1970*b*) both show that this conclusion is not now justified on the basis of the most recent information. Essentially the differences in palaeoinclination observed between this region and stable Europe are within the limits of accuracy of the palaeomagnetic method as further data have now shown. Therefore it is not necessary to postulate a major shear zone in the Mediterranean on the basis of the palaeomagnetic data, so that concepts such as the 'Tethys Twist' of Van Hilten (1964) are not now a necessity. Only the differences in declination are significant so that rotations without translations are all that are required. This does not mean that any shear has not taken place, they are just not required by the palaeomagnetic results.

The data from Pannonia also show that rotations have occurred in the Carpathian Mountains of Czechoslovakia (Kotasek & Krs, 1965) within the Tethys mobile belt. Results from the Choč nappe suggest a post-Permian rotation of 45° with respect to neighbouring regions of the Carpathians. This rotation is probably a less fundamental movement than the other instances outlined above (Hospers & Van Andel, 1969).

7.2.4 *The Far East*

Mesozoic mobile belts occupy large areas of the eastern U.S.S.R. and China. The Verkhoyansk Mountains separate the Siberian platform in the west from the Kolyma platform to the east. This mobile belt was formed in Lower Cretaceous times from a geosyncline of late Palaeozoic and Mesozoic sediments. In addition there is a Mesozoic mobile belt to the south of the Aldan massif of eastern Transbaikalia and Amur and closely connected with it is the Mesozoic mobile belt of Sikhote Alin (see fig. 133 below). Hamilton (1970) regards these various orogenic belts of the structural chaos of Asia south and east of the Siberian platform as having formed as small continental plates that were swept against each other and into Asia. The orogenic belts thus represent sites of collision zones on plate tectonic theory if Hamilton's interpretation is correct. If the North Atlantic opened along the region of the mid-Atlantic ridge and separated Europe from North America, then one might expect this tensional feature to be compensated by a compressional zone on the other side of the globe. Indeed Wilson (1965 a) proposed that the mid-Atlantic ridge transformed into the Verkhoyansk Mountains. The data at present available from the Far East make it possible to test these ideas.

The Permian to Cretaceous palaeomagnetic poles from Siberia, China, the Kolyma block and Sikhote Alin region are all compared in fig. 132. Although there are only Jurassic and Cretaceous results from China, they are similar to the Siberian data and suggest that the Chinese and Siberian platforms were already part of the same plate in Jurassic times. The apparent polar-wander paths for Kolyma block and the Sikhote Alin region differ widely from one another and from the Siberian path during the Permian. The three paths approach one another during the Triassic, and converge during the Jurassic, becoming tightly grouped during the Cretaceous. This feature provides strong support for the collision theory and suggests that both the Verkhoyansk and Sikhote Alin Mountains represent the sites of collision zones at which these regions became welded onto Asia by the Early Cretaceous.

The Cainozoic mobile belts of the Far East occupy the extreme northeastern part of Asia including the Kamchatka peninsular and the island arcs of the Kuriles, Japan and Sakhalin. Kawai *et al.* (1961) proposed a Cretaceous–Tertiary geotectonic movement on the basis of palaeomagnetic results, which showed that the main island of Japan had been bent to form the present bow-shaped structure. The deformation of the Japanese arc had previously been postulated by Carey (1958). Kawai *et al.* (1969) have summarized the most recent information from the northeast part of Japan,

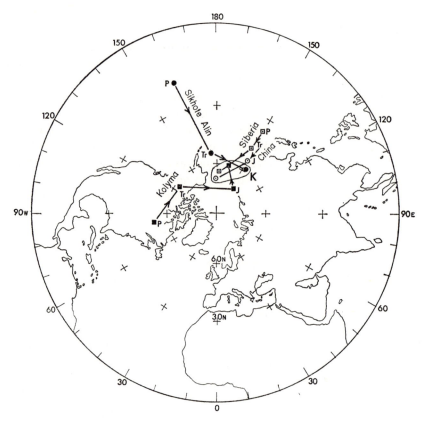

FIGURE 132 Comparison of the apparent polar-wander paths for the various regions of the Far East from the Permian to the Cretaceous (cf. fig. 118, §6.2.4). The Cretaceous poles are circled and indicate the time when the paths converge. Polar stereographic projection.

including K–Ar ages. All this information is summarized in the appendix and shows the following main features.

Cainozoic volcanic rocks from the northeast and southwest parts of Japan give palaeomagnetic results which are in general agreement during the Tertiary (JA11.1–3). On the other hand, Cretaceous results from various regions differ. Those from southwest Japan (JA10.1 and 10.6) and Kamchatka (JA10.4) have easterly declinations, whereas those from northeast Japan (JA10.2) and Sakhalin (JA10.5) show westerly declinations. These declinations are indicated on fig. 133 for the various regions, and they show that differences in declination of about 70° are involved. The present angle between the axes of the arms of the main island of Japan is about 60°, so the palaeomagnetic results are in general agreement with the bending

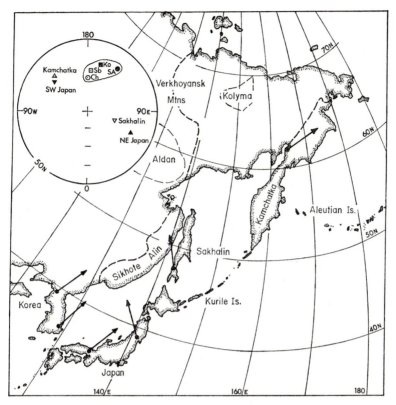

FIGURE 133   Cretaceous palaeomagnetic data for the Far East. The palaeomagnetic directions (declinations) are indicated for various sampling regions in Korea,
Japan, Sakhalin and Kamchatka and show that northeast Japan and Sakhalin have
rotated with respect to southwest Japan and Kamchatka. The inset shows the
Cretaceous palaeomagnetic poles plotted on a stereographic projection from 50N.
The ringed poles correspond to those in fig. 132 for the four regions of the Asian
mainland.

hypothesis within the limits of accuracy. They therefore suggest that both
northeast Japan and Sakhalin have been rotated with respect to southwest
Japan and Kamchatka. On the basis of one rather poor result (JA10.3) with
slightly younger K–Ar age, Kawai *et al.* (1969) suggested the main bending
took place between 120 and 90 My ago. This seems a somewhat doubtful
conclusion to make on the basis of the rather limited information, but the
bending must certainly be post 120 My ago and is probably Late Cretaceous
to Early Tertiary in age.

The Cretaceous results from the Cainozoic mobile belt of Japan–
Sakhalin–Kamchatka can be compared with those from the other regions

251

of Asia (Siberia, China, Kolyma and Sikhote Alin). The inset of fig. 133 compares the Cretaceous palaeomagnetic poles for the various regions. The southwest Japan and Kamchatka poles agree well, but deviate from the group of four for the various regions of Asia. The latter also differ from the poles for the bent regions of northeast Japan and Sakhalin, as might be expected. These differences provide evidence that Kamchatka and southwest Japan have moved relative to the eastern part of Asia since the Cretaceous. The interpretation of this movement depends largely on the origin of the basins west of the Japan and Kurile–Kamchatka island arc. Beloussov & Ruditch (1961) take a somewhat classical view that their origin is essentially due to crustal subsidence, but Karig (1971) ascribes their origin to crustal extension on a plate tectonic model. On this model the 'interarc basins' of the western Pacific were formed by extensional rifting within an older 'frontal arc'. As a result, the trench–frontal arc complexes migrate away from the Asian continent with creation of new basins with oceanic crust on the convex sides of the frontal arcs. The palaeomagnetic data support this model rather than the classical one since relative motion is involved.

The palaeomagnetic data for the Far East, whilst providing clear indications that the various portions of eastern Asia have been the result of successive collisions of continental plates with Asia proper, also demonstrate that the formation of island arc complexes off continental margins can lead to rather complicated palaeomagnetic situations. This suggests that one might have to be rather wary when dealing with data from regions whose geological history has involved these complexities.

7.3 SUPERCONTINENTS

7.3.1 *Euramerica*

After the first analyses of results from the British Isles (Creer *et al.*, 1954) and North America (Runcorn, 1956*a*) in terms of apparent polar-wander paths, Runcorn (1956*a*) attempted to represent both the North American and British results by a single polar-wander path. However, Runcorn (1956*b*) showed that the North American path was systematically west of the European one and that this was statistically significant. Thus started the first palaeomagnetic argument in favour of continental drift. The argument, being statistically based, was disputed but soon the data from the southern hemisphere (§7.3.2) confirmed this interpretation in an entirely convincing manner. Successive analyses of the latest information (Runcorn, 1962, 1965) from North America and Europe have confirmed this significant difference between their polar-wander paths. The two paths are compared in fig. 134*a* using the data of tables 16 and 29. With the amount of data now available there is no doubt that the paths are significantly different. The North American curve lies to the west of the European curve until the Late Triassic.

Wells & Verhoogen (1967) showed that if the method of reconstruction of the Atlantic Ocean used by Bullard *et al.* (1965) was applied to the late Palaeozoic palaeomagnetic poles of the surrounding continents, then there was a significant improvement in the grouping of the poles for the Permian but not for the Carboniferous. The analysis was based largely on the summary data given in Irving (1964). Hospers & Van Andel (1968) noted that these summary data failed to establish beyond doubt whether Europe and North America have common or significantly different apparent polar-wander curves. They therefore analysed the most reliable (cleaned) data for the two continents and showed that the differences between the two paths were significant. They further showed that if the method of reconstruction of the North Atlantic used by Bullard *et al.* (1965) was applied to the palaeomagnetic poles (rotation about a pole at 88.5N, 27.7E) the Carboniferous, Permian and Triassic poles need to be rotated through 40° to make corresponding poles coincide. Bullard *et al.* (1965) used a rotation of 38° to close the North Atlantic between Europe and North America, so that the agreement between the two methods is excellent.

The method of reconstruction of the North Atlantic used by Bullard *et al.* (1965) can be applied to the apparent polar-wander paths of fig. 134*a*. If North America and its polar-wander path are rotated through an angle of 38° about a pole at 88.5N, 27.7E so as to close up the North Atlantic ocean then this results in the arrangement of fig. 134*b*. Agreement between the paths from the Silurian to the Triassic is virtually exact. Note that the Triassic pole for Europe is closest to the Lower Triassic pole for North America. This is consistent with the fact that twelve of the fourteen European results are from Lower and Middle Triassic rocks. The Jurassic of Europe coincides with the Upper Triassic of North America, but the European results are from rocks of questionable Jurassic age from the Crimea; they might be Upper Triassic or Lower Jurassic. The two paths diverge after the Triassic as might be expected because current views date the most recent opening of the North Atlantic Ocean between Europe and North America as post-Triassic (see §5.2.2; Funnell & Smith, 1968; Pitman *et al.*, 1971; Williams & McKenzie, 1971). The limited Cambrian results are brought into closer agreement by the method of reconstruction in fig. 134*b*, but the lower Palaeozoic poles from the Caledonian and Appalachian orogenic belts (shown in fig. 134*b*) still remain widely separated. These results are consistent with views concerning the existence of

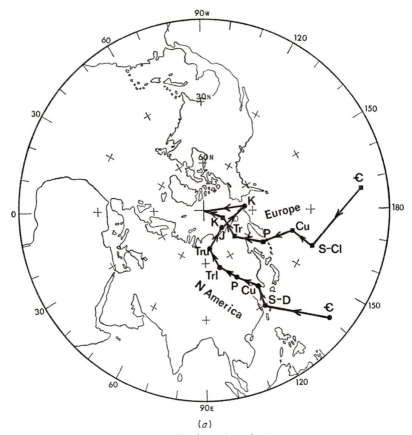

FIGURE 134. For legend see facing page.

a proto-Atlantic Ocean (J. T. Wilson, 1966; Dewey, 1969). Deutsch (1969) has proposed a palaeomagnetic scheme for testing these views, and whilst, as he suggests, a tenfold increase of data may be required, it represents a challenging application of the palaeomagnetic technique.

A single Euramerican landmass, which included North America and Europe (west of the Urals) probably only existed from the Silurian through to the Triassic. The proximity of the Cambrian poles on the reconstruction suggests however that any proto-Atlantic was not a very wide ocean. Whether or not Euramerica and Gondwanaland were themselves part of a single unit (Pangaea) during any of this period is discussed in §7.3.3. The regions to the east of the Urals were only added to Europe during the late Palaeozoic and early Mesozoic, whilst the Kolyma block and other regions of the far east were only added to Asia during the late Mesozoic after the opening of the North Atlantic had commenced.

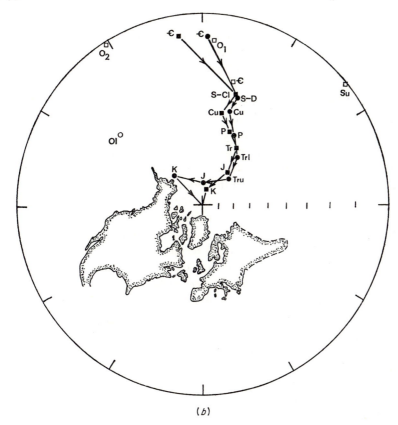

(b)

FIGURE 134 (a) Comparison of the apparent polar-wander paths for North America (circles, table 16) and Europe (squares, table 29). Polar stereographic projection. (b) The two polar-wander paths after rotation to the Bullard *et al.* (1965) fit of the North Atlantic. Solid symbols refer to the essentially stable regions of the two continents, whereas the open symbols refer to the regions of the Caledonian and Appalachian orogenic belts giving deviating results. Polar stereographic projection.

7.3.2 *Gondwanaland*

The observations made on Mesozoic rocks from the southern continents and in particular Australia, provided the most compelling evidence in favour of the continental drift hypothesis (Irving, 1957c, 1958a, b, 1959; Creer *et al.*, 1958). Not only were the Mesozoic poles from these continents widely divergent but also the late Palaeozoic, Mesozoic and early Tertiary poles for Australia departed markedly from those for North America and Europe. Furthermore Irving (1957c, 1958a) showed that the data were more compatible with reconstructions proposed on geological grounds (du Toit, 1937; Carey, 1958) than with the present distribution of continents. With the improvement in palaeomagnetic techniques the divergences between the Mesozoic poles for the southern continents were further emphasized (Irving *et al.*, 1963; Gough *et al.*, 1964). With the publication of data for a number of geological periods it became possible to attempt to use the method outlined in §7.1 and illustrated in fig. 127. This was undertaken successively by Creer (1964, 1965), Briden (1967), McElhinny (1967*b*) and Creer (1967*c*, 1968*a*). The procedure is to slide transparent spherical shells, on each of which is drawn a continent and its apparent polar-wander path, over the surface of the globe until the paths match in the best possible way. Unfortunately only data for the Mesozoic were available for Antarctica and India, and it was not possible to derive any palaeomagnetically unique reconstruction. The results, however, appeared to provide general support for du Toit's (1937) reconstruction of Gondwanaland.

Briden (1967) first pointed out that, on a reconstruction of Gondwanaland, the palaeomagnetic poles tended to fall into a number of groups rather than to lie on a continuous apparent polar-wander path common to all the continents. These groups of palaeomagnetic poles represented varying lengths of time which Briden termed *quasi-static intervals*. During these intervals the pole remained essentially stationary whilst between them the pole made a rapid transition from one group to the next. These transitions were called *drift episodes* after Irving (1966). In order to determine a unique palaeomagnetically based reconstruction, it is necessary to have data for each continent from at least two of these quasi-static intervals. The determination of the first lower Palaeozoic poles for Antarctica (Zijderveld, 1968) and India (McElhinny, 1970*b*) and more reliable ones for Australia (McElhinny & Luck, 1970*a*) completed these requirements. This enabled McElhinny & Luck (1970*b*) to propose a reconstruction that was based upon palaeomagnetic results alone. A feature of this reconstruction was a gap between Australia and Antarctica compared with the usual morphological fit (Sproll & Dietz, 1969; Smith & Hallam, 1970). This arose from the proposal originally made by McElhinny & Opdyke (1968) that African palaeomagnetic results show the same rapid polar shift during the Carboniferous as was found by Irving (1966) for Australia, and that these two shifts could be matched on a reconstruction of Gondwanaland. Lower Palaeozoic poles from Australia provided an impressive fit between the African and Australian polar-wander paths throughout the Palaeozoic with these two continents in a fixed relative position and appeared to provide strong support for this proposition (McElhinny & Luck, 1970*a*). New Permo-Carboniferous results from South America (Embleton, 1970)

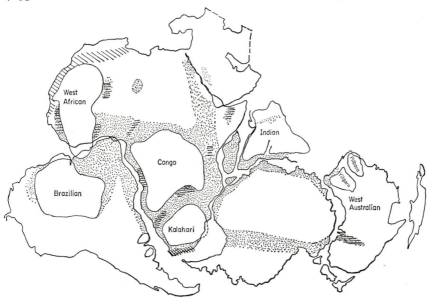

FIGURE 135   Computer reconstruction of Smith & Hallam (1970) of Gondwanaland, showing the ancient cratons and Precambrian shields (named regions) of age
approximately 2000 My or more. Late Precambrian to Cambrian geosynclines and
metamorphic belts whose isotopic ages peak around 500 My are stippled. Horizontal
lines indicate the areas where essentially flat lying Late Precambrian to Cambrian
sediments occur. Diagonal lines indicate the younger late Palaeozoic fold belts of
the Cape and the Mauritinides. Equal area projection with geological information
by courtesy of A. R. Crawford.

now suggest that a Carboniferous polar shift of the same magnitude is not
to be found in the palaeomagnetic record of South America. The most recent work on Silurian to Lower Carboniferous formations in southeast
Australia (Luck, 1971) substantially modifies the simple initial picture
obtained by Irving (1966). It now appears to be necessary to abandon the
McElhinny & Opdyke (1968) proposal. This is not to say that the two
Permo-Carboniferous poles from Africa should be combined, as will be
shown.

The reason for the failure of the apparent matching of the polar-wander
paths of Australia and Africa to produce a viable reconstruction of their
relative positions lies in the fact that more than one plate was involved.
As was noted in §7.1, attempts to match the polar-wander paths of two
plates in relative motion can lead to erroneous conclusions. Therefore the
approach that will be adopted here is to compare the palaeomagnetic data
for Gondwanaland with what appears to be the most reasonably based
fit on both morphological and geological criteria (Smith & Hallam, 1970).

Before doing so, however, it is worth considering the basic overall geological features that are relevant to the Phanerozoic plate tectonics of Gondwanaland. These are illustrated in fig. 135.

The main features of the geology of the reconstruction is that there are two main age provinces, comprising the ancient cratons of age roughly 2000 My or more and their surrounding younger metamorphic belts or geosynclines whose ages tend to cluster around 450 to 650 My (Cahen & Snelling, 1966; Clifford, 1968, 1970; Crawford, 1969; Hurley & Rand, 1969). In general these latter regions in the western part of Gondwanaland (largely based on Africa) have Late Precambrian to Cambrian sediments and metamorphic ages peaking at 500 My. A major question that needs to be asked and answered is what is the interpretation of this 500 My old event in terms of plate tectonic models? Dewey & Horsfield (1970) note that large areas of these belts consist of regenerated earlier continental crust, and that this is difficult to interpret in terms of these zones representing the sites of contracted oceans, although this cannot be excluded. Does this mean therefore that Gondwanaland only formed as a unit during the Late Precambrian or early Palaeozoic? If this is so, it might not be valid to compare early Palaeozoic results from parts of Gondwanaland in matching apparent polar-wander paths.

In the eastern part of Gondwanaland, the Late Precambrian to Cambrian geosyncline crossing Antarctica (Hamilton, 1967) joins with the similar Adelaide geosyncline of South Australia (Brown *et al.*, 1968). Oliver (1964) and Hamilton (1967) have indicated that orogenic belts may have crossed the whole of Antarctica, becoming progressively younger westwards. Thus the Tasman geosyncline of eastern Australia corresponds to Hamilton's Palaeozoic geosyncline lying west of the Transantarctic Mountains. A still younger Mesozoic geosyncline runs down the Antarctic peninsula. It has already been pointed out in §6.3.3 that the geological history of eastern Australia could be quite complex in plate tectonic terms. There is evidence that the region might be made up of a succession of island arcs accreted onto Australia through continent–island arc collision mechanisms. Oversby (1971) suggests that continent–oceanic plate interaction has produced an easterly progression of trenches and island arcs from the latest Proterozoic through to the late Palaeozoic. If any of these processes also involved some of the ideas of Karig (1971) and Packham & Falvey (1971) with respect to the marginal basins of the western Pacific Ocean today, then there is every possibility that the most complex palaeomagnetic data will be derived from this region. The palaeomagnetism of the Japanese arc (§7.2.4) serves as an example in this respect. If a similar history is also envisaged for the corresponding parts of Antarctica, this also neeeds to be borne in mind.

Of some interest is the Mauritinide zone of north-west Africa. This major zone has been ascribed to the Hercynian (Sougy, 1962), and geochronological data suggest two events recorded by ages in the 300–350 My and 200–250 My ranges (Clifford, 1970). All of these geological characteristics of Gondwanaland need to be borne in mind when analysing the palaeomagnetic results from the various regions of this supercontinent.

The palaeomagnetic poles for Gondwanaland are plotted on the Smith–Hallam reconstruction in fig. 136a. In the region of northwest Africa there is a group of lower Palaeozoic poles, which cover the period from the very latest Precambrian to the Ordovician. They correspond to the oldest of Briden's (1967) quasi-static intervals. They form only a loose group, a factor which may be related to the different ages covered by the group (e.g. the Ordovician of Antarctica lies to the east of the others, whilst African results include the latest Precambrian). Their general grouping in this region suggests Gondwanaland probably was a unit in the lower Palaeozoic or at least a group of closely situated subcontinents. Two mid-Palaeozoic poles from Africa and South America fall in the region of southern Africa and form the succeeding quasi-static interval. The poles in this group range from Lower Devonian to Lower Carboniferous, but stratigraphic evidence from north and south Africa suggests the shift between these two quasi-static positions occurred very rapidly in the Late Ordovician and close to the Ordovician–Silurian boundary (§6.3.2). Note that the Silurian, Devonian and Lower Carboniferous poles from southeast Australia do not fall in this group in southern Africa. Their anomalous position provides evidence that one or more plates may have been involved. Siluro–Devonian poles are derived from rocks in an established stratigraphical sequence in the Canberra–Yass region, whereas the Lower Carboniferous rocks are from the Hunter Valley to the northeast. These would refer to separate island arc systems on Oversby's (1971) scheme and therefore refer to different plates. Neither of these regions need refer to Gondwanaland at all. A comparison of Gondwanaland with Euramerica in §7.3.3 further suggests these poles are anomalous.

The third quasi-static interval is for the Late Carboniferous and Permian. These poles lie in Antarctica on the reconstruction and form a group consisting of the Australian and South American poles and the *combined* Permo-Carboniferous African pole. The Indian Permo-Carboniferous pole does not agree with these, but forms a group to the east with the Arabian Permian pole and the Lower Permian (Songwe–Ketewaka) pole for Africa. The reason for this more easterly group of poles is not clear, but it may be significant that they are all derived from one region of Gondwanaland–India, East Africa and Arabia. One explanation therefore is that this

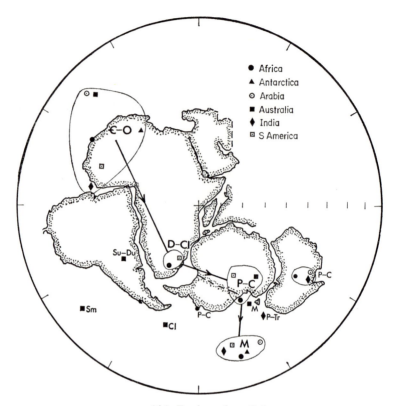

FIGURE 136(*a*). For legend see facing page.

region moved with respect to the rest of Gondwanaland in the Late Permian or Early Triassic. Note that Irving's (1966) mid-Carboniferous rapid polar shift does not now refer to Australia as a whole but only to a small plate east of the Hunter Valley region from which both Lower and Upper Carboniferous results are derived. This plate only formed part of Australia during the Late Carboniferous and subsequently.

Mesozoic poles form a group situated some 25° to the south of the Permo-Carboniferous group. The shift between these two groups occurs within the Early Triassic. This Mesozoic group is made up of Triassic poles for South America and India, all Mesozoic poles for Africa and Arabia and Jurassic poles for Antarctica. The Mesozoic poles for Australia, however, diverge from this group. This puzzling inconsistency has been noted by several authors (Irving & Robertson, 1969; Creer, 1970*a*; McElhinny, 1970*c*), because sea-floor spreading data indicate a Tertiary separation of Australia and Antarctica (Le Pichon & Heirtzler, 1968). Yet the Jurassic

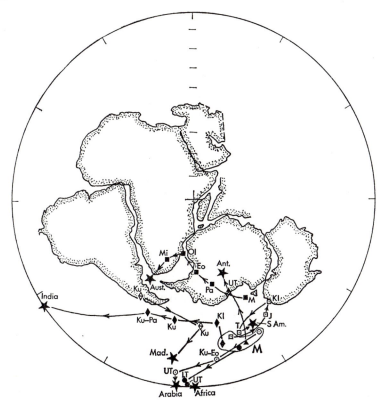

FIGURE 136   (*a*) Cambrian to Mesozoic (M) palaeomagnetic poles for Gondwanaland plotted on the reconstruction of Smith & Hallam (1970). Equal area projection.
(*b*) Mesozoic to Present palaeomagnetic poles using the same symbols as in (*a*) with
the addition of data for Madagascar (open diamonds). The stars represent the
position of the present south pole with respect to the continent in each case, and is
the endpoint of each polar-wander path. Equal area projection.

results for Australia and Antarctica are derived from the Tasmanian and
Ferrar dolerites, which not only have the same age (McDougall, 1963) but
also belong to the same geochemical province (Compston *et al.*, 1968). The
results from these rocks are quite incompatible with the usual morphological fit of Australia and Antarctica. There are possible geological explanations in terms of plate tectonics, because the results of much oil exploration
on the Australian continental shelves have revealed that the initial break
between Australia and Antarctica must have occurred much earlier than
sea-floor spreading data suggest. Griffiths (1971) shows that the presence
of Upper Jurassic sediments and the intrusion of the Tasmanian and
Ferrar dolerites are consistent with an initial break about 150 My ago. The

261

initial rifting did not continue beyond the Red Sea stage, so that the final separation of Australia and Antarctica took place over the past 50 My as indicated by the magnetic anomaly pattern. The Mesozoic data for Australia, however, indicate that the history of eastern Australia during the Mesozoic is probably more complicated than just an initial separation because the palaeomagnetic data are incompatible with the Australia–Antarctica fit with or without a small rift. Some motion of the eastern part of Australia is required, a possibility already being countenanced by Griffiths (1971) in suggesting the motion of a separate Tasmanian plate during the early rifting stage.

It is somewhat ironic now that whereas the results from southeast Australia initially provided the most convincing palaeomagnetic evidence for continental drift, results from this region now seem to indicate that southeast Australia has had a much more complicated history than simply having been part of Gondwanaland before its break-up. By the same token Mesozoic results from West Antarctica (§6.3.5) were shown to be incompatible with the region(s) having been attached to East Antarctica during the Mesozoic. This also has some geological support (Hamilton, 1967; Schopf, 1969). Divergences between the palaeomagnetic poles for the Mesozoic start to occur during the Jurassic, which essentially dates the initial break-up of Gondwanaland. The Mesozoic to Recent results are all plotted in fig. 136*b* and show the manner in which the poles diverge. Their relationship to the break-up and dispersal of the Gondwanic continents will be discussed in the next section.

7.3.3 *Pangaea and its dispersal*

The reconstruction of the continents around the Atlantic Ocean of Bullard *et al.* (1965) can be combined with the reconstruction of Gondwanaland of Smith & Hallam (1970) to provide a reconstruction of Pangaea based upon morphological fits of the continental shelves. This can then be compared with the palaeomagnetic results to see whether or not they conform with this reconstruction. The period mean poles are plotted in fig. 137. They are most compatible with the Pangaea reconstruction during the Silurian to Early Carboniferous, when the poles group in the region of southern Africa and for the Late Triassic. There are three different Permo-Carboniferous groups labelled, 1, 2 and 3 in fig. 137. Each of these groups refers to a different region of Pangaea. The westerly group (1) is from Euramerica, but includes one of the East African Permo-Carboniferous poles (Galula). The central group (2) is for South America, Australia and the combined African Permo-Carboniferous poles, whilst the easterly

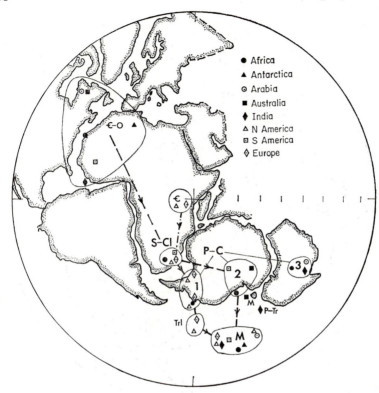

FIGURE 137   Computer reconstruction of Pangaea using the fit of the Atlantic Ocean from Bullard *et al.* (1965) and the fit of the Gondwanic continents from Smith & Hallam (1970). Equal area projection. Period mean palaeomagnetic poles are plotted from the Cambrian through to the Mesozoic. There are three Permo-Carboniferous groups of poles labelled 1, 2 and 3.

group (3) includes India, Arabia and the East African pole from Songwe–Ketewaka. The reason for this grouping is not clear, but two possible explanations can be offered. The first involves the existence of relative movement between parts of Pangaea during the late Palaeozoic. Such movements do not involve the break-up of the supercontinent, but a re-arrangement of parts of it. The evidence for such movement must be found in the Hercynian activity in Europe, which extended down to the Mauri-tinide zone of northwest Africa. The alternate explanation involves the existence of non-dipole fields during the Permo-Triassic (Briden *et al.*, 1971*b*). This is a less satisfactory explanation because it suggests a break-down of the axial geocentric dipole hypothesis, but it has the merit of at least being capable of being tested at least in principle. Only more detailed work will decide which of the explanations is correct.

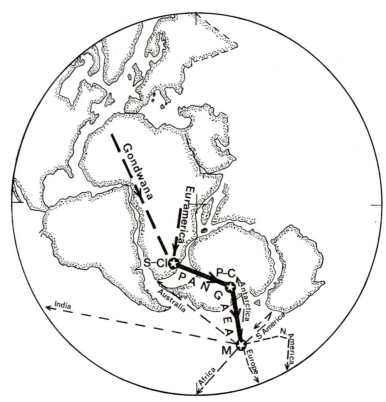

FIGURE 138   Generalized version of the overall results in fig. 137 showing the path of the south pole during the Phanerozoic. The two continents of Gondwanaland and Euramerica combine into Pangaea approximately during the Silurian and there then exists a common polar-wander path until various times during the Mesozoic (M). The paths for the constituent parts of Pangaea then diverge as shown.

The Phanerozoic data can be summarized in the form of the generalized diagram of fig. 138. The pre-Silurian results from Euramerica and Gondwanaland show the existence of two separate supercontinents separated by an ocean during the early Palaeozoic. The polar-wander paths converge into a single combined path for Pangaea from the Silurian through to the late Triassic. The pole was initially situated in southern Africa, but this did not give rise to the Great Gondwana glaciation until the Early Carboniferous. During the Carboniferous the pole moved southeastwards with respect to Pangaea so that by the Early Permian the centres of glaciation had moved towards Australia. Glaciation ceased when the pole moved further south during the Late Permian, when it occupied its Mesozoic position shown. At various times during the Mesozoic the present con

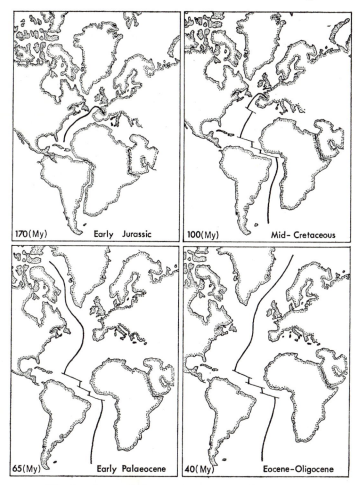

FIGURE 139   The break-up of the continents around the Atlantic Ocean after Francheteau (1970) showing their relative positions at four different times. The location of the mid-Atlantic ridge at each stage is indicated.

stituent parts of Pangaea broke off and drifted to their present positions. The most spectacular of these movements is represented by the northward motion of India discussed below.

The dispersal of the constituent parts of Pangaea is best discussed with reference to the magnetic anomaly patterns in the Atlantic and Indian Oceans. Francheteau (1970) has produced a model for the opening of the Atlantic Ocean by finite rotations of six rigid plates (North America, Greenland, Europe, Spain, Africa and South America) about various poles of rotation for each pair of plates. The application of Francheteau's model

265

leads to the relative positions of the continents at four times (170, 100, 65 and 40 My) as shown in fig. 139. These positions are all compatible with palaeomagnetic information.

The first break in the North Atlantic appears to be dated now as Late Triassic in the time range 200–180 My (Funnell & Smith, 1968; Pitman *et al.*, 1971; McElhinny & Burek, 1971). Williams & McKenzie (1971) suggest that rifting between Iberia and Newfoundland was shortly afterwards accompanied by the extension of the spreading centre into what is now the Bay of Biscay. This is the situation depicted for the Early Jurassic (170 My) in fig. 139, and corresponds roughly to the so-called quiet zone fit of North America and Africa proposed by Drake *et al.* (1968) to avoid the overlap of Mexico, Honduras and Nicaragua with South America on the fit of Bullard *et al.* (1965). The arguments for the quiet zone fit are not entirely convincing and a perfectly plausible plate tectonic model for the evolution of the Caribbean–Gulf of Mexico region has been suggested (Freeland & Dietz, 1971) which avoids this overlap. However, none of the aspects of this model have been depicted in fig. 139.

The opening of the South Atlantic is regarded as having started about 130–140 My ago (Jurassic–Cretaceous boundary), about 60 My later than the North Atlantic. The earliest marine deposits on the western margin of Africa are Upper Jurassic beds in northern Senegal overlain by Lower Cretaceous beds associated with evaporites. These latter sediments are also found in the Congo and Gabon basins (Funnell & Smith, 1968). The oldest marine sediments from the tropical Atlantic off South America are Upper Jurassic to Lower Cretaceous (Fox *et al.*, 1970) whilst those on the South Atlantic date from the Lower Cretaceous (Martin, 1961; Reyment, 1965). The results of Maxwell *et al.* (1970) from deep-sea drilling indicate that the South Atlantic cannot be older than about 130 My. Palaeomagnetic comparisons of the polar-wander paths of Africa and South America (Valencio & Vilas, 1970) show that these diverge between the Late Triassic and Early Cretaceous. Thus by the mid-Cretaceous (100 My) the opening of the South Atlantic is already under way as illustrated in fig. 139. Note how the Mediterranean region is what amounts to a shear zone between Europe and Africa as North Africa slides past. The rotation of Corsica, Sardinia and Italy are well under way.

The rotation of Spain and the formation of the Bay of Biscay occurred largely in the Late Jurassic and Early Cretaceous and was completed by about 80 My, at which time the spreading extended northwest into the Labrador Sea (Williams & McKenzie, 1971). Geological evidence supports the view that the opening of the Labrador Sea occurred in the late Mesozoic (Nafe & Drake, 1969; Kay, 1969). Greenland thus started to separate from North America so that by the Early Palaeocene (65 My) the western Mediterranean activity had ceased and Africa is located quite far to the south of the region (fig. 139). By 40 My ago the opening of the Labrador Sea was completed and the mid-Atlantic ridge then extended north between Greenland and Norway. This last phase in the opening of the Atlantic is regarded as having commenced in the early Tertiary (Vine, 1966; Kay, 1969). By Eocene–Oligocene times (40 my) the separation of Greenland and Europe has caused Europe and Africa to approach one another again (fig. 139). The zone of collision occurs along the Caucasus starting in the mid-Tertiary, and the meeting of the Middle East region and southeast Europe is the cause of the complications seen in the eastern Mediterranean today.

The magnetic lineations in the Indian Ocean have been extensively analysed by McKenzie & Sclater (1971). The anomalies south of India can be traced back to 75 My ago from which the Late Cretaceous (75 My) and Oligocene (36 My) reconstructions of fig. 140 were determined. The magnetic lineations provide no clue as to how to reconstruct Gondwanaland from the Late Cretaceous positions. McElhinny (1970c) has however examined the palaeomagnetic poles for the mid-Cretaceous and made some suggestions as to how this might be done. The earliest marine sediments along the East African coast during the Mesozoic range from Upper Triassic–Lower Jurassic in North Kenya and North Madagascar to Lower–Middle Jurassic in Tanzania and South Madagascar (Arkell, 1956). Recently marine Upper Jurassic sediments have been reported from the coast of South Africa (Dingle & Klinger, 1971). Faunal similarities between this region and the Majunga basin of northwest Madagascar suggest a marine connection between the two regions in early Upper Jurassic times. Thus east Gondwanaland comprising Antarctica, Australia, India and Madagascar must have separated from west Gondwanaland by Middle Jurassic times. The initial rifting was presumably related to the intrusion of the Karroo dolerites of Early to Middle Jurassic age (McDougall, 1963).

By the mid-Cretaceous (100 My) east Gondwanaland must have moved well southwards, because Antarctica was already near the south pole at that time and has remained there since. Mid-Cretaceous results from India suggest this subcontinent was about 45s at that time. If the motion of India suggested in fig. 140 is extrapolated backwards to 100 My, it is easy to imagine it occupying its Gondwanic position adjacent to Antarctica and lying at the correct latitude. McElhinny (1970c) related extrusion of the 100–105 My Rajmahal traps to the initial break between Antarctica and India, consistent with the first appearance of marine deposits of Ceno-manian age along the coast of southeast India transgressing Precambrian or

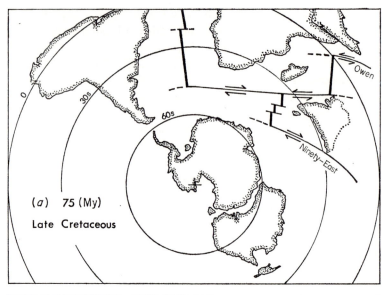

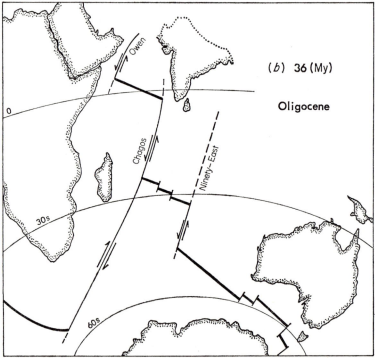

FIGURE 140 The positions of the Gondwanic continents during their dispersal at 75 and 36 My ago as deduced from magnetic anomalies in the Indian Ocean after McKenzie & Sclater (1971). Palaeolatitude lines are shown on an equal area projection.

Gondwana formations (Wadia, 1953). In fig. 136b it is worth noting that the Late Cretaceous poles for Madagascar lie in the same region as Late Cretaceous poles for India. This led McElhinny (1970c) to suppose that the proximity of India and Madagascar was maintained at least up to the Late Cretaceous, say 70–80 My ago. However, the sequence of the Madagascan poles in fig. 136b is in the opposite sense to those from India during the Late Cretaceous, so the proposition is largely speculation.

At the time of extrusion of the Deccan traps around 65 My ago (Wellman & McElhinny, 1970) India was still situated at about 30s. McElhinny (1968c) noted that this required that India move northwards through 50° of latitude to its present position of 20N in about 50 My, or at an average rate of 10 cm y^{-1} or more. This observation has been dramatically confirmed by McKenzie & Sclater (1971) who have shown from the magnetic lineations south of India that India moved away from Antarctica at about 18 cm y^{-1} for 20 My between 75 and 55 My ago. This rapid motion then virtually ceased between 55 and 10 My ago. During this time spreading started along the southeast branch of the Indian Ocean ridge system with the separation of Australia from Antarctica about 45 My ago. Present-day spreading directions became established later in the Oligocene and shortly afterwards produced the continental breaks which are now the Gulf of Aden, the Red Sea and East African Rift Valleys. The main uplift of the Himalayas took place during this period, and not during the rapid motion of India northward during the Late Cretaceous and early Tertiary.

In figs. 136b and 138 the long polar-wander paths between the Mesozoic and the present day for both India and Australia are the records of their northward drift in each case. The dispersal of these continents essentially formed the Indian Ocean of today, but on the reconstruction of Pangaea (figs. 135 to 138), this region was previously occupied by the Tethys Sea and there existed a bay between India and Australia which Dietz & Holden (1970) have termed the Sinus Australis. The break-up of Pangaea and the motion of India and Australia northwards (fig. 140) shows that what is now the Wharton Basin lying between the ninety east ridge and Western Australia might be old (pre-Mesozoic) oceanic crust, which formed part of the Sinus Australis (Dietz & Holden, 1971). However Veevers *et al.* (1971) have shown that the Perth Basin of southwest Australia contains dominantly non-marine sequences through the Permian, Triassic and Jurassic, and that deep-sea sedimentation along the length of Western Australia only commenced during the Cretaceous. Therefore it appears that there must have been some missing piece of Gondwanaland occupying the Sinus Australis, and the Wharton Basin may not necessarily represent a remnant of old oceanic crust as Dietz & Holden (1971) suppose.

7.4 SOME PROBLEMS

7.4.1 *The remagnetization hypothesis*

Creer (1968 *b*) has argued that unless the traditionally accepted reconstructions of Wegener (1929) and du Toit (1937) are completely invalid, then the vast majority of palaeomagnetic poles computed from the palaeomagnetism of European and North American lower Palaeozoic rocks are anomalous. In order to account for these supposed anomalous results Creer introduced the remagnetization hypothesis. This hypothesis arose from the discovery of an apparently anomalous Lower Devonian pole from palaeomagnetic studies on the Midland Valley lavas of Scotland (Stubbs, 1958) and from thermal demagnetization experiments on the Lower Old Red Sandstone of the Anglo-Welsh Cuvette (Chamalaun & Creer, 1964). More recent work on the Midland Valley lavas using cleaning techniques (Creer & Embleton, 1967; McMurry, 1970) have tended to confirm the original result. Arguing that the Midland Valley lavas were the only Lower Devonian igneous rocks studied in either the U.S.S.R., Europe or the U.S.A., Creer (1968 *b*) accepted the pole from these rocks as being records of the true Lower Devonian pole for Europe. The previously accepted Devonian pole position for Europe (Irving, 1964) was regarded as being a record of late Palaeozoic remagnetization in the rocks studied.

In order to explain why lower Palaeozoic rocks from parts of Laurasia were remagnetized and rocks of the same age from Gondwanaland were not, Creer (1968 *b*) pointed to the different latitude belts occupied by each of these supercontinents in the late Palaeozoic. Laurasia occupied low latitudes, whilst Gondwanaland occupied high latitudes. It is supposed that many of the lower Palaeozoic rock formations which subsequently occupied low latitudes in the late Palaeozoic and used for palaeomagnetic studies contain iron oxides of secondary origin. These could be formed by the dehydration and oxidation of iron hydroxides and oxyhydroxides to form magnetite, maghaemite, haematite and other magnetic minerals (Van Houten, 1961; see also §2.3.5). Lateritization, one of the relevant chemical reactions, is confined to the tropical zone at the present time as is illustrated in fig. 141 showing the distribution of lateritic soils and the Terra Rossa. Creer (1968 *b*) therefore argues that the broad zone of low latitudes across Laurasia in the Permo-Carboniferous produced an environment favourable for the formation of redbeds on a wide scale and never since repeated. At the same time Siberia occupied high latitudes, so Creer supposed that the true early Palaeozoic pole should be found from the magnetization of Siberian rocks of this age. A consequence of this

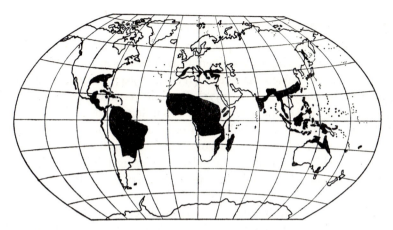

FIGURE 141   The present distribution of lateritic soils and the
Terra Rossa. From Creer (1968*b*), *Nature*, **219**, 246–50.

supposition is of course that Siberia and western Europe were still a single
unit during the early Palaeozoic.

Whilst it is true that some formations have suffered remagnetization at
later dates, these cases can largely be related to orogenic events (e.g. the
Hercynian) and not to the remagnetization hypothesis. Creer's (1968*b*)
reasoning seems to be circular, because it starts from the *assumption* of
Pangaea during the early Palaeozoic and is an attempt to try to fit the
palaeomagnetic results to this configuration. The application of plate
tectonic models to the geology of the Caledonian and Appalachian orogenic
belts (Dewey, 1969; Bird & Dewey, 1970) and the Uralides (Hamilton,
1970) and the possible existence of a proto-Atlantic during the early
Palaeozoic (J. T. Wilson, 1966) now show, on geological grounds, that
the assumptions made by Creer (1968*b*) are probably invalid. If geological arguments to account for discrepancies between pole positions are
to be allowed to prevail, then we must now abandon Creer's (1968*b*)
remagnetization hypothesis once and for all. This is not to say that remagnetization never occurs, but that it should require some geological
evidence, including the exclusion of plate tectonic arguments, to explain
the apparently anomalous result. There has been far too liberal a use of the
general hypothesis that the 'magnetic' age of rock formations is likely to be
significantly younger than its stratigraphic age. The hypothesis has far
too many degrees of freedom to be scientifically attractive.

7.4.2 *Polar-wander versus plate tectonics*

Whether differences between palaeomagnetic poles from Palaeozoic rocks reflect polar-wandering or continental drift has been the subject of some debate amongst palaeomagnetists (Irving, 1964; Briden, 1967; Creer, 1970a; McElhinny & Briden, 1971). Possible versions of polar-wandering include wandering of the geomagnetic axis independent of the rotation axis or else a net shift of the whole earth or of a whole outer shell relative to the rotation axis. There has been some confusion in the literature as to the manner in which the concept of polar-wandering might or might not fit into the general theory of plate tectonics. Irving & Robertson (1969) suggested a possible test for polar-wandering on the basis of palaeomagnetic data. They noted that the apparent polar-wander curves for North America and Europe followed a north–south direction but that the continents have separated in an east–west direction. They proposed that this north–south component represented polar-wandering. This test fails, however, because the importance of relative motion has been ignored. The earth's surface is covered with a mosaic of plates, all of which are in relative motion, so that neither the plates themselves *nor their margins* define an absolute frame of reference. This point has been admirably demonstrated by Francheteau & Sclater (1970), who used the poles of rotation determined by Le Pichon (1968) to show that neither ridges nor trenches form such a frame, since they are all in relative motion. Since the end result of palaeomagnetic investigations is the deduction of the relative motion between a plate and the rotational axis (the palaeomagnetic pole), this axis together with some arbitrarily chosen meridian become the reference frame for analysing plate motions. Polar-wandering therefore becomes an unnecessary concept in palaeomagnetism and plate tectonics. This point is one essentially made originally by Munk & MacDonald (1960). They showed that if the problem is posed as one consisting of both polar-wandering and continental drift, then there is no unique solution. There are $2n$ equations with $2(n+1)$ unknowns where n is the number of continental blocks.

In plate tectonic terms one might define polar-wandering quantitatively as the vector sum of the horizontal displacements of all points on the earth's surface. A formal expression of this has been made by McKenzie (1972) as follows. If the angular velocity of the pole P relative to a plate A is $_A\mathbf{\Omega}_P$, the velocity and direction of polar motion $_A\mathbf{v}_P$ is

$$_A\mathbf{v}_P = a(_A\mathbf{\Omega}_P \times \mathbf{a}_z),\tag{7.6}$$

where $\mathbf{a}_z$ is a unit vector along the rotational axis and a is the radius of the

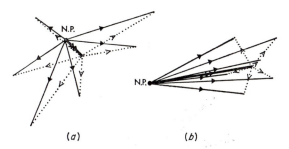

(a)   (b)

FIGURE 142   Polar-wandering as a concept in plate tectonics. The thin solid lines with arrows represent the vectors $_nV_P$ of the north pole (N.P.) for each of six plates imagined to cover the earth's surface (see (7.7)). The thicker line with two arrows shows the vector V_m of (7.9) and the dotted lines $_nV_P - V_m$. Polar-wandering is useful when V_m is large (the rate of relative motion of all plates relative to the pole is very much reduced by a particular choice of polar-wandering direction and velocity). This condition is satisfied by (b) but not by (a). After McKenzie, *The nature of the solid earth.* © McGraw-Hill Book Company, 1972.

earth. To allow for the varying sizes of plates it is convenient to define a new vector $_AV_P$ by

$$_AV_P = A_A \cdot {_A\boldsymbol{v}_P}, \tag{7.7}$$

where A_A is the total area of plate A. Polar-wandering really only becomes useful if the rate of relative motion of all plates relative to the pole is very much reduced by a particular choice of polar-wandering direction and velocity. Mathematically this can be expressed in the following way

$$\sum_{n=1}^{N} |_nV_P| \gg \sum_{n=1}^{N} |(_nV_P - V_m)|, \tag{7.8}$$

where

$$V_m = \frac{1}{N} \sum_{n=1}^{N} {_nV_P} \tag{7.9}$$

and represents the polar-wandering vector. Two situations are illustrated in fig. 142, where the thin solid lines with arrows are the vectors $_1V_P \ldots {_NV_P}$ drawn from the north pole to represent the apparent polar-wandering directions of all N plates which cover the earth's surface. The condition of (7.8) is satisfied by the situation shown in fig. 142b, but not by that shown in fig. 142a.

It might be argued that polar-wandering is demonstrated whenever V_m in (7.9) has some value other than zero, but this can only be decided when the relative motions between all the plates and the pole have been determined. This is extremely difficult to do for past geological eras, and becomes impossible if any purely oceanic plates, of which there are now no trace, covered the earth's surface. Even if some supercontinent can be reconstructed with a common polar-wander path, this does not demon-

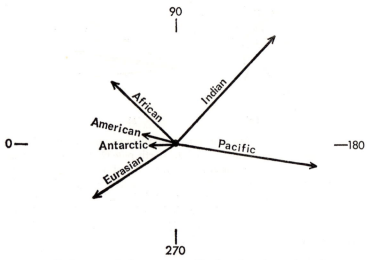

FIGURE 143   Estimates of the vectors $_nV_P$ for the six major plates presently covering the earth's surface for the past 50 My. The situation corresponds to that described by fig. 142a. Note the importance of the data from the purely oceanic Pacific plate.

strate the existence of polar-wandering as Creer (1970a) postulates, because only one plate is involved. It only requires that the vector sums of the polar motions of the other plates be such that V_m in (7.9) be approximately equal to zero for polar-wandering to be excluded. An attempt has been made to combine sea-floor spreading data with palaeomagnetic results to try to detect any polar-wandering component during the Tertiary (McElhinny & Wellman, 1969). Although a polar-wandering component of 10° to 15° since the Eocene was thought to have been detected, the analysis involved roughly only half the earth's surface and is therefore incomplete. The palaeomagnetic information for the six major plates currently covering the earth's surface can be represented by six vectors as in fig. 142 to describe the magnitude and direction of polar motion over the past 50 My. This is shown in fig. 143, from which it will be seen that the situation during most of the Tertiary corresponds to that of fig. 142a, and polar-wandering is therefore excluded. Note however the importance of the vector representing the Pacific plate. Lack of information for this purely oceanic plate could easily have led to the conclusion from the remaining five plates that the situation is more nearly that of fig. 142b.

If the rate of polar-wandering is rapid compared with plate motions then the situation shown in fig. 142b applies according to (7.8). Palaeomagnetic data provide one possible instance when this might have occurred. This is the rapid shift of the pole across Africa during the Late Ordovician (§6.3.1 and §7.3.1). The rate of polar movement is estimated to be at least 30 cm y^{-1}, since a polar shift of 50–60° is observed over a period of not greater than 20 My. This implies either very rapid plate motions or else rapid polar-wandering. If the latter is the case, then this rapid polar shift should have the same magnitude and direction in all the apparent polar-wander paths for all plates. McElhinny & Briden (1971) have shown not only that there are differences in the magnitude of this shift when observed from different plates, but also that attempts to explain the data exclusively in terms of polar-wandering lead to geologically and geometrically untenable conclusions. There is at present therefore no palaeomagnetic evidence that makes it necessary to invoke the concept of polar-wandering.

The problem of the feasibility of polar-wandering has received some attention over the past decade or more by Gold (1955), Inglis (1957), Munk & MacDonald (1960), MacDonald (1963), McKenzie (1966, 1972) and Goldreich & Toomre (1969). The argument essentially revolves around estimates of the viscosity of the lower mantle. Because of the equatorial bulge, the earth rotates about its maximum principal axis of inertia, which remains fixed in space and with respect to fixed surface features. Suppose however that as a result of plate tectonics, continental crust is redistributed over the surface of the globe and mountain belts are formed. Small changes in the axis of the principal moment of intertia would be produced and the pole would move to a new position. Gold (1955) has shown that if a continent the size of South America were suddenly raised by 30 m, the change in moment of inertia would cause the pole to migrate by 0.01°. If the earth is capable of yielding by flow, the equatorial bulge will migrate towards the new equator until the rotational axis and the axis of principal moment of inertia coincide again. The rate of migration of the pole will depend upon how fast the equilibrium figure of the earth can be re-established. Gold suggested that a decay time of about ten years observed for the Chandler wobble would be applicable to the re-adjustment of the equatorial bulge. Taking a more realistic 3 m uplift of South America, this would cause 10° of polar-wandering in 10^5 years. If however the earth behaves as a rigid body and is not capable of yielding to flow, this sort of polar-wandering is excluded. The crucial factor therefore is the viscosity of the lower mantle, because this will determine the time-delay in attainment of the equilibrium bulge at the equator.

If the earth is in hydrostatic equilibrium it should be an ellipsoid of revolution flattened at the poles. Satellite measurements of the earth's gravitational field show that the observed flattening is greater than that to be expected if the earth were in hydrostatic equilibrium by about 0.5 per cent. The present equatorial bulge is about 200 m larger than it should be. The excess over the hydrostatically expected value is highly significant and is termed the *non-hydrostatic bulge*. Munk & MacDonald (1960) and MacDonald (1963) suggested that the non-hydrostatic bulge represented a relic of the earth's faster rate of rotation. As the rotation of the earth slows down due to tidal friction, there is a lag in the attainment of hydrostatic equilibrium due to the high viscosity of the lower mantle. The present equatorial bulge corresponds to the hydrostatic bulge of 2×10^7 years ago. On this interpretation this then represents the time-constant for the delay in attaining the earth's equilibrium figure. The viscosity of the lower mantle is then calculated to be about 10^{26} poise. McKenzie (1966) has calculated a value of 6×10^{26} poise using the same argument. Returning to Gold's calculation for polar-wandering, if the decay time is 10^7 years instead of 10 years as he assumed, the pole will migrate $10°$ in only 10^{11} years rather than 10^5 years. Polar-wandering would be virtually excluded over geological time. For any reasonable rate of polar-wandering, say $10°$ in 10^7 or 10^8 years, a relaxation time of 10^3 to 10^4 years is required, or a lower mantle viscosity of 10^{22} to 10^{23} poise, not very different from that of the upper mantle as deduced from the uplift of Fennoscandia (McConnell, 1965) or the drying out of Lake Bonneville (Crittenden, 1963).

Goldreich & Toomre (1969) have objected to this interpretation of the non-hydrostatic bulge and take an entirely different view. They propose that the pole of rotation follows the *non-hydrostatic* principal moment of inertia. They therefore subtract the hydrostatic part of the equatorial bulge and obtain the non-hydrostatic moments of inertia $A' < B' < C'$. This is a triaxial ellipsoid such that

$$\frac{C' - B'}{C' - A'} \approx 0.5, \qquad (7.10)$$

and the axis of maximum principal non-hydrostatic moment of inertia lies along the polar axis of rotation. Suppose the earth evolves by the gradual redistribution (or decay or manufacture) of density inhomogeneities within the earth, then according to Goldreich & Toomre (1969), the observed axial ratio of moments of inertia is that to be expected in such a randomly evolving spheroid. For the bulge to follow the pole of rotation the decay time must be low enough and consequently Goldreich & Toomre place an upper limit of about 10^{24} poise for the viscosity of the lower mantle.

Theoretical calculations of the viscosity of the lower mantle provide conflicting viewpoints. Anderson & O'Connell (1967) suppose that viscosity or creep rate, and seismic anelasticity are both probably defect controlled and would therefore be expected to be similar functions of depth through their temperature and pressure dependence. On their models the average viscosity of the lower mantle is about 10^{23} poise, giving a relaxation time of about 3000 years for the non-equilibrium bulge of the earth. Gordon (1967) however objects to the supposition of a direct relationship between the viscosity of the mantle and the attenuation of seismic waves. McKenzie (1967) has also looked at the theoretical aspects of creep processes in the mantle. He suggests that recovery and recrystallization creep occur in the upper mantle and diffusion creep occurs in the lower mantle. The viscosity derived by this method gives a value of 10^{26} poise or more for the lower mantle and agrees well with that derived from the fossil bulge interpretation of the non-hydrostatic bulge (Munk & MacDonald, 1960; MacDonald, 1963; McKenzie, 1966).

At the present time therefore the problem of the feasibility of polar wandering of the type discussed by Goldreich & Toomre (1969) remains unsolved and awaits independent estimates of the viscosity of the lower mantle. However, the poles might still wander through the movement of an outer shell, or in plate tectonic terms through the slip of a lithospheric shell. The feasibility of this form of polar-wandering must await future developments in plate theory and the mechanisms of plate motions.

7.4.3 *Expanding earth*

It has been suggested by a number of authors that the earth has expanded during geological time. Carey (1958) and Hilgenberg (1962) both suggested an approximately exponential increase of the earth's radius with time so that the earth's surface area would have nearly doubled since the Palaeozoic. Much smaller expansion rates have been suggested by Egyed (1957), Heezen (1960), Wilson (1960) and Van Hilten (1963). Egyed (1960) was the first to point out that palaeomagnetic data may be used to test the hypothesis of earth expansion. In his first method, the *palaeomeridian method*, he pointed out that if two widely separated palaeomagnetic sampling sites are available for the same geological period and which lie on the same palaeomeridian, the palaeoradius (R_a) may be calculated from the equation

$$R_a = \frac{d}{\lambda_1 - \lambda_2}, \tag{7.11}$$

where d is the present separation of the sampling sites and λ_1 and λ_2 are their palaeolatitudes calculated from the usual dipole equation (§1.2.4). The method is illustrated in fig. 144, and only applies of course to measurements from a crustal block which has retained its primitive dimensions on the expanded surface.

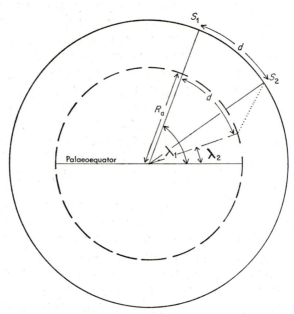

FIGURE 144 The palaeomeridian method for testing the expanding earth hypothesis. Two sampling sites S_1 and S_2 lying on a palaeomeridian are separated by a distance d. If their palaeolatitudes from the dipole equation are λ_1 and λ_2, the ancient radius R_a is determined from (7.11).

Cox & Doell (1961) applied this method to Permian palaeomagnetic data from Eurasia, using sampling sites separated by almost 5000 km. They obtained a value of $R_a = 0.99$ of the present radius, so that large expansion rates of Carey (1958) and Hilgenberg (1962) were certainly very unlikely. Egyed (1961) later produced a *triangulation method*, generally applicable to cases where the sampling sites did not lie on the same palaeomeridian. Van Andel & Hospers (1968 a) however have shown that Egyed's general method was not readily applied to the palaeomagnetic data and so produced a revised triangulation method. Ward (1963) has also produced a generally applicable method, in which the procedure is to vary the palaeoradius, recalculating the poles for each sampling site until the minimum pole dispersion is achieved. The method is termed the *minimum dispersion method*. A modified version of this method has also been suggested by Van Hilten (1968).

Van Andel & Hospers (1968 b) have reviewed the results of the three general methods applied to Permian data for Europe and Siberia and to Triassic and Cretaceous data for Europe. These results are summarized in table 31, which shows that the values of palaeoradius determined by the

TABLE 31 *Values of the earth's palaeoradius R_a expressed as a fraction of the present radius as determined from palaeomagnetic data by different authors. From Van Andel & Hospers (1968b)*

Geographical period	Region	Authors	R_a
Permian	Europe	Ward (1963)	0.94
		Van Hilten (1968)	1.00
		Van Andel & Hospers (1968a)	1.03
	Siberia	Van Hilten (1968)	0.76
		Van Andel & Hospers (1968a)	0.70
Triassic	Europe	Ward (1963)	0.99
		Van Andel & Hospers (1968a)	0.89
Cretaceous	Europe	Van Hilten (1968)	0.97
		Van Andel & Hospers (1968a)	0.99

different methods agree within a few per cent. Again the hypotheses of fast earth expansion put forward by Carey (1958) and Hilgenberg (1962) can be rejected at the ninety-five per cent confidence level, but the method cannot resolve slow expansion rates of the order of 1 mm y^{-1}.

7.4.4 *The Precambrian*

Palaeomagnetic data from Precambrian rocks are too limited to draw any serious conclusions about ancient continental configurations at the present time. Data of any consequence are available only from the North American Precambrian and the southern African shield. There is some indication from the North American results (fig. 111) that Briden's (1967) scheme of quasi-static intervals interspersed with rapid shifts could apply to the Pre-cambrian as well as to the Phanerozoic. The periods from 2060–1600 My and 1300–1150 My could both represent quasi-static intervals, whilst in the time range of approximately 2080–2060 My a rapid polar shift of 40° is indicated. This rapid shift might be related to the difference of 80° which separates the African poles between 2200 and 1950 My (fig. 120). It is worth noting that about 120° of polar movement is required with respect to southern Africa between 1950 and 1750 My, during a quasi-static period observed in North America. This provides considerable support for the interpretation of the apparent polar-wander curves in terms of plate motions rather than polar-wandering as was argued in §7.4.2.

The interpretation of Precambrian palaeomagnetic results will need to go hand in hand with geological interpretations in terms of plate tectonics. Even at this stage it might be debatable whether all the North American

TABLE 32   *Apparent polar-wandering rates over geological time*

Region	Period (My)	Angular shift	Rate per My
All continents	600–0	120°–180°	0.2°–0.3°
Southern Africa	2600–1300	400°	0.3°
North America	2500–600	340°	0.2°
Baltic shield and Scotland	1800–600	360°	0.3°
Siberia	~ 1500–600	280°	0.3°

Precambrian results refer to a single plate for the whole of geological time. Already there are suggestions from the Canadian shield that the boundary between the Churchill and Superior structural provinces represents a suture between two protocontinents (Gibb & Walcott, 1971). This suggests that the mechanisms of plate tectonics were active during the period 2400 to 1650 My ago. Within the Rhodesian craton, probably the oldest cratonic nucleus of Africa, the mafic volcanic rocks and serpentine forming the lower parts of the Sebakwian and Bulawayan sequences are overlain by more andesitic volcanics and coarse clastic sediments (Swift, 1961). Dewey & Horsfield (1970) suggest that the andesites and coarse clastic sediments represent island arcs built on a Sebakwian/Bulawayan ocean floor older than 3000 My. Likewise in the Yellowknife region of the Great Slave province of Canada, theoleitic and calc-alkaline volcanics about 2600 My old have been suggested to belong to an old island arc complex (Folinsbee *et al.*, 1968).

If, as now seems likely, the processes of plate tectonics have been a feature of all geological time, the present Precambrian apparent polar-wandering information allows certain speculations to be made about the amount of lithosphere that has been created during the earth's history. The average rates of apparent polar-wandering observed in Phanerozoic and Pre-cambrian investigations are summarized in table 32. They show that an average rate of 0.2°–0.3° per My or about 3 cm y^{-1} is a reasonable average throughout geological time. It was demonstrated in §7.1 that in any given instance apparent polar-wandering is not necessarily related to plate motions in any simple way, but might be a large magnification or diminu-tion of the actual motion of a plate. Over very long periods such as 10^9 years or so, one might expect that these magnifications and reductions would tend to average out so that the *average* rates of apparent polar wandering as calculated in table 32 could be a rough estimate of *average* plate motions over geological time. Rates of 3 cm y^{-1} are not uncommon spreading rates observed at the present time, so it could be reasonable to extrapolate this figure back in time. For a conservative estimate assume this to be a double spreading rate and that the present length of active ridge systems of 40000 km has also persisted throughout geological time. For a lithospheric plate of 70 km thickness, this means that about 80 km³ of lithosphere is created per annum or about 3×10^{11} km³ in 4×10^9 years. This is equivalent to the volume of about 600 km thickness of the upper mantle, so that the entire volume of the mantle cannot have been completely cycled through lithospheric production and consumption throughout the earth's history.

Appendix

The results which conform with the criteria outlined in §3.5 are listed by regions. Only the minimum information is provided. Geographical co-ordinates of each rock unit are provided either as a mean position, or if the area is large as a range. The rock unit name is marked with a dagger if its location is in a mobile belt adjoining two regions, in which case it is not clear whether this result may refer to the region as a whole. A footnote generally indicates the belt concerned.

Geological periods are specified by the symbols: Precambrian Pc, Cambrian Є, Ordovician O, Silurian S, Devonian D, Carboniferous C, Permian P, Triassic Tr, Jurassic J, Cretaceous K, Tertiary T and Quaternary Q. For North America the Carboniferous is indicated as the Mississipian Mi, and Pennsylvanian Pn. Radiometric ages, when available, are given in brackets, and subdivisions of the Cambrian to Cretaceous are indicated by l, m and u. Subdivisions of the Tertiary are indicated as Tpa (Palaeocene), Te (Eocene), To (Oligocene), Tm (Miocene), Tp (Pliocene). The Quaternary is subdivided as Qp (Pleistocene) and Qr (Recent).

The number of sites (Si) and samples (Sa) used in each study are given as a possible guide to the relative importance of each. The stability tests carried out are indicated by the symbols: alternating field demagnetization a, thermal demagnetization t, chemical demagnetization h, fold test f, congolomerate test g, baked contact test c, presence of reversals r. Where magnetic, thermal or chemical cleaning was carried out the bold symbols a, t or h respectively are used. In a few cases where little or no stability evidence is available, but consistent results were obtained from a wide area, then the consistency test (cons) is indicated.

The pole position given is that which falls in the northern hemisphere, and the polarity (P) is indicated as normal (N), reversed (R) or mixed (M). Only mixed polarities are indicated for Precambrian studies. More detailed information (including the original references in each case) can be found by reference to the appropriate entry in the pole lists (numbers 1–7, Irving, 1960–5; numbers 8–12, McElhinny, 1968–72) published in the Geophysical Journal (GJ) and given in the last column. Entries to the Russian list (Khramov & Sholpo, 1967) are prefixed by the letter R.

Pole no.	Rock unit	Age (My)	Si (Sa)	Stab.	Pole Posn	P	GJ
		NORTH AMERICA					
(a) Matachewan–Stillwater trend (2080–~ 2500 My)							
NAI.1	Stillwater complex, Montana 45.4N, 110W	Pc (2500)	9 (96)	*a*	62N, 112E	—	12/166
NAI.2	Soudan iron formation 46.5N, 87.6W	Pc (< 2600)	(8)	*a, f*	39N, 117E	—	9/153
NAI.3	Matachewan dykes (SW magn.) 48N, 81W	Pc (2485) 25	(78)	*a*	37N, 59E	—	8/174
NAI.4	Matachewan dykes (NE magn.) 48N, 81W		(30)	*a*	63N, 61E	—	8/175
NAI.5	Matachewan dykes (SW magn.) 48N, 79W	Pc (2485)	5 (43)	*a*	45N, 81E	—	8/181
NAI.6	Abana dykes (Mat. NE magn.) 48N, 79W	Pc (2485)	(31)	*a*	54N, 71E	—	8/182
NAI.7	Older Abitibi dykes 48N, 78.5W	Pc (> 1230)	2 (13)	*a*	24N, 107E	—	8/172
NAI.8	Older Abitibi dykes 48N, 78.5W	Pc (> 1230)	4 (14)	*a*	21N, 58E	—	8/173
NAI.9	Nispissing diabase 47.4N, 79.7W	Pc (2180)	11 (40)	*a*	19N, 92E	—	12/164
NAI.10	Indian Harbour dyke swarm 54.4N, 57.1W	Pc (2080)	12 (12)	*a, t, r*	6N, 63E	M	†(1)
b) Animikie–Sudbury trend (1600–2060 My)							
NAI.11	Cobalt Group sediments 47N, 79W	Pc (2180–2650)	(8)	*a, c*	22N, 263E	—	9/159
NAI.12	Wind River dykes 42.5N, 108.8W	Pc (1880–2060)	4 (23)	*a, t*	43N, 239E	—	†(2)
NAI.13	Gunflint iron formation 48.5N, 89.2W	Pc (1600–1850)	(20)	*a, r*	28N, 266E	M	10/152
NAI.14	Sudbury irruptive (1) 46.6N, 81.2W	Pc (1704)	(47)	*a*	39N, 261E	—	5/92
NAI.15	Sudbury irruptive (2) 46.6N, 81.2W	Pc (1704)	(47)	*a*	53N, 245E	—	5/92
NAI.16	Marathon dykes 49N, 86W	Pc (1810)	5 (9)	*a*	29N, 213E	—	8/176
NAI.17	Molson dykes 55N, 96W	Pc (1445)	4 (10)	*a*	36N, 251E	—	8/177
c) Elsonian trend (1300–1475 My)							
NAI.18	Younger Abitibi dykes 48N, 78.5W	Pc (1230)	14 (55)	*a*	32N, 228E	—	8/170
NAI.19	Younger Abitibi dykes 48N, 78.5W	Pc (1230)	10 (83)	*a*	27N, 226E	—	8/171
NAI.20	Younger Abitibi dykes 48N, 79W	Pc (1230)	5 (35)	*a*	17N, 228E	—	8/180
NAI.21	Arbuckle granites 34N, 97W	Pc (1320–1400)	(15)	*a*	17N, 210E	—	10/166
NAI.22	Croker Island complex 46N, 82W	Pc (1445)	19 (23)	*a*	5N, 217E	—	11/90
NAI.23	Michikamau anorthosite 54.5N, 64W	Pc (1400)	6 (29)	*a*	1N, 34E	—	10/191
NAI.24	St Francois igneous rocks 37.6N, 90.7W	Pc (1300–1400)	11 (49)	*a*	1N, 38E	—	8/159
NAI.25	St Francois igneous rocks 37.5N, 90.5W	Pc (1300–1400)	5 (95)	*a*	1N, 37E	—	8/160
NAI.26	Sherman granite 41N, 105.4W	Pc (1410)	5 (14)	*a, r*	8N, 29E	M	9/144
NAI.27	Grand Canyon Series combined (1/137–138, 3/83–84, 86–87, 8/163, 168–169)	Pc (1100–1400)	9 (149)	*r*, cons	9N, 29E	M	—
AI.28	Belt Series combined (3/77–82, 5/89–91, 8/161–162)	Pc (1100–1325)	11 (216)	*a, r*, cons	10N, 27E	M	—
d) MacKenzie trend (1150–1300 My)							
AI.29	MacKenzie dykes 65N, 112W	Pc (1315)	23 (34)	*a*	4N, 183E	—	8/179
AI.30	MacKenzie dykes 54.7 – 65.8N, 82.2 – 117.6W	Pc (1300)	14 (63)	*a*	4N, 193E	—	11/89
AI.31	Sudbury dykes 46.5N, 81W	Pc (1285)	(13)	*a, c*	8N, 17E	—	7/78
AI.32	Sudbury dykes 47N, 82W	Pc (1285)	22 (37)	*a*	4N, 13E	—	8/178
AI.33	Sudbury dykes 46.4N, 81.5W	Pc (1285)	13 (53)	*a*	2N, 9E	—	9/143
AI.34	Muskox gabbro 66N, 113W	Pc (1150)	8 (19)	*a*	4N, 185E	—	8/183
AI.35	Coppermine lavas 67.5N, 116W	Pc (1150)	24 (48)	*a*	1N, 184E	—	8/184

Pole no.	Rock unit	Age (My)	Si (Sa)	Stab.	Pole Posn	P	GJ
(e) Keweenawan trend (1000–1150 My)							
NA1.36	Mungford basalts 57.8N, 62.0W	Pc (948)	16 (56)	*a, t*	49N, 217E	—	†(1)
NA1.37	Purcell System 49N, 114W	Pc (1000–1150)	7 (101)	*r*	18N, 33E	M	7/79
NA1.38	Nonesuch shale and Freda sandstone 47N, 88.5W	Pc (1046)	(68)	*f*	9N, 169E	—	1/132
NA1.39	Nonesuch shale and Freda sandstone 47N, 88.5W	Pc (1046)	5 (—)	*t*	7N, 186E	—	10/184
NA1.40	Boulter intrusive 45N, 77.5W	Pc (1000–1150)	(8)	*a*	42N, 203E	—	5/93
NA1.41	Thanet and Umfraville intrusives 45N, 78W	Pc (1000–1150)	(18)	*a*	1N, 338E	—	5/94
NA1.42	Tudor intrusive 44.5N, 77.5W	Pc (1000–1150)	(11)	*a*	42N, 149E	—	5/95
NA1.43	Central Arizona diabase 37N, 111W	Pc (1140–1150)	8 (62)	*a, t*	27N, 179E	—	†(3)
NA1.44	Allard Lake anorthosite 51N, 63W	Pc (1000)	4 (102)	*a, t, r*	39N, 320E	M	9/142
NA1.45	Copper Harbour lavas and sediments 47N, 88.5W	Pc (1046–1200)	(25)	*g*	30N, 176E	—	1/133
NA1.46	Copper Harbour lavas 47N, 88.5W	Pc (1046–1200)	6 (—)	*a, t*	13N, 176E	—	10/183
NA1.47	Portage Lake lavas 47N, 88.5W	Pc (1200)	14 (18)	*a, t*	24N, 183E	—	10/168
NA1.48	Portage Lake lavas 47.2N, 88.4W	Pc (1200)	29 (—)	*a*	27N, 181E	—	12/158
NA1.49	Grand Portage lavas 47.9N, 89.7W	Pc (1000–1150)	11 (52)	*a*	44N, 197E	—	12/159
NA1.50	Ironwood flows 46.5N, 90.1W	Pc (1000–1150)	14 (61)	*a*	29N, 232E	—	12/160
NA1.51	Beaver Bay complex 47.0N, 91.5W	Pc (1000)	29 (—)	*a*	28N, 190E	—	11/87
NA1.52	Keweenawan gabbro, Minnesota 48.0N, 90.4W	Pc (1000–1150)	7 (—)	*a*	48N, 183E	—	11/88
NA1.53	Keweenawan gabbro, Wisconsin 46.3N, 91W	Pc (1000–1150)	14 (67)	*a*	36N, 185E	—	12/156
NA1.54	Duluth anorthosite 47N, 92W	Pc (1115)	12 (—)	*a*	42N, 191E	—	12/150
NA1.55	Duluth layered series 47N, 92W	Pc (1115)	24 (—)	*a*	36N, 190E	—	12/151
NA1.56	Duluth gabbro 48N, 90W	Pc (1115)	10 (—)	*a*	32N, 200E	—	12/152
NA1.57	Duluth gabbro and Logan diabase 48N, 90W	Pc (1115)	13 (—)	*a*	42N, 204E	—	12/153
NA1.58	Logan diabase 48N, 90W	Pc (1115)	8 (—)	*a*	26N, 194E	—	12/154
NA1.59	Endion Sill 47N, 92W	Pc (1115)	14 (—)	*a*	32N, 191E	—	12/155
NA1.60	Lester River Sill 47N, 92W	Pc (1115)	20 (—)	*a*	35N, 199E	—	12/157
NA1.61	Pikes Peak granite 39N, 105W	Pc (1020–1040)	4 (10)	*a, t*	6N, 181E	—	12/161
NA1.62	Logan Sills, 49N, 89.5W	Pc (1060)	30 (—)	*a*	48N, 218E	—	12/167
NA1.63	Osler Group Volcanics 48.6N, 88W	Pc (1100)	12 (24)	*a, f*	49N, 203E	—	12/168
NA1.64	Alona Bay Volcanics 47N, 84.8W	Pc (1100)	(11)	*a, f*	47N, 233E	—	12/169
NA1.65	Mamainse Point Volcanics 47N, 84.8W	Pc (1076)	(52)	*a, f*	37N, 180E	—	12/170
NA1.66	Mamainse Point Volcanics 47N, 84.8W	Pc (1076)	(20)	*a, f*	54N, 212E	—	12/171
NA1.67	Michipicoten Island Volcanics 47.7N, 85.5W	Pc (1100)	(28)	*a, f*	29N, 179E	—	12/172
NA1.68	Cape Gargantua Volcanics 47.6N, 85.0W	Pc (1100)	(8)	*a, f*	34N, 180E	—	12/173
NA1.69	Cape Gargantua Volcanics 47.6N, 85.0W	Pc (1100)	(9)	*a, f*	72N, 223E	—	12/174
NA1.70	North Shore Volcanics 47.8N, 89.9W	Pc (1115)	(14)	*a, f*	47N, 200E	—	12/175
NA1.71	North Shore Volcanics 47.5N, 91.0W	Pc (1115)	(54)	*a, f*	32N, 188E	—	12/176
NA2.1	Bradore formation, Newfoundland 51N, 58W	€l	(10)	*a*	9N, 149E	R	8/147
NA2.2	Ratcliffe Brook formation 45N, 67W	€l	(12)	*a*	10N, 124E	R	8/148
NA2.3	Lodore formation 41N, 110W	€	(11)	*r*	23N, 6E	M	3/76
NA2.4	Wichita granites 35N, 99W	€ (525)	18 (120)	*a, t, r*	2N, 147E	M	10/148
NA3.1	Wabana Group, Newfoundland 47.6N, 53.0W	Ol	11 (116)	*a, t*	28N, 192E	R	†(4)

Pole no.	Rock unit	Age (My)	Si (Sa)	Stab.	Pole Posn	P	GJ
NA4.1	Rosehill formation 39.5N, 79W	Sm	5 (35)	*f*	20N, 136E	N	1/117
NA4.2	Bloomsburg redbeds 39–41N, 76.5 – 78.5W	Su	9 (—)	*t, f*	32N, 102E	N	9/125
NA5.1	Clam Bank Group, Newfoundland 48N, 59W	Dl	(18)	*a*	28N, 146E	N	8/123
NA5.2	Perry formation Volcanics 45N, 67W	Du	(16)	*a*	26N, 109E	R	8/120
NA5.3	Perry formation sediments 45N, 67W	Du	(36)	*a*	35N, 121E	R	8/121
NA5.4	Perry lavas, Maine 45N, 68W	Du	(32)	*a*	24N, 128E	R	9/120
NA5.5	Perry formation 45N, 67W	Du	8 (55)	*t, f*	32N, 118E	R	10/126
NA6.1	Barnett formation 31N, 99W	Mi	9 (68)	*r*	41N, 135E	M	1/106-7
NA6.2	Codroy Group, Newfoundland 48N, 59W	Mi	(32)	*a*	30N, 127E	R	8/117
NA6.3	Mauch Chunk formation	Mi (u)	20 (96)	*t, f, r*	43N, 127E	M	10/119
NA6.4	Maringouin formation 45.6N, 64.8W	Mi (u)	8 (46)	*t, f*	34N, 117E	R	10/120
NA6.5	Pre-Pictou Sandstone 48N, 66W	Mi (u)–Pn (l)	(8)	*a*	24N, 133E	R	8/113
NA6.6	Hopewell Group 47.7N, 64.5W	Mi (u)–Pn (l)	15 (67)	*t, h, f, r*	34N, 118E	M	11/78
NA6.7	Minturn formation 39.6N, 106.4W	Pn (l)	10 (—)	*a, f, r*	39N, 105E	M	10/115
NA6.8	Cumberland Group 45.7N, 64.5W	Pn (l)	10 (48)	*t, f*	36N, 125E	R	11/76
NA6.9	Pictou Group 46N, 64W	Pn	(18)	*t*	41N, 132E	R	8/101
NA6.10	Hurley Creek formation 46N, 66W	Pn (u)	5 (19)	*t*	39N, 125E	R	9/98
NA6.11	Prince Edward Island 46N, 64W	Pn (u)–Pl	3 (154)	*a*	40N, 126E	R	8/95-7
NA6.12	Prince Edward Island 46.5N, 63.7W	Pn (u)–Pl	17 (58)	*t*	42N, 133E	R	8/99
NA6.13	Bonaventure formation 48N, 65W	Pn (u)–Pl	11 (21)	*t*	38N, 133E	R	8/100
NA7.1	Dunkard Series 39.5N, 81W	Pl	9 (57)	*a, r*	44N, 122E	M	8/88
NA7.2	Maroon formation 39.6N, 106.6W	Pn (u)–P	36 (—)	*a, f*	33N, 126E	R	10/105
NA7.3	Fountain and Lykins formations 40.2N, 105.3W	Pn (u)–P	27 (—)	*a, c, f*	48N, 119E	R	10/106
NA7.4	Basic Sill, P.E. Island 46N, 64W	Pu	(12)	*a*	52N, 113E	R	9/65
NA7.5	Toroweap formation 36.1N, 112.2W	P	(11)	*t*	47N, 103E	R	11/65
NA7.6	Halgaito tongue 36.1N, 112.2W	P	(14)	*t*	49N, 117E	R	11/67
NA8.1	Moenkopi formation 36–41N, 109–112W	Trl	8 (92)	*r*	57N, 107E	M	3/37-44
NA8.2	Chugwater formation 41–45N, 106–109W	Trl	10 (98)	*r*	48N, 112E	M	3/45-54
NA8.3	Manicouagan structure 52N, 68W	Trl (225)	11 (44)	*a*	60N, 88E	N	9/52
NA8.4	Manicouagan structure 52N, 68W	Trl (225)	6 (—)	*t*	57N, 89E	N	9/51
NA8.5	Chugwater formation 43N, 108.5W	Trl	(163)	*r*	58N, 117E	M	9/53-4
NA8.6	Upper Maroon formation 39.6N, 106.6W	Trl	11 (—)	*a, f, r*	56N, 100E	M	10/96
NA8.7	Hoskinnini tongue 36.1N, 112.2W	Trl	(18)	*t*	50N, 121E	R	11/58
NA8.8	Moenkopi formation 38.6N, 108.9W	Trl	12 (318)	*a, t, r*	57N, 89E	M	11/59
NA9.9	Connecticut Valley Rocks 42N, 73W	Tru	(12)	*r*	54N, 86E	M	1/67
NA8.10	Newark Group 40.5N, 74.9W	Tru (190)	29 (78)	*a*	63N, 108E	N	5/34
NA8.11	Massachusetts lavas 42N, 72.5W	Tru	5 (16)	*a*	55N, 88E	N	5/35
NA8.12	Diabase dyke, Nova Scotia 44N, 66W	Tru (197)	2 (11)	*a*	69N, 98E	N	8/68
NA8.13	Diabase, Pennsylvania 40N, 76.5W	Tru	20 (95)	*a*	62N, 105E	N	8/69
NA8.14	North Mountain basalt 45N, 64W	Tru (200)	17 (28)	*a*	66N, 113E	N	9/49
NA8.15	Intrusives and sediments 36–45N, 65–80W	Tru	50 (387)	*r*	68N, 91E	M	9/50
NA8.16	Igneous rocks, Connecticut Valley 41.5N, 72.7W	Tru (193)	7 (313)	*a*	65N, 87E	N	10/88
NA8.17	North Mountain basalt 44.9N, 65.4W	Tru (200)	25 (40)	*a, t, g*	73N, 104E	N	10/89

Pole no.	Rock unit	Age (My)	Si (Sa)	Stab.	Pole Posn	P	GJ
NA8.18	Grand Manan Island lavas 44.5N, 66.5W	Tru	4 (8)	*a*	80N, 100E	N	10/90
NA8.19	Mistatin Lake Volcanics 55.9N, 63.4W	Tru (202)	10 (73)	*a*	86N, 118E	R	11/44
NA9.1	White Mountain Volcanics 44N, 71W	Jl (180)	12 (130)	*a, r*	85N, 126E	M	8/61
NA9.2	Mesozoic dykes, Appalachians 33.5 – 41.5N, 72–83.5W	Jl (?)	74 (121)	*a*	66N, 145E	N	9/47
NA9.3	Franciscan formation‡ 38N, 122.5W	Ju–Kl	25 (127)	*a, t, f*	29N, 316E	N	10/68
NA10.1	Guadeloupe Mountains 37.5N, 120W	Ju–Kl (136)	4 (—)	*a, r*	43N, 171E	M	9/42
NA10.2	Bucks Batholith 39.9N, 121.3W	Ju–Kl (129–142)	9 (—)	*a, r*	58N, 195E	M	9/43
NA10.3	Isachen diabase 78.7N, 103.7N	K	10 (20)	*a*	69N, 180E	N	7/25
NA10.4	Sierra Nevada Plutons 38N, 120W	Ku (83–90)	14 (80)	*a, t*	64N, 195E	N	8/48
NA10.5	Mt Ascutney gabbro 43.5N, 72.5W	Kl (130)	2 (24)	*a*	64N, 187E	R	8/52
NA10.6	Elkhorn Mt Volcanics 46N, 112W	Ku	8 (—)	*a, f, r*	70N, 190E	M	9/36
NA10.7	Mesaverde Group 41N, 109W	Ku	45 (—)	*t*	65N, 198E	N	11/35
NA10.8	Monteregian Hills 45.3N, 72.8W	Kl (100–122)	32 (147)	*a, r*	71N, 190E	M	11/36
NA10.9	Magnet Cove and Potash Sulphur Spring Complexes 34.5N, 92.8W	Kl–u (98)	19 (—)	*a*	65N, 187E	N	11/37
NA11.1	Siletz River Volcanics‡ 45N, 123.5W	Te	8 (50)	*t, f, r*	37N, 311E	M	1/36
NA11.2	Beaverhead Valley Volcanics 45N, 113W	Te	4 (—)	*a*	66N, 239E	N	9/29
NA11.3	Intrusion and baked zone 40N, 105.3W	Tpa–e	3 (15)	*a, c*	68N, 189E	N	10/46
NA11.4	Cape Dyer basalts 66.7N, 61.3W	Tpa (58)	5 (24)	*a, t*	83N, 305E	N	†(5)
NA11.5	Spanish Peaks dyke swarm 37.4 N, 105W	Te–o	6 (39)	*a*	81N, 211E	N	11/25
NA11.6	Buck Hill Volcanic Series 29.3N, 103.3W	To (27–36)	20 (—)	*a, r*	81N, 89E	M	12/45
NA11.7	Basalts, Golden 40.8N, 105.2W	Tpa (59)	2 (15)	*a*	72N, 64E	R	11/30–
NA12.1	B.C. and Yukon basalts 61N, 134W	Tm–p	4 (46)	*r*	85N, 150E	M	1/27
NA12.2	Columbia River basalts 44.5N, 119.6W	Tm (14–21)	433 (911)	*a, r*	86N, 26E	M	8/28
NA12.3	Lovejoy basalt 40N, 121W	Tm (24)	13 (158)	*a, t, r*	76N, 74E	M	8/27
NA12.4	Abert Rim lavas 42.6N, 120.2W	Tm	16 (98)	*a, c*	80N, 62E	R	8/26
NA12.5	Lavas and baked sediments 35.2N, 111.6W	Tp (4)	24 (111)	*a, t, c, r*	85N, 109E	M	9/13
NA12.6	Beaverhead Valley Volcanics 45N, 113W	Tm–p	2 (—)	*a, r*	72N, 251E	M	9/20
NA12.7	Cariboo Plateau basalts 51.8N, 121.8W	Tm (10–14)	48 (251)	*a, r*	84N, 220E	M	11/17
NA12.8	Gabbro plugs 51.5N, 121.2W	Tm (10–14)	17 (101)	*a, r*	85N, 213E	M	11/18
NA12.9	Nevada ignimbrites 38.5N, 115W	Tm (18–30)	8 (98)	*a, t, r*	73N, 142E	M	12/37
NA12.10	Wrangell Volcanics 62N, 143W	Tp	10 (—)	*a*	48N, 115E	N	12/23
NA12.11	Wrangell Volcanics 62N, 143W	Tp	9 (—)	*a*	43N, 109E	N	12/24
NA13.1	Wrangell Volcanics 62N, 143W	Tp–Qp	9 (—)	*a*	85N, 173E	N	12/18
NA13.2	Wrangell Volcanics 62N, 143W	Tp–Qp	11 (—)	*a*	51N, 248E	N	12/19
NA13.3	Suttle Lake lavas, Oregon 44N, 122W	Qp	19 (—)	*a*	88N, 272E	R	12/16
NA13.4	Mt Griggs Volcanics, Alaska 58N, 155W	Q	5 (—)	*a*	77N, 151E	N	12/8
NA13.5	Mt Edgecumbe lavas, Alaska 62N, 143W	Q	5 (—)	*a*	75N, 183E	N	12/7

†(1) Murthy & Deutsch (1972). †(2) Spall (1971*b*). †(3) Helsley & Spall (1972).
†(4) Deutsch & Rao (1970). †(5) Deutsch *et al.* (1971). ‡ Sites located in western cordillera region.

Pole no.	Rock unit	Age (My)	Si (Sa)	Stab.	Pole Posn	P	GJ
		GREENLAND					
GR8 1	Kapp Biot sediments 72N, 23W	Trm–u	(44) *r*		68N, 160E	M	4/14
GR11.1	Volcanics 70N, 22–28W	Te (50–60)	28 (157) *a*		63N, 174E	R	9/27
		SPITSBERGEN					
SN9.1	Diabase dyke 78.3N, 16.2E	Ju (149)	(8) *a*		58N, 178E	R	10/69
SN9.2	Dolerites 78.3 – 78.7N, 14.3 – 16.4E	Ju–Kl (110–149)	7 (37) *t*		53N, 358E	R	10/67
		WESTERN EUROPE					
WE1.1	Upper Dala Volcanics, Sweden 61.5N, 14E	Pc (1405)	3 (27) *a*		27N, 174E	—	10/192
WE1.2	Jotnian basalts, Sweden 61N, 13.5E	Pc (1405)	2 (16) *a*		32N, 186E	—	10/193
WE1.3	Late Jotnian dykes, Sweden 61N, 14E	Pc (1405)	4 (27) *a*		23N, 178E	—	10/194
WE1.4	Hyperite dolerite dykes, Sweden 58N, 14.5E	Pc (1550)	5 (70) *a*		12N, 46E	—	10/195
WE1.5	Hame dykes, Finland 61.4N, 24.8E	Pc (1640–1500)	(30) *a, t*		67N, 72E	—	10/196
WE1.6	Jotnian dolerites, Finland 61.2N, 22E	Pc (1330)	18 (18) *a, t*		2N, 158E	—	8/185
WE1.7	Vaasa dolerites, Finland 63N, 20.9E	Pc (1330)	15 (15) *a, t*		7N, 164E	—	8/186
WE1.8	Dalarna igneous rocks, Sweden 61N, 13E	Pc (1200)	(9) *a*		10N, 295E	—	12/177
WE1.9	Longmyndian, England 53N, 3W	Pc	12 (40) *a, r*		2N, 240E	M	1/128
WE1.10	Lower Torridonian, Scotland 58N, 6W	Pc (935)	13 (32) *f*		35N, 242E	—	1/127
WE1.11	Upper Torridonian, Scotland 58N, 6W	Pc (760)	81 (205) *t, f, r*		6N, 43E	M	1/126
WE2.1	Caerfai Series, Wales 51.8N, 5.0W	€l	16 (91) *a, f*		26N, 169E	R	†(1)
WE2.2	Hartshill quartzite, England 52.5N, 1.5W	€l	(11) *a*		18N, 165E	R	5/83
WE3.1	Borrowdale Volcanics 54.7N, 3.0W	Ol	27 (152) *a, f*		7N, 177E	N	†(2)
WE3.2	Builth lavas 52.2N, 3.5W	Ol	15 (68) *a, t*		16N, 179E	R	†(3)
WE3.3	Volcanics, U.K. 54N, 4W	Ol	3 (28) *r, a, t*		13N, 165E	M	9/129–31
WE3.4	Killary Harbour ignimbrites, Eire 53.6N, 9.7W	Om	14 (—) *a, t, f, g*		2N, 212E	R	11/80 †(4)
WE3.5	Builth intrusives 52.2N, 3.5W	Ou	6 (18) *a, t, r*		2N, 182E	M	†(3)
WE4.1	Ringerike Sandstone, Norway 60N, 10E	Su	10 (21) *t*		21N, 159E	N	10/129
WE4.2	Arrochar Complex, Scotland 56.2N, 4.8W	Su (418)	6 (47) *a, t, c*		8N, 144E	R	12/139
WE4.3	Garabal Hill–Glen Fyne Complex 56.3N, 4.8W	Su–Dl (415)	5 (17) *a, t*		5N, 146E	N	12/138
WE5.1	Roragen Sandstones, Norway 62.5N, 11.9E	Dl	8 (13) *t, r*		19N, 160E	M	8/124
WE5.2	Old Red Sandstone, Wales 52N, 3W	Dl	(35) *t, f*		3N, 298E	N	8/126
WE5.3	O.R.S. lavas, Scotland 57N, 2W	Dl	16 (—) *r*		10N, 321E	M	8/125
WE5.4	O.R.S. lavas, Scotland 57N, 2W	Dl (395)	10 (124) *a, r*		1N, 121E	M	9/124
WE5.5	O.R.S. lavas, Scotland 57N, 2W	Dl	26 (112) *a, f, r*		1N, 325E	M	12/137
WE5.6	Kvamshesten O.R.S. Norway 61.4N, 5.6E	Dm	(41) *a, t, r*		22N, 170E	M	12/134
WE6.1	Kinghorn lavas, Scotland 56N, 3W	Cl	21 (43) *a, t, r*		17N, 167E	M	1/99–101
WE6.2	Derbyshire lavas 53N, 1.5W	Cl	4 (8) *t, r*		10N, 147E	M	1/102
WE6.3	Carboniferous lavas, Scotland 56N, 3W	Cl	17 (—) *a, t, c, r*		18N, 161E	M	8/118

Appendix

Pole no.	Rock unit	Age (My)	Si (Sa)	Stab.	Pole Posn	P	GJ
WE6.4	Tideswelldale rocks 53.5N, 1.5W	Cl	(13)	c	9N, 162E	R	1/97
WE6.5	Carboniferous Limestone 54N, 3W	Cl	(14)	r	43N, 119E	M	5/47
WE6.6	Lancashire sediments 54N, 3W	C	7 (24)	r	34N, 160E	M	1/103
WE6.7	Derbyshire sediments 53N, 1.5W	C	6 (24)	r	33N, 148E	M	1/103
WE6.8	Midland Sills 52.5N, 2W	Cu	10 (18)	c, r	28N, 157E	M	1/96
WE6.9	Whin Sill 55N, 2W	Cu (281)	34 (102)	a, d	37N, 169E	R	2/36
WE6.10	Whin Sill, 55.5N, 1.7W	Cu (295)	5 (24)	a, t	44N, 159E	R	11/77
WE6.11	Inner Sudetic Basin 50.6N, 16.1E	Cu	76 (104)	a, t	39N, 177E	R	10/111
WE6.12	Plzen Basin sediments 49.8N, 13.3E	Cu	65 (65)	t	31N, 152E	R	10/112
WE6.13	Kladuo-Rakovink Basin sediments 50N, 14E	Cu	73 (73)	t	36N, 154E	R	10/113
WE6.14	Lower Silesian Volcanics, Poland 50.5N, 16.5E	Cu	8 (33)	a	43N, 174E	R	11/71
WE6.15	Mt Billinger Sill, Sweden 58.5N, 14E	Cu (287)	9 (80)	a	31N, 174E	R	10/107
WE6.16	Mt Hunnenberg Sill, Sweden 58.5N, 12.5E	Cu (–Pl) (279)	3 (33)	a	38N, 177E	R	10/108
WE6.17	Skane dolerites, Sweden 55.5N, 13.5E	Cu (–Pl)	8 (48)	a	37N, 174E	R	10/109
WE6.18	Ny-Hellesund diabase, Norway 58N, 7.8E	Cu (–Pl)	10 (37)	a, t	39N, 161E	R	12/119
WE7.1	Igneous complex, Oslo 59.7N, 10.4E	Pl (270)	27 (494)	a	47N, 157E	R	5/37
WE7.2	Exeter lavas, England 51N, 4W	Pl (280)	5 (30)	a, f	50N, 149E	R	9/89
WE7.3	Exeter lavas, England 51N, 4W	Pl (280)	22 (66)	a, t, c	46N, 165E	R	9/90
WE7.4	Lower Silesian Volcanics 51N, 15.5E	Pl	14 (60)	a	43N, 175E	R	11/70
WE7.5	Volcanics, Krakow 50N, 20E	Pl	11 (41)	a	43N, 165E	R	8/87
WE7.6	Esterel igneous rocks, France 43.5N, 6.8E	P	3 (22)	a, t	50N, 146E	R	1/78–8
WE7.7	Esterel igneous and sediments 43.5N, 6.8E	P	(14)	a, t, f	47N, 145E	R	1/81
WE7.8	Nahe igneous rocks, Germany 50N, 8E	P	4 (18)	a, t	40N, 168E	R	2/37
WE7.9	Nahe igneous rocks, Germany 50N, 8E	P	5 (75)	a	46N, 167E	R	5/36
WE7.10	Sandstones, Czechoslovakia 49.5N, 16.6E	P	(15)	d, a, t	36N, 150E	R	8/83
WE7.11	Malmedy congolmerate, Belgium 50.3N, 6E	P	7 (26)	a	46N, 166E	R	9/76
WE7.12	Nideck–Domen Volcanics, France 53.9N, 5.7E	P	9 (37)	t	41N, 169E	R	7/35
WE7.13	Dome de Barrot redbeds, France 44N, 6.8E	Pu	30 (114)	a, t	46N, 148E	R	12/110
WE8.1	Vosges Sandstone, France 48.5N, 7E	Trl	10 (—)	r	44N, 151E	M	1/63, 4/7
WE8.2	Keuper Marls, England 53N, 2W	Tru	9 (43)	a, t, f, r	43N, 131E	M	1/64
WE8.3	Keuper Marls, England 55.6N, 5.3W	Tru	(35)	a, r	44N, 134E	M	†(5)
WE8.4	Ladinian Volcanics, Germany 46.4N, 11.7E	Trm	2 (8)	a	46N, 155E	N	10/91
WE9.1	Midford and Cotswold Sands 51N, 2.5W	Jl	(37)	r	36N, 50E	M	2/23–4
WE10.1	Iron grit, England 51N, 0.5E	Kl	(21)	r	87N, 349E	M	5/15–6
WE10.2	Iron grit, England 51N, 0.5E	Kl	(120)	a, t, r	84N, 11E	M	8/51
WE10.3	Sandstones, Czechoslovakia 50.2N, 14.6E	Ku	(12)	d, a, t	87N, 347E	N	8/46

Pole no.	Rock unit	Age (My)	Si (Sa)	Stab.	Pole Posn	P	GJ
WE11.1	Skye lavas and intrusives 57.4N, 6.3W	Te–o	53 (174)	*a, t*	74N, 157E	R	2/13
WE11.2	Ardnamurchan gabbro 56.7N, 6.2W	Te–o	(15)	*a, t*	69N, 165E	R	2/14
WE11.3	Rum gabbro 57.0N, 6.4W	Te–o	(8)	*a, t*	69N, 171E	R	2/15
WE11.4	Mull intrusives 56.4N, 5.8W	Te–o	5 (50)	*r*	72N, 133E	M	1/31
WE11.5	Arran dykes 55.5N, 7.2W	Te–o	7 (77)	*c, r*	78N, 149E	M	1/32
WE11.6	Antrim igneous suite 55.1N, 6.4W	Te–o	89 (163)	*c, t*	78N, 145E	R	†(5)
WE11.7	Lundy dykes 51.2N, 4.7W	Te–o	8 (—)	*c*	75N, 130E	R	1/33
WE11.8	Rheinland Platz igneous rocks 50.6N, 7.5E	To–m	22 (53)	*a, r*	70N, 108E	M	6/11–13
WE11.9	Tertiary dykes, Scotland 56N, 5W	Te–o	9 (9)	*c, t, r*	79N, 128E	M	8/23
WE11.10	Tertiary dykes, Scotland 55N, 4W	Te–o (34–57)	21 (84)	*a, t, c, r*	73N, 197E	M	8/33
WE11.11	Tertiary dykes, U.K. 52.8–54.5N, 4.7–1.1W	Te	11 (54)	*a, c*	77N, 211E	R	11/28
WE11.12	Northern Ireland igneous rocks 55N, 6W	Te	54 (225)	*a, c*	70N, 163E	M	12/49
WE11.13	Gergovie intrusives, France 45.8N, 3.1E	To	9 (—)	*c*	72N, 115E	R	1/34
WE12.1	Cantal and Limagne lavas 45.5N, 3E	Tm	5 (—)	*t, r*	57N, 236E	M	3/21–2
WE12.2	Vogelsberg intrusives and lavas 50.5N, 9E	Tm	(25)	*r*	86N, 168E	M	4/4
WE12.3	Vogelsberg basalt 50.5N, 9.5E	Tm	42 (200)	*r*	76N, 160E	M	3/23
WE12.4	Suevites, Nordlinger Ries 49.9N, 10.5E	Tm (15)	12 (111)	*a*	78N, 143E	R	8/30
WE12.5	Kaiserstuhl volcanics 48.1N, 10.6E	Tm (16)	8 (—)	*a*	71N, 171E	R	10/25
WE12.6	Lausitz Volcanics 51N, 14.7E	To–m	27 (167)	*a, r*	75N, 125E	M	10/35
WE12.7	Lower Silesian basalts, Poland 51N, 16E	Tm–p	65 (390)	*a, r*	82N, 184E	M	11/16
WE12.8	Gottingen Volcanics 51.4N, 9.8E	Tm–p	15 (155)	*a, r*	83N, 163E	M	7/7
WE12.9	Habitschtswald tuffs and lavas 51.3N, 9.4E	Tm–p	4 (11)	*a, r*	87N, 219E	M	7/8
WE12.10	Coiron lavas, France 45N, 4E	Tm–p	36 (220)	*a, r*	80N, 143E	M	12/30
WE13.1	Auvergne lavas, France 45.1N, 3.5E	Tp–Qp	6 (—)	*c*	78N, 93E	R	1/21
WE13.2	Postglacial varves, Sweden 63.1N, 17.7E	Qr	29 (150)	*a*	89N, 156E	N	1/3
WE13.3	Massif central lavas, France 43.4–45.8N, 2.8–4.3E	Tp–Qp	10 (73)	*a, r*	78N, 188E	M	12/17
WE13.4	Massif central lavas, France 43.4–45.8N, 2.8–4.3E	Qp–r	31 (219)	*a*	86N, 212E	N	12/9
WE13.5	Plateau du Velay basalts, France 45N, 3.8E	Qp	28 (81)	*a, r*	80N, 132E	M	11/12

†(1) Briden *et al.* (1971a).   †(2) Morris (1972).   †(3) Piper (1972).
†(4) Deutsch (personal communication).   †(5) Irving (1964).

Note. Cambrian and Ordovician results reported from Czechoslovakia (Bohemian Massif) are inconsistent n publications of Bucha (1961, 1965) and Andreeva *et al.* (1965) and have not been included in the list above or western Europe.

SOUTHERN EUROPE

2) *Spain/Portugal*

P4.1	Atienza igneous rocks (159, +19) 41N, 3W	Su	6 (33)	*a*	36N, 203E	R	9/126
P4.2	Almaden igneous rocks (131, +22) 39N, 5W	Su	2 (10)	*a*	21N, 228E	R	9/127
P7.1	Bucaco redbeds (149, +11) 40.3N, 8.4W	Cu–Pl	4 (17)	*a, t, f*	36N, 212E	R	11/72

Appendix

Pole no.	Rock unit	Age (My)	Si (Sa)	Stab.	Pole Posn	P	GJ
SP7.2	Viar redbeds (151, +02) 37.5N, 5.9W	Cu–Pl	3 (8)	*a, t*	42N, 216E	R	11/73
SP7.3	Viar dykes and sills (156, +10) 37.5N, 5.9W	Cu–Pl	3 (14)	*a, t*	41N, 208E	R	11/74
SP7.4	Andesites, Huesca Province (152, −22) 43N, 1W	P (–Tr)	(14)	*a*	51N, 227E	R	7/36
SP7.5	Sediments, Spain (158, −18) 42N, 1.5E	P	2 (13)	*a*	52N, 219E	R	9/78
SP7.6	Igneous rocks, Spain (170, −03) 42N, 1.5E	Pl	10 (41)	*a*	49N, 197E	R	9/80
SP8.1	Alcazar sediments (0, +23) 39N, 3W	Trl	2 (39)	*a*	63N, 178E	N	9/61
SP8.2	Gerralda redbeds (350, +18) 42.9N, 1.3W	Tr	5 (95)	*a, t, f*	55N, 196E	N	11/54
SP8.3	Volcanics, Pyrenees‡ (29, +47) 43N, 1.3E	Tru	7 (26)	*a, f*	62N, 114E	N	10/79
SP9.1	Volcanics, Pyrenees‡ (16, +39) 43N, 1.3E	Jl	8 (26)	*a, f*	65N, 143E	N	10/78
SP10.1	Sintra granite (359, +44) 38.8N, 9.5W	Ku	8 (25)	*a*	76N, 174E	N	11/32
SP11.1	Monchique syenite (182, −37) 37.3N, 8.5W	Tpa–e	2 (8)	*a*	73N, 166E	R	11/29
SP11.2	Lisbon Volcanics (347, +37) 38.8N, 9.2W	Te	12 (39)	*a*	69N, 207E	N	10/42
SP11.3	Lisbon Volcanics (352, +40) 39N, 9W	Te	6 (17)	*a, t, f*	72N, 196E	N	10/43
SP13.1	Gerona Volcanics (356, +54) 42.2N, 2.5E	Qr	7 (69)	*a*	81N, 204E	N	10/5
(b) Alps/Italy							
IT6.1	Pramollo sediments‡ (24, +40) 46N, 11E	Cu	(9)	*a*	60N, 142E	N	9/91
IT7.1	Bolzano Porphyry, Merano (164, −11) 46.6N, 11.2E	Pl	28 (51)	*a*	45N, 214E	R	6/49
IT7.2	Bolzano Porphyry, Val-di-Non (150, −31) 46.5N, 11.4E	Pl	10 (33)	*a*	51N, 241E	R	6/50
IT7.3	Bolzano Porphyry, General (150, −20)	Pl	39 (152)	*a, t, f*	46N. 236E	R	†(1)
IT7.4	Bolzano Porphyry, Lagorai (143, −15) 46N, 11E	Pl	9 (33)	*a*	40N, 242E	R	9/82
IT7.5	Chiusa, Cavalese, Posina (161, −20) 46N, 11E	Pl	3 (21)	*a*	51N, 223E	R	9/81, 83–4
IT7.6	East Lombardy Volcanics (135, −21) 45.8N, 10.2E	Pl	5 (43)	*a*	38N, 252E	R	11/69
IT7.7	Valgardena redbeds (148, −13) 46.1 − 46.5N, 11.1 − 12.4E	Pm–u	4 (8)	*t*	42N, 237E	R	12/111
IT7.8	Lugano Porphyries (144, −17) 45.9N, 8.8E	P	4 (11)	*a*	42N, 240E	R	8/81
IT7.9	Staro and Camparmo Volcanics (332, +32) 45N, 11E	Pu	4 (17)	*a, r*	53N, 241E	M	9/67
IT7.10	San Martino and Cartiano sediments (332, +24) 45N, 11E	Pm–u	2 (15)	*a, r*	49N, 235E	M	9/68
IT7.11	Vallee du Guil Volcanics (147, −32) 44.5N, 7E	P	(8)	*a*	51N, 243E	R	11/68

Pole no.	Rock unit	Age (My)	Si (Sa)	Stab.	Pole Posn	P	GJ
IT7.12	Coccau sediments‡ (35, +24) 46N, 12E	Pm–u	(9)	*a*	45N, *140*E	N	9/77
IT7.13	Rhyolites, French Alps‡ (110, −76) 44.1N, 7.3E	Pu	3 (8)	*a*	47N, *329*E	R	10/97
IT7.14	Rhyolites, French Alps‡ (62, −52) 44.1N, 7.3E	Pu	4 (14)	*a*	5N, *319*E	R	10/98
IT8.1	Schio Volcanics and sediments (330, +40) 45N, 11E	Trl	5 (43)	*a, r*	57N, 249E	M	9/60
IT8.2	Porphyries, dykes, tuffs (329, +25) 46.4N, 11.7E	Trm–u	12 (67)	*a*	48N, 240E	N	12/100
IT8.3	Predazzo dykes (336, +43) 46.3N, 11.7E	Tr	2 (9)	*a*	61N, 242E	N	12/101
IT8.4	Valle di Scalve and Schio Intrusives (336, +36) 46N, 11E	Trm	2 (12)	*a, r*	57N, 240E	M	9/55, 11/48
IT8.5	Tarvis and Dobratsch intrusives‡ (23, +46) 46N, 11E	Trm	2 (36)	*a*	65N, *136*E	N	9/56–7
IT11.1	Priabona basalts (353, +50) 45N, 11E	Te	(29)	*a*	75N, 214E	N	9/28
IT11.2	Marostica basalts (184, −61) 45N, 11E	To	(16)	*a*	86N, 145E	R	9/26

(c) Corsica/Sardinia

Pole no.	Rock unit	Age (My)	Si (Sa)	Stab.	Pole Posn	P	GJ
CS6.1	Ota gabbrodiorite, Corsica (189, −13) 42.2N, 8.8E	C?	2 (14)	*a, t*	54N, 173E	R	9/119
CS7.1	Lavas and tuffs, Corsica (141, +9) 42.3N, 8.6E	(Cu–) Pl	15 (15)	*a*	32N, 239E	R	9/88
CS7.2	Volcanics and dykes, Corsica (177, −14) 42.3N, 8.5E	Cu–Pl	37 (—)	*a, t*	55N, 194E	R	11/75
CS7.3	Red Sandstones, Sardinia (110, −16) 41N, 9E	P	2 (15)	*a, t, f*	20N, 271E	R	12/112
CS7.4	Ignimbrites, Sardinia (142, −02) 41N, 9E	P	6 (29)	*a*	38N, 239E	R	12/113
CS11.1	Alghero trachyandesites, Sardinia (322, +42) 40.5N, 8.5E	To	10 (59)	*a, r*	54N, 263E	M	11/23

(d) Pannonia

Pole no.	Rock unit	Age (My)	Si (Sa)	Stab.	Pole Posn	P	GJ
PN7.1	Kosice region, Czechoslovakia (35, +17) 49N, 21E	P–Tr	(—)	*a, t, r*	40N, 153E	M	†(2)
PN7.2	Spisska Nova Ves, Czechoslovakia (28, +14) 49, 21E	P–Tr	(—)	*a, t, r*	42N, 162E	M	†(2)
PN7.3	Choc Nappe, Czechoslovakia‡ (73, +20) 49N, 21E	P–Tr	(—)	*a, t, r*	19N, *116*E	M	†(2)
PN8.1	Werfenian strata, Slovakia (6, +19) 48.9N, 19.2E	Trl	(21)	*d, a, t*	51N, 190E	R	8/75
PN10.1	Mecsek Mountains, Hungary (334, +47) 46.1N, 18.4E	Kl	11 (94)	*a, r*	62N, 256E	M	12/78
PN10.2	Mecsek Mountains, Hungary 46.1N, 18.5E	Kl	7 (27)	*a, r*	82N, 160E	M	12/79
PN11.1	Andesites, Bulgaria (19, +61) 41N, 25E	To	3 (19)	*a, t, r*	76N, 105E	M	9/25
PN12.1	East Slovak Volcanics (359, +64) 48N, 21E	Tm	33 (170)	*a, r*	87N, 217E	M	9/24

Pole no.	Rock unit	Age (My)	Si (Sa)	Stab.	Pole Posn	P	GJ
PN12.2	Matra Cserhat Volcanics, Hungary (0, +55) 47.9N, 20E	Tm	9 (72)	*a, r*	78N, 201E	M	12/38
PN12.3	Matra Cserhat Volcanics, Hungary 47.9N, 19.8E	Tm	16 (83)	*a, r*	73N, 194E	M	12/41
PN12.4	Cserhat Volcanics, Hungary 48N, 19.5E	Tm	4 (28)	*a*	84N, 184E	R	12/42
PN12.5	Zempleni Volcanics, Hungary (352, +66) 48.4N, 21.6E	Tm	10 (96)	*a, r*	85N, 299E	M	12/39
PN12.6	Komlo Laccolith, Hungary (21, +63) 46.1N, 18.3E	Tm	(18)	*a*	75N, 106E	N	12/40
PN12.7	Wzar Andesite dykes (192, −73) 49.5N, 20.5E	Tm	15 (52)	*a, c*	79N, 54E	R	11/21
PN12.8	Wzar Mountain andesites (198, −80) 49.4N, 20.3E	Tm–p	(16)	*r*	67N, 36E	M	10/23
PN12.9	Central Slovak Volcanics (11, +63) 48.5N, 19E	Tm–p	94 (436)	*a, r*	81N, 132E	M	10/24
PN12.10	Almopias Complex, Greece (195, −66) 41.1N, 22E	Tp	5 (15)	*a, t, r*	77N, 70E	M	11/14
PN13.1	Balaton basalts, Hungary (9, +60) 47N, 17.5E	Tp–Qp	13 (136)	*a, r*	82N, 145E	M	12/21
PN13.2	Basalts, Hungary 46.9–48.1N, 17.5 − 19.9E	Tp–Qp	15 (156)	*a, r*	88N, 97E	M	12/22

†(1) Zijderveld *et al.* (1970b). †(2) Kotasek & Krs (1965), values read from diagram.
‡ Sites located within the Tethys Mobile Belt–Pyrenees, Alpine Front etc.

Note. Numbers in brackets after each rock unit are the mean declination and inclination (D_m, I_m).

RUSSIAN PLATFORM AND SOUTHERN U.S.S.R

Pole no.	Rock unit	Age (My)	Si (Sa)	Stab.	Pole Posn	P	GJ
RP1.1	Katav River Suite, South Urals 53N, 57E	Pc (1000)	(16)	*a*	17N, 224E	—	R/297
RP2.1	Asha River Suite, South Urals 53.5N, 57E	€l (570)	(83)	*a, r, d*	2N, 181E	M	R/282
RP2.2	Asha River Suite, South Urals 54.5N, 57E	€l	(50)	*a, r, d*	0N, 182E	M	R/283
RP2.3	Asha River Suite, South Urals 54.5N, 57E	€l	(16)	*a, r, d*	9N, 202E	M	R/284
RP2.4	Asha River Suite, South Urals 52.5N, 57E	€l	(33)	*a, r, d*	20N, 191E	M	R/285
RP3.1	Obolid sandstones, Tosna and Narva Rivers 59.5N, 29E	Ol	2 (16)	*a*	38N, 150E	R	R/254–5
RP3.2	Limestones, River Popovka 60N, 30E	Ol	(11)	*a*	34N, 135E	R	R/253
RP3.3	Diabases, North Serginski Region, Urals‡ 57N, 60E	Om–u	(134)	*t, d*	16N, 184E	R	R/252
RP4.1	Asha River Suite, South Urals 54N, 57E	Ou–Sl	2 (28)	*a*	8N, 146E	R	R/249–51
RP4.2	Asha River Suite, South Urals 53.6N, 56.5E	Ou–Sl	(110)	*a*	20N, 151E	R	R/250
RP4.3	Dolomites, River Belaya, South Urals 52.5N, 57E	Sl	(9)	*a*	38N, 166E	R	R/237
RP5.1	Zhedian stage, River Dniestr 49N, 25.5E	Dl	(20)	*a*	43N, 159E	R	R/230
RP5.2	Zhedian stage, River Dniestr 49N, 25.5E	Dl	(97)	*a*	42N, 160E	R	R/229
RP5.3	Zhedian stage, River Dniestr 49N, 25.5E	Dl	(63)	*a*	39N, 159E	R	R/228

Pole no.	Rock unit	Age (My)	Si (Sa)	Stab.	Pole Posn	P	GJ
RP5.4	Zhedian stage, Volyn' 49N, 25E	Dl	(64)	d, r	31N, 167E	M	R/227
RP5.5	Redbeds, Volyn' 49N, 25E	Dl	(32)	d	36N, 168E	R	R/226
RP5.6	Coblentzian stage, River Dniestr 49N, 25E	Dl	(80)	d	41N, 166E	R	R/224
RP5.7	Red bauxites, South Urals‡ 55N, 58.5E	Du	(35)	d	20N, 164E	R	R/210
RP5.8	Bauxites, Pashnia‡ 57N, 57E	Du	(127)	a, d	35N, 181E	R	R/209
RP5.9	Red loams, River Sias 60N, 33E	Du	(50)	a, d	28N, 159E	R	R/207
RP5.10	Redbeds, Lake Il'men 57N, 31E	Du	(49)	a, d, r	29N, 164E	M	R/206
RP5.11	Redbeds, Rivers Mda, Lininka and Msta 58N, 33E	Du	(43)	a, d, r	34N, 158E	M	R/205
RP5.12	Redbeds, Rivers Lininka and Mda 59N, 34E	Du	(8)	a, d	32N, 159E	R	R/204
RP5.13	Zilayir River Suite‡ 54N, 59E	Du	(80)	a, t, r	40N, 179E	M	R/203
RP6.1	Sediments, Azov Sea Area 48N, 38E	Cl	(10)	a	21N, 198E	R	R/199
RP6.2	Limestones, South Urals 53N, 57E	Cl	(13)	a	19N, 147E	R	{ 9/116 \ R/198 }
RP6.3	Limestones, Donbass 48N, 38E	Cl	(25)	a	27N, 167E	R	R/196
RP6.4	Sediments, Donbass 48N, 38E	Cl (m)	(42)	a, r	38N, 178E	M	R/195
RP6.5	Porphyrites, South Urals 52N, 59E	Cl	(145)	a, t, d	20N, 163E	R	{ 9/115 \ R/194 }
RP6.6	Redbeds, River Vyterega 61N, 37E	Cl (u)	(13)	a, d	46N, 152E	R	{ 9/113 \ R/191 }
RP6.7	Tula horizon, Nebol'chi 59N, 34E	Cl (u)	(25)	r, d	41N, 158E	M	{ 9/112 \ R/190 }
RP6.8	Oka strata, Boksitogorsk City 59N, 34E	Cl (u)	(20)	a, d	49N, 164E	R	R/189
RP6.9	Ok-Serpukhov strata, Tikhvin City 59N, 34E	Cl (u)	(38)	d	45N, 156E	R	{ 9/111 \ R/188 }
RP6.10	Suite 'E', Donbass 48N, 38E	Cl	(18)	a, r, d	24N, 173E	M	R/183
RP6.11	Suite 'E' Donbass 48N, 38E	Cl	(9)	r, d	20N, 168E	M	R/182
RP6.12	Suite 'E' Donbass 48N, 38E	Cl	(18)	r, d	26N, 173E	M	R/181
RP6.13	Chistiakovskaya Suite, Donbass 48N, 38E	Cm	(16)	d	32N, 183E	R	R/177
RP6.14	Lr Nagol'chi Suite, Donbass 48N, 38E	Cm	(13)	a, r, d	32N, 177E	M	R/176
RP6.15	Upr Nagol'chi Suite, Donbass 48N 38E	Cm	(26)	a, r, d	19N, 166E	M	R/175
RP6.16	Sediments, Donbass 48N, 38E	Cm	(12)	d	29N, 183E	R	R/174
RP6.17	Smolianinov Suite, Donbass 48N, 41E	Cm	(9)	a, r, d	28N, 208E	M	R/173
RP6.18	Belaya Kalitva Suite, Donbass 48N, 38E	Cm	(35)	a, r, d	29N, 173E	M	R/172
RP6.19	Belaya Kalitva Suite, Donbass 48N, 41E	Cm	(10)	a, r, d	28N, 181E	M	R/171
RP6.20	Kamenskoye Suite, Donbass 48N, 41E	Cm	(12)	a, r, d	35N, 195E	M	R/170
RP6.21	Kamenskoye Suite, Donbass 48N, 38E	Cm	(36)	a, d	17N, 168E	R	R/169
RP6.22	Diamond Suite, Donbass 48N, 38E	Cm	(20)	a, d	26N, 157E	R	R/168
RP6.23	Lisichansk Suite, Donbass 48N, 41E	Cm	(20)	a, r, d	36N, 195E	M	R/167
RP6.24	Vereya Horizon, Shatsk City 54N, 42E	Cm	(14)	a, d	31N, 168E	R	{ 9/104 \ R/166 }
RP6.25	Vereya Horizon, Vereya City 55N, 36E	Cm	(17)	a	32N, 166E	R	R/164
RP6.26	Vereya Horizon, Serpukhov City 55N, 38E	Cm	(20)	a, d	29N, 155E	R	{ 9/103 \ R/165 }
RP6.27	Kashira River Horizon, Rzhev City 56N, 34E	Cm	(40)	a	33N, 158E	R	{ 9/101 \ R/162 }
RP6.28	Kashira River Horizon, Ozery City 55N, 39E	Cm	(15)	a	34N, 170E	R	{ 9/102 \ R/163 }

Appendix

Pole no.	Rock unit	Age (My)	Si (Sa)	Stab.	Pole Posn	P	GJ
RP6.29	Isayev Suite, Donbass 48N, 41E	Cu	(23)	a, d	25N, 186E	R	9/95, R/157
RP6.30	Kasimovian Stage, Noginsk 56N, 39E	Cu	(43)	a, d	42N, 169E	R	R/155
RP6.31	Kasimovian Stage, Voskresensk City 55N, 39E	Cu	(28)	a, d	41N, 167E	R	R/156
RP6.32	Avilov Suite, Donbass 48N, 38E	Cu	(93)	d	40N, 182E	R	9/94, R/154
RP6.33	Avilov Suite, Donbass, 49N 38E	Cu	(93)	d	40N, 184E	R	R/153
RP6.34	Araucarite Suite, Donbass 49N, 38E	Cu	(137)	d	44N, 171E	R	R/151
RP6.35	Araucarite Suite, Donbass, 48N, 38E	Cu	(129)	d	38N, 170E	R	9/92, R/152
RP6.36	Gzhelian Stage, Moscow 55N, 38E	Cu	(41)	a, d	46N, 177E	R	R/149
RP6.37	Gzhelian Stage, Moscow 56N, 38E	Cu	(33)	d	42N, 167E	R	9/93, R/148
RP7.1	Cupriferous Sandstone, Donbass 48N, 38E	Pl	(92)	r	43N, 160E	M	R/144
RP7.2	Cupriferous Sandstone, Donbass 48N, 38E	Pl	(62)	a, t	40N, 152E	R	R/143
RP7.3	Cupriferous Suite, Donbass 48N, 38E	Pl	(107)	d, r	46N, 166E	M	R/142
RP7.4	Cupriferous Suite, Donbass 48N, 38E	Pl	(61)	d	36N, 159E	R	R/141
RP7.5	Asselian Stage, Donbass 49N, 38E	Pl	(27)	d	46N, 170E	R	R/140
RP7.6	Redclays, Kama and Belaya Rivers 56N, 55E	Pu	(31)	d	40N, 168E	R	9/75, R/134
RP7.7	Redclays, Kama River 57N, 55E	Pu	(15)	d	40N, 167E	R	9/74, R/133
RP7.8	Dronovskaya Suite, Donbass 49N, 38E	Pu	(19)	d	38N, 159E	R	R/132
RP7.9	Dronovskaya Suite, Donbass 48N, 38E	Pu	(10)	d	40N, 158E	R	R/131
RP7.10	Red clays, Buguruslan 54N, 52E	Pu	(20)	d	44N, 169E	R	R/129
RP7.11	Red clays, River Kama 57N, 54E	Pu	(35)	d	39N, 167E	R	9/73, R/128
RP7.12	Sukhona Suite, River Sukhona 61N, 46E	Pu	(8)	d, r	40N, 160E	M	9/69, R/127
RP.713	River Sukhona 61N, 46E	Pu	(49)	d, r	48N, 165E	M	9/69, R/125
RP7.14	Red clays, River Viatka 59N, 51E	Pu	(29)	d, r	43N, 162E	M	9/70, R/124
RP7.15	Red clays, Buguruslan 54N, 52E	Pu	(38)	d, r	49N, 167E	M	R/123
RP7.16	Sarma Suite, River Donguz 52.5N, 55E	Pu	(12)	d	61N, 152E	N	R/122
RP7.17	Lake Inder redbeds 48.5N, 52E	Pu	(35)	r	52N, 152E	M	R/121
RP8.1	Red clays, Vetluga Stage 59N, 51E	Trl	(10)	d	50N, 174E	R	R/112
RP8.2	Buzuluk Suite, Buzuluk 53N, 52E	Trl	(9)	d	54N, 164E	R	R/111
RP8.3	Buzuluk Suite, River Donguz 52.5N, 55E	Trl	(19)	a	45N, 152E	N	R/110
RP8.4	Buzuluk Suite 52.5N, 51E	Trl	(10)	a, r	51N, 164E	M	R/109
RP8.5	Donguz Suite, River Donguz 52.5N, 55E	Trl	(42)	a	62N, 125E	N	R/107
RP8.6	Lake Bashkunchak 48N, 47E	Trl	(48)	d	52N, 150E	N	R/106
RP8.7	Lake Inder 48.5N, 52E	Trl	(17)	d	48N, 153E	N	R/105
RP8.8	Serebriansk Suite, Donbass 48N, 38E	Trm	(26)	r	60N, 135E	M	R/91
RP8.9	Serebriansk Suite, Donbass 48N, 38E	Trm	(53)	d, r	49N, 152E	M	R/92
RP8.10	Yushatyr Suite, River Yushatyr 52.5N, 55E	Trm–u	(18)	a	35N, 158E	N	R/89
RP9.1	Igneous rocks, Crimea 45N, 35E	Jm–u?	(12)	a	60N, 115E	N	R/82
RP9.2	Igneous rocks, Crimea 45N, 35E	Jm–u?	(36)	a, d	50N, 135E	N	R/83

Pole no.	Rock unit	Age (My)	Si (Sa)	Stab.	Pole Posn	P	GJ
RP9.3	Igneous rocks, Crimea 45N, 35E	Jm–u?	(35)	*a, d*	65N, 165E	N	R/84
RP9.4	Igneous rocks, Crimea 45N, 35E	Jm–u?	(107)	*a, d*	82N, 150E	N	R/85
RP10.1	Bol'shoi Balkan Range 39N, 55E	Kl	(22)	*r*	60N, 167E	M	R/76
RP10.2	sw Spurs of the Hissar Range 38N, 67E	Kl	(16)	*a*	72N, 164E	R	8/53 R/75
RP11.1	Sediments, Rakhmatur 36N, 61E	Tpa	(47)	*a, r*	69N, 192E	M	R/70
RP11.2	Sediments, Shor-Gaudan 38N, 59E	Te	(132)	*a*	67N, 225E	N	R/69
RP11.3	Sediments, Turkmenia 38N, 58E	To	(19)	*a*	69N, 202E	N	R/68
RP11.4	Sediments, West Turkmenia 41N, 55E	To	(26)	*r*	60N, 158E	M	R/65
RP12.1	Maikop Suite, Kerch' Pen 45N, 36E	Tm	(115)	*a*	83N, 191E	R	R/63
RP12.2	Karagaudan Suite, Turkmenia 38N, 58.5E	Tm	(145)	*a, r*	73N, 209E	M	R/61
RP12.3	Karagaudan Suite, Turkmenia 37N, 60E	Tm	(42)	*a*	71N, 164E	N	R/62
RP12.4	Maikop clays, Kerch' Pen 45N, 36E	Tm	(16)	*t*	66N, 225E	R	R/60
RP12.5	Turkmenia, Uzek-Dag 39N, 55.5E	Tm	(32)	*d*	64N, 224E	R	R/59
RP12.6	Sarmatian clays, Keliata 39N, 58E	Tm	(59)	*a, r*	70N, 213E	M	R/57
RP12.7	Sarmatian clays, Ilanly 39N, 56E	Tm	(27)	*d*	81N, 204E	N	R/58
RP12.8	Kazanchai Suite, Kaksar-Bulak 40N, 53E	Tp	(48)	*a, t, r*	76N, 165E	M	R/54
RP12.9	Kazanchai Suite, Kuru-Gaudan 38N, 59E	Tp	(93)	*a*	78N, 204E	R	R/52
RP12.10	Kazanchai Suite, Kuru-Gaudan 38N, 59E	Tp	(98)	*a*	75N, 228E	R	R/53
RP12.11	Kerch' Pen sediments 47N, 35E	Tp	(13)	*t*	63N, 69E	R	R/51
RP12.12	Red clays, West Turkmenia 39N, 53E	Tp	(119)	*d*	71N, 199E	N	R/25
RP12.12	Red clays, West Turkmenia 39N, 53E	Tp	(166)	*d*	74N, 194E	R	R/26
RP12.14	Red clays, West Turkmenia 39N, 53E	Tp	(84)	*d*	71N, 207E	N	R/27
RP12.15	Red clays, West Turkmenia 39N, 54E	Tp	(16)	*d*	75N, 228E	R	R/28
RP12.16	Red clays, West Turkmenia 39N, 55E	Tp	(36)	*d*	74N, 175E	N	R/29
RP12.17	Red clays, West Turkmenia 39N, 55E	Tp	(43)	*d*	73N, 249E	R	R/30
RP12.18	Sediments, Kerch' Pen 45N, 36E	Tp	(15)	*t*	75N, 170E	N	R/24
RP12.19	Sediments, Crimea 44N, 33E	Tp	(23)	*t*	77N, 65E	N	R/21
RP12.20	Sediments, River Kuban 45N, 40E	Tp	(8)	*t*	71N, 73E	N	R/20
RP12.21	Kaksar Bulak, Turkmenia 40N, 53E	Tp	(257)	*a, t*	71N, 157E	N	R/19
RP12.22	Sediments, Turkmenia 38N, 56E	Tp	(79)	*a, t*	71N, 146E	R	R/18
RP12.23	Sediments, Turkmenia 38N, 59E	Tp	(64)	*a*	76N, 217E	N	R/17
RP12.24	Sediments, Turkmenia 38N, 59E	Tp	(45)	*a*	81N, 229E	R	R/16
RP12.25	Sediments, Turkmenia 38N, 59E	Tp	(34)	*a, t*	68N, 190E	R	R/15
RP13.1	Sediments, Turkmenia 39N, 55E	Tp–Qp	(104)	*d*	77N, 213E	R	R/14
RP13.2	Azov Region sediments 47N, 39E	Qp	(12)	*t*	79N, 53E	N	R/13
RP13.3	River Kuban sediments 45N, 40E	Qp	(8)	*t*	81N, 44E	N	R/12
RP13.4	Baku Stage, West Turkmenia 39N, 53E	Qp–r	(80)	*d*	81N, 188E	N	R/10
RP13.5	Azov Region sediments 47N, 39E	Qp–r	(10)	*d*	79N, 63E	N	R/9
RP13.6	Kerch' Pen sediments 45N, 36E	Qr	(13)	*t*	80N, 56E	N	R/8
RP13.7	Odessa City sediments 47N, 30E	Qr	(9)	*t*	87N, 322E	N	R/7
RP13.8	River Dnieper sediments 50N, 30E	Qr	(19)	*t*	87N, 91E	N	R/6
RP13.9	Azov Sea sediments 47N, 37E	Qr	(10)	*t*	80N, 34E	N	R/5
RP13.10	Moldavia sediments 45N, 29E	Qr	(12)	*t*	82N, 62E	N	R/4
RP13.11	Moldavia sediments 45N, 28E	Qr	(27)	*t*	78N, 54E	N	R/3

‡ Sites located within the Urals.

295

Pole no.	Rock unit	Age (My)	Si (Sa)	Stab.	Pole Posn	P	GJ
		SIBERIA AND TRANSURALS					
SB1.1	Maimakan Suite 59N, 135E	Pc (Lr Sn.)	(16)	a	30N, 23E	—	R/302
SB1.2	Malgin Creek Suite 59N, 134E	Pc (Lr Sn.)	(104)	a, r	19N, 35E	M	R/301
SB1.3	Cipanda Suite 59N, 134E	Pc (Lr Sn.)	(21)	a	24N, 28E	—	R/300
SB1.4	Bol'shaya Lakhanda River Suite 59N, 134E	Pc (Lr Sn.)	(88)	a, r	25N, 26E	M	R/229
SB1.5	Sukhopit Series, Gorbilok Suite 59N, 98E	Pc (> 1300)	(24)	d	13N, 98E	—	R/295 10/182
SB1.6	Sukhopit Series, Uderei Suite 59N, 95E	Pc (1270–1300)	(77)	d, r	20N, 102E	M	R/294 10/181
SB1.7	Sukhopit Series, Pogoryin Suite 59N, 95?	Pc (1120–1150)	(38)	d, r	20N, 106E	M	R/293 10/180
SB1.8	Tingusik Series, Potoskuy Suite 59N, 98E	Pc (1050–1100)	(35)	d, r	25N, 115E	M	10/179
SB1.9	Tingusik Series, Shuntar Suite 59N, 95E	Pc (930–950)	(25)	d, r	18N, 116E	M	R/292 10/178
SB1.10	Tingusik Series, Kirgitey Suite 59N, 95E	Pc (< 930)	(99)	d, r	17N, 110E	M	R/291 10/177
SB1.11	Oslyansk Series, Lr Angara Suite 59N, 95E	Pc (< 930)	(9)	d	10N, 142E	—	R/290 10/176
SB1.12	Kokin Creek Suite 59N, 93E	Pc (1140)	(22)	a, d	37N, 257E	—	R/298
SB1.13	Burovaya Suite, Lr Tunguska River 66N, 89E	Pc (925)	(39)	a, t	40N, 298E	—	R/296
SB1.14	Haematite ores, Lr Angara Suite 58N, 95E	Pc (745–925)	(40)	a, d, r	17N, 269E	M	R/289
SB1.15	Karagasski Suite, Sayan Region 55N, 98E	Pc (Upr Sn.)	(23)	a, d, r	22N, 321E	M	R/288
SB1.16	Izluch'ye Suite 67N, 87E	Pc (–Є)	(38)	a	24N, 356E	—	R/287
SB2.1	Lena Stage, Chara River Suite 60.5N, 134E	Єl	(88)	a	25N, 26E	R	R/281
SB2.2	Lena Stage, Sub-Redrock Suite 60.5N, 120.5E	Єl	(21)	a	54N, 12E	N	R/280
SB2.3	Lena Stage, Sukharian Suite 67N, 87E	Єl	(22)	a, r	19N, 350E	M	R/279
SB2.4	Ust'-Botoma Suite 61.5N, 129E	Єm	(12)	a	38N, 8E	R	R/277
SB2.5	Chaya River Suite, River Maiya 60N, 135E	Єm	(101)	a	46N, 346E	R	R/276
SB2.6	Ust' Maiya Suite, River Maiya 60N, 135E	Єm	(182)	a	46N, 343E	R	R/275
SB2.7	Maiyan Stage, River Aldan 61N, 135E	Єm	(16)	a	50N, 328E	R	R/274
SB2.8	Maiyan Stage, River Sukharikha 67N, 87E	Єm	(58)	a, t, r	31N, 332E	M	R/273
SB2.9	Maiyan Stage, River Koliumbe 68N, 88E	Єm	(27)	a, t	36N, 315E	R	R/272
SB2.10	River Koliumbe 68N, 88E	Єu	(53)	a, t	36N, 314E	R	R/271 11/84
SB2.11	Upper Lena Suite, River Lena 55N, 106E	Єu	(167)	a, r, d	36N, 312E	M	R/270 11/83
SB2.12	Upper Lena Suite, River Nepa 59N, 106.5E	Єu	(18)	a, r	38N, 308E	M	R/269
SB2.13	Upper Lena Suite, River Nepa 59N, 106.5E	Єu	(52)	a, r	34N, 300E	M	R/268
SB2.14	Upper Lena Suite, River Nepa 59N, 106.5E	Єu	(43)	a, r	34N, 302E	M	R/267
SB2.15	Upper Lena Suite, Angara River 58N, 97E	Єu	(159)	r, d	29N, 296E	M	R/266 11/82
SB2.16	Upper Lena Suite, Angara and Oka Rivers 54N, 102E	Єu	(28)	a, r	41N, 306E	M	R/265 10/144
SB2.17	Upper Lena Suite, Upper Lena River 54N, 106E	Єu	(35)	a, r	33N, 310E	M	R/264 10/141

Pole no.	Rock unit	Age (My)	Si (Sa)	Stab.	Pole Posn	P	GJ
SB2.18	Upper Lena Suite, Upper Lena River 58N, 108E	Єu	(39)	a, r	36N, 299E	M	R/263, 10/142
SB2.19	Upper Lena Suite, Upper Lena River 57N, 107E	Єu	(39)	a, r	41N, 307E	M	R/262, 10/143
SB2.20	Upper Lena Suite, Middle Lena 61N, 116E	Єu	(47)	r	37N, 296E	M	R/261
SB2.21	Verkholensk redbeds 54.3N, 104.6E	Єu	17 (202)	d, *a* t, r	34N, 312E	M	10/145
SB2.22	Evenkiy redbeds 58N, 97E	Єu	(300)	d, a, t, r	37N, 320E	M	11/81
SB3.1	River Kureika 67N, 88E	Ol	(28)	a, t	26N, 324E	R	R/260
SB3.2	River Kuliumbe, Chunya Stage 68N, 88E	Ol	(42)	a, t	19N, 324E	R	R/259
SB3.3	Chunya Stage, Ilim River 57N, 104E	Ol	(19)	d, a, t	42N, 311E	R	R/258, 10/139
SB3.4	Ust' Kutian Stage, Upper Lena River 57N, 107E	Ol	(22)	d, a, t, r	41N, 308E	M	R/257, 10/138
SB3.5	Ust' Kutian Stage, Angara River 58N, 97E	Ol	(101)	d, a, r	19N, 300E	M	R/256
SB3.6	Krivaya Luka Stage, River Lena 58N, 106E	Om	(33)	a, r	22N, 312E	M	R/247
SB3.7	Krivaya Luka Stage, Kuliumbe River 68N, 88E	Om	(90)	a	18N, 328E	R	R/248
SB3.8	Krivaya Luka Stage, River Lena 60N, 118E	Om	(20)	d, a, t, r	27N, 314E	M	R/246, 10/137
SB3.9	Krivaya Luka Stage, Upper River Lena 58N, 108E	Om	(31)	d, a, t, r	23N, 308E	M	R/245, 10/136
SB3.10	Mangazeyan Stage, River Lena 60N, 118E	Om	(27)	d, a, t, r	22N, 314E	M	R/244, 10/135
SB3.11	Mangazeyan Stage, Upper River Lena 58N, 108E	Om	(20)	d, a, t, r	23N, 308E	M	R/243, 10/134
SB3.12	Dolbor Stage, Lena River 60N, 118E	Ou	(49)	d, a, t, r	21N, 307E	M	R/242, 10/133
SB3.13	Mararovo Suite, Dolbor Stage 58N, 108E	Ou	(78)	d, a, t, r	25N, 308E	M	R/241, 10/132
SB3.14	Dolbor Stage, Bratsk Suite 57N, 103E	Ou	(133)	d, a, t	20N, 300E	N	R/240, 10/131
SB3.15	Dolbor Stage, River Niuya 61N, 116E	Om–u	(20)	r	19N, 292E	M	R/239
S4.1	Llandovery Stage, River Lena 61N, 116E	Sl	(29)	a	2N, 98E	N	R/238
S4.2	River Kureika‡ 67N, 88E	Sl	(16)	a	41N, 210E	N	R/236
S4.3	River Kuliumbe 68N, 88E	Sl	(53)	a, t, r	23N, 152E	M	R/235
S4.4	Ludlow Stage, River Kuliumbe 68N, 88E	Su	(16)	a	34N, 132E	R	R/234
S4.5	River Kureika‡ 67N, 88E	Su	(21)	a	34N, 212E	N	R/233
S4.6	Chergak Suite, Tuva 52N, 94E	Su	(42)	a, d	13N, 132E	N	R/232
S4.7	Porphyrites, Rivers Is and Vyya‡ 59N, 60E	Su–Dl	(241)	a, d	8N, 154E	N	R/231
S5.1	Kureika and Zubovian Suites, R. Kuliumbe 68N, 88E	Dl	(38)	a, d	29N, 156E	R	R/225
S5.2	B'yerdar Suite, Tuva 52N, 94E	Dl	(23)	a, d	4N, 336E	N	R/223
S5.3	Byskarian Series, Rybinsk Depression 55N, 93E	Dl–m	(500)	a, r	31N, 154E	M	R/222, 10/127
S5.4	River Kuliumbe 68N, 88E	Dm	(37)	a, t, r	40N, 147E	M	R/219
S5.5	River Kureika‡ 67N, 88E	Dm	(68)	a, t, r	38N, 203E	M	R/218
S5.6	River Kureika‡ 67N, 88E	Dm	(22)	a	50N, 178E	N	R/217
S5.7	Central Kazakstan‡ 48N, 74E	Dm	(22)	a, d	43N, 200E	R	R/216

Pole no.	Rock unit	Age (My)	Si (Sa)	Stab.	Pole Posn	P	GJ
SB5.8	Rybinsk Depression 53N, 95E	Dm	(162)	*a, d, r*	32N, 162E	M	R/215, 10/128
SB5.9	Red clays, Krasnoyarsk 56N, 93E	Dm	(18)	*a*	19N, 136E	R	R/214
SB5.10	Red bauxites, North Urals‡ 60N, 60E	Dm	(144)	*a, d*	31N, 182E	R	R/221
SB5.11	Red aleurolites, Krasnoyarsk 56N, 93E	Dm–u	(28)	*a*	23N, 150E	R	R/213
SB5.12	Redbeds, North Kazakhstan‡ 52N, 68E	Dm–u	(78)	*a*	34N, 101E	R	R/212
SB5.13	Kokhai, Tuba and Oidanovo Suites 55N, 90E	Du	(16)	*d*	8N, 147E	R	R/211
SB5.14	Rybinsk Depression redbeds 56N, 94E	Du	(204)	*a, d, r*	24N, 151E	M	R/208
SB5.15	Limestones, North Kazakhstan‡ 50N, 66E	Du	(8)	*a*	56N, 193E	R	R/202
SB6.1	Redbeds, North Kazakhstan 52N, 68E	Cl	(117)	*a*	48N, 114E	R	R/201
SB6.2	Minusinsk Depression 54N, 91E	Cl	(107)	*a, d, r*	34N, 132E	M	R/200, 10/123
SB6.3	River Kuliumbe 66N, 89E	Cl	(14)	*a*	39N, 146E	R	R/197, 9/114
SB6.4	Altai and Bystrianskaya Suites, Minusinsk 54N, 92E	Cl	(38)	*d*	25N, 134E	R	R/193a, 10/124
SB6.5	Minusinsk Depression 54N, 92E	Cl	(19)	*d*	7N, 158E	R	R/193, 10/122
SB6.6	Tumashe Suite, River Angara 58N, 103E	Cl?	(19)	*a, d*	53N, 121E	R	R/192
SB6.7	Red Sandstone, River Miass, Urals‡ 55.3N, 61.5E	Cl	(19)	*a, d*	45N, 199E	R	R/187, 9/110
SB6.8	River Bagariak, Urals‡ 56.2N, 61.8E	Cl	(20)	*a, d*	28N, 229E	R	R/186, 9/109
SB6.9	Valer'yanoskaya, Suite, River Tobol, Urals‡ 52.6N, 62.5E	Cl	(23)	*a, d*	16N, 184E	R	R/185
SB6.10	Ankanzh and Aktal Suites, Tuva 51N, 94E	Cm	(41)	*d*	3N, 160E	R	R/184, 10/118
SB6.11	Vladimirovskaya Suite, North Kazakhstan‡ 50.5N, 68E	Cm	(28)	*a*	42N, 160E	R	R/180
SB6.12	Vladimirovskaya Suite, North Kazakhstan‡ 52N, 68E	Cm	(23)	*a*	55N, 162E	R	R/179, 9/99
SB6.13	Kata River Suite, Angara and Tunguska Rivers 59N, 105E	Cm–u	(61)	*a, d, r*	47N, 125E	R	R/178
SB6.14	Moscovian Stage, River Bagariak 56.4N, 61.9	Cm	(44)	*a, d*	22N, 160E	R	R/161, 9/100
SB6.15	Upper Tomsk Suite, Kuznets Basin 55N, 88E	Cm	(18)	*d*	13N, 146E	R	R/160
SB6.16	Ostrog Suite, Kuznets Basin 55N, 88E	Cm–u	(14)	*d*	1N, 319E	R	R/159, 10/117
SB6.17	Lower Balakhonika Suite, Kuznets Basin 54N, 88E	Cm–u	(30)	*d*	20N, 163E	R	R/158, 10/116
SB6.18	Red clays, River Tobol, Urals‡ 52.5N, 62.5E	Cu	(79)	*a, d*	59N, 198E	R	R/150, 9/96
SB6.19	Upper Balakhonika Suite, Kuznets Basin 54N, 88E	Cu–Pl	(16)	*d*	8N, 287E	N	R/147, 10/103
SB7.1	Barren measures, Minusinsk Depression 53N, 92E	P	(18)	*d, r*	37N, 155E	M	9/85, R/146
SB7.2	Belyi Yar Suite, Yenesei-Abakan trough 53N, 91E	P	(62)	*a, d, r*	47N, 151E	M	R/145
SB7.3	Kuznets Suite, Kuznets Basin 54N, 87E	Pl	(35)	*d*	3N, 136E	R	R/137

Pole no.	Rock unit	Age (My)	Si (Sa)	Stab.	Pole Posn	P	GJ
sb7.4	Il'yinskian Suite, Kuznets Basin 54N, 87E	P	(40)	d	1N, 288E	N	{R/136 {10/99
sb7.5	Yerunakovo Suite, Kuznets Basin 54N, 87E	Pu	(10)	d	21N, 136E	N	{R/135 {10/100
sb7.6	Dolerites, River Kuliumbe 64N, 88E	P–Tr	(50)	a, r	26N, 146E	M	R/120
sb7.7	East Taimyr Sandstones‡ 76N, 111E	Pu–Trl	(60)	a	19N, 139E	R	R8/78
sb8.1	Tuffogenic Suite, River Tunguska 63N, 107E	Trl	(79)	a, d	52N, 125E	N	R/119
sb8.2	Tuffogenic Suite, River Angara 59N, 103E	Trl	(65)	a, d	53N, 115E	N	R/118
sb8.3	Intrusives, River Viliui 63N, 112E	Trl	(25)	?	54N, 145E	N	R/117
sb8.4	Tuffs, River Viliui 64N, 112E	Trl	(11)	d	52N, 162E	N	R/116
sb8.5	Dykes, River Viliui 64N, 112E	Trl	(114)	d	61N, 142E	N	R/115
sb8.6	Basalts, Central Kazakhstan‡ 48N, 80E	Trl	(21)	a	25N, 135E	R	R/113
sb8.7	East Taimyr Sandstones‡ 75N, 108E	Trl	(22)	d	41N, 168E	R	R/104
sb8.8	East Taimyr Sandstones‡ 76N, 111E	Trl	(160)	a, f	9N, 129E	R	8/76
sb8.9	Khannamakit Suite, Tungus Depression 68N, 91E	Trl?	(36)	d	43N, 153E	N	R/103
sb8.10	Ultrabasic intrusions, River Kotui 71N, 103E	Trl?	(194)	d	33N, 141E	R	R/102
sb8.11	Dykes, River Maimecha 71N, 101E	Trl?	(58)	d	30N, 150E	R	R/101
sb8.12	Ayan Suite basalts, Tungus Depression 70N, 96E	Trl	(120)	d	53N, 146E	N	R/100
sb8.13	Tuffs, River Maimecha 71N, 101E	Trl	(35)	d	39N, 180E	R	R/99
sb8.14	River Maimecha 71N, 101E	Trl	(230)	d	45N, 183E	R	R/98
sb8.15	Basalts, River Maimecha 71N, 101E	Trl	(136)	d, r	36N, 158E	M	R/97
sb8.16	Kogotok Suite, River Maimecha 71N, 101E	Trl	(92)	d, r	41N, 161E	M	R/96
sb8.17	Tuffaceous Suite, River Pyshma 58N, 62E	Trl–m	(63)	a, t, r, d	52N, 150E	M	R/95
sb8.18	Basalts and dolerites, River Bichur 58N, 62E	Trl–m	(450)	a, t, d	42N, 153E	R	R/94
sb8.19	Basalts and dolerites, River Sinara 57N, 62E	Trl–m	(198)	a, t, d	48N, 136E	N	R/93
sb8.20	East Taimyr Sandstones‡ 76E, 111E	Trm	(98)	a	4N, 303E	R	8/70
sb9.1	Petropavlosk Suite, River Dzhida 51N, 105E	Jl–m	(53)	d, r	77N, 132E	M	R/86
sb9.2	Lower coal bed, Kuznets Basin 54N, 88E	Jl–m	(10)	a, t, d	73N, 214E	N	12/89
sb9.3	Upper coal bed, Kuznets Basin 54N, 88E	Jl–m	(10)	a, t, d	52N, 244E	R	12/90
sb9.4	East Taimyr Sandstones‡ 76N, 111E	Jm–u	(—)	a	22N, 162E	R	8/55
b10.1	Sediments Anabar Gulf 73N, 113E	Kl	(—)	a, t, d, r	63N, 174E	M	†
b10.2	East Ferghana Cretaceous 41N, 73E	Kl (–u)	(50)	a, d	65N, 178E	N	{R/74 {10/53
b10.3	East Ferghana Cretaceous 41N, 73E	Kl (–u)	(41)	a, d	65N, 178E	N	{R/74 {10/53
b10.4	East Ferghana Cretaceous 41N, 73E	Ku	(56)	a, d	65N, 176E	N	{R/74 {10/53
b10.5	East Ferghana Cretaceous 41N, 73E	Ku	(52)	d	75N, 332E	N	{R/73 {10/53
b10.6	Ileskaya and Simonovaya formations 56N, 92E	Ku	(93)	a	74N, 18E	N	8/45
b11.1	Liulinov Suite 67N, 74E	Te–o	(14)	a	57N, 152E	N	R/67

Pole no.	Rock unit	Age (My)	Si (Sa)	Stab.	Pole Posn	P	GJ
SB12.1	Kompasskii Bor 60N, 83E	Tm	(15)	*t, d*	68N, 223E	R	R/64
SB12.2	Plateau basalts, Tunka Depression 53N, 103E	Tm–p	(75)	*a, d*	64N, 245E	R	R/55
SB13.1	Clays, Lower Tunguska River 66N, 88E	Qp–r	(21)	*a*	84N, 227E	N	R/1

† Pospelova *et al.* (1968).
‡ Sites located within the Urals, Taimyr Peninsula, Kazakhstan and River Kureika sections.

VERKHOYANSK–SIKHOTE ALIN

Pole no.	Rock unit	Age (My)	Si (Sa)	Stab.	Pole Posn	P	GJ
VS6.1	Umkuveyen Volcanics and sediments† 62.5N, 166E	Du–Pl	4 (17)	*r*, cons	39N, *151*E	M	12/120, 130–2
VS7.1	Yuzagol Suite, Vladivostok 43N, 132E	Pl	(48)	*a, d*	19N, 200E	N	{R/139, 10/102}
VS7.2	Kazulinian Suite, Vladivostok 43N, 132E	Pu	(29)	*a*	18N, 196E	R	{R/138, 10/102}
VS7.3	Khivach sediments 63N, 159.3E	Pu	(12)	*a*	52N, 279E	N	12/108
VS7.4	Nerskaya Suite, West Verkhoyan Region† 66N, 128E	Pu	(18)	*a*	*15*N, 74E	N	R/130
VS8.1	Lower Kelter Suite, West Verkhoyan Region† 64N, 130E	Trl	(15)	*a*	49N, *141*E	N	R/114
VS8.2	Khivach Limestones and argillites 63N, 159.3E	Tr	(20)	*a, r*	59N, 234E	M	12/104
VS8.3	Khivach argillites 63N, 159.3E	Tr	(15)	*a*	52N, 258E	N	12/103
VS8.4	Finish sediments 65.3N, 159.1E	Tru	(42)	*a, r*	76N, 184E	M	12/99
VS8.5	Khivach argilites 63N, 159.3E	Tru	(22)	*a, r*	58N, 235E	M	12/98
VS8.6	Finish sediments 65.3N, 159.1E	Tru	(71)	*a, r*	64N, 240E	M	12/97
VS8.7	Sandstones, Vladivostok 43N, 132E	Trm	(19)	*a*	54N, 186E	N	{R/90, 10/92}
VS8.8	Begidzhan Suite, West Verkhoyan Region† 64N, 130E	Trm–u	(21)	*a*	*32*N, *151*E	N	R/88
VS9.1	Sandstones, West Verkhoyan Region† 66N, 125E	Jl	(36)	*a, r*	62N, 45E	M	R/87
VS9.2	Finish sediments 65.3N, 159.1E	Jl	(26)	*a, r*	80N, 305E	M	12/92
VS9.3	Start sediments 64.0N, 158.3E	Jl	(70)	*a, r*	79N, 202E	M	12/91
VS9.4	Khivach andesites and met. sediments 63.0N, 159.3E	Jm	(12)	*a*	64N, 103E	N	12/88
VS9.5	Viliga sediments 61.5N, 156.0E	Jm	(12)	*a, t, r*	75N, 150E	M	12/87
VS9.6	Start sediments 64.0N, 158.3E	Jm	(24)	*a, r*	40N, 140E	M	12/86
VS9.7	Viliga sediments 61.5N, 156.0E	Jm	(81)	*a, r*	62N, 122E	M	12/85
VS9.8	Basalts and tuffs 66.5N, 166.3E	Ju–Kl	(13)	*a, r, c*	70N, 195E	M	12/81
VS10.1	Suchan Suite, Vladivostok 43N, 132E	Kl	(25)	*a, r, d*	58N, 146E	M	{R/77, 10/65}
VS10.2	Pezhenka met. sediments 66.8N, 166.3E	Kl	(—)	*a*	58N, 140E	N	12/77
VS10.3	Viliga met. sediments 61.8N, 156.0E	Kl	(20)	*t*	60N, 150E	N	12/76
VS10.4	Viliga met. sediments and dykes, East wing 61.8N, 156.0E	Kl–m	(—)	*a, t*	70N, 171E	N	12/68
VS10.5	Viliga met. sediments and dykes, West wing 61.8N, 156.0E	Kl–m	(—)	*a, t*	46N, 186E	N	12/69
VS10.6	Viliga and Khivach met. sediments and dykes 62N, 158E	Kl–u	2 (8)	*a, t*	56N, 176E	N	12/66, 70
VS10.7	Umkuveyen met. seds 65.3N, 166.0E	Ku	6 (—)	*a*	66N, 166E	N	12/55

Pole no.	Rock unit	Age (My)	Si (Sa)	Stab.	Pole Posn	P	GJ
VS12.1	Khabarovsk andesites and basalts 49N, 140E	Tp	(75)	a, r	83N, 246E	M	R/45
VS12.2	Andesites and basalts, Maritime region 43N, 131R	Tp	(102)	a, d	70N, 30E	R	R/46
VS12.3	Andesites and basalts, Maritime region 43N, 131E	Tp	(220)	a, d	75N, 182E	N	R/47
VS12.4	Andesites and basalts, Maritime region 43N, 131E	Tp	(344)	a, d	84N, 22E	R	R/48
VS12.5	Andesites and basalts, Maritime region 43N, 131E	Tp	(216)	a, d	84N, 282E	N	R/49

† Sites located in the Verkhoyansk Mountains, Umkeveyen depression etc.

CHINA AND KOREA

Pole no.	Rock unit	Age (My)	Si (Sa)	Stab.	Pole Posn	P	GJ
CH1.1	Xuiung and Lianto Series, China 30N, 118E	Pc (Lr Sn.)	4 (—)	a, cons	40N, 96E	—	10/162–5
CH2.1	Sandstones, North Korea 39N, 126E	Є1–m	2 (23)	d, a, t	11N, 317E	N	10/146–7
CH9.1	Xichuan redbeds, China 25–29N, 102–105E	J	5 (—)	cons	55N, 149E	N	10/72–6
CH10.1	Canton, Yunnan & Hubei redbeds, China 23–29N, 102–114E	K	7 (—)	cons	61N, 161E	N	10/51–2 10/57–61
CH10.2	Sandstones, North Korea (26, +67) 39N, 126E	Ku	(8)	d, a, t	69N, 182E	N	10/55
CH10.3	Igneous rocks, South Korea (20, +53) 36N, 129E	K	5 (35)	a	73N, 220E	N	8/49
CH11.1	Hunnan redbeds, China 27N, 112.7E	Te–m	4 (—)	d	75N, 38E	N	10/36–8 10/44
CH13.1	Basalts, South Korea 36–38N, 126–129E	Tp–Qp	5 (35)	r	66N, 237E	M	10/16

Note. Numbers in brackets after Cretaceous rock units from Korea are the declination and inclination (D_m, I_m).

JAPAN–SAKHALIN–KAMCHATKA

Pole no.	Rock unit	Age (My)	Si (Sa)	Stab.	Pole Posn	P	GJ
JA10.1	Cretaceous intrusives, SW Japan (30, +47) 35N, 134E	K	7 (63)	cons	63N, 230E	N	5/13
JA10.2	Granitic intrusives NE Japan (321, +55)	Kl (101–119)	9 (74)	cons	59N, 62E	N	†(1)
JA10.3	Granitic intrusives NE Japan (51, +53)	Ku (84)	1 (10)	?	49N, 219E	N	†(1)
JA10.4	Penzhinskaya Bay, Kamchatka (61, +75) 61.7N, 164.0E	Kl–u	(10)	t	61N, 225E	N	12/65
JA10.5	Sakhalin Island sediments (338, +68) 47.3N, 142.4E	Ku	(—)	a, r	75N, 78E	M	12/53
JA10.6	Quartz diorite, SW Japan (239, −61) 35N, 134.5E	K–T	6 (62)	a, c, r	44N, 194E	M	8/35
JA11.1	Cenozoic rocks, SW Japan (9, +50) 35N, 134E	T	8 (182)	r, cons	81N, 251E	M	5/11
JA11.2	Cenozoic rocks, NE Japan (10, +50) 37N, 140E	T	5 (71)	cons	80N, 264E	N	5/12
JA11.3	Volcanics, SW Japan (33, +53) 35N, 134E	Te–o	10 (158)	a, t, r	62N, 221E	M	10/40

Appendix

Pole no.	Rock unit	Age (My)	Si (Sa)	Stab.	Pole Posn	P	GJ
JA12.1	Igneous rocks, Japan	Tm	7 (52)	*r*	73N, 216E	M	2/9–10
JA12.2	Volcanics, Japan	Tm–p	7 (—)	*r*	79N, 13E	M	†(2)
JA12.3	Intrusives of Fossa Magna, Japan 36.3N, 138.6E	Tm	12 (63)	*a*	85N, 351E	N	7/13
JA12.4	Kita-Matsuura basalts, Japan 33.3N, 129.8E	Tm–p (7–11)	13 (65)	*a, r*	89N, 131E	M	10/22
JA12.5	Kyushu Volcanics, Japan 33N, 130E	Tm–p	15 (69)	*a, r*	74N, 270E	M	12/32–3
JA12.6	Volcanics, Japan	Tp	4 (22)	*r*	80N, 271E	M	2/7–8
JA12.7	Enrei Formation, Japan, 36N, 138E	Tp	27 (—)	*a*	77N, 352E	R	6/27
JA12.8	Sakhalin andesites and basalts 49N, 142E	Tp	(95)	*a, d*	77N, 132E	N	R/43
JA12.9	Sakhalin andesites and basalts 49N, 142E	Tp	(107)	*a, d*	82N, 232E	R	R/44
JA12.10	Kamchatka andesites and basalts 56N, 159E	Tp	(42)	*a, d*	80N, 329E	R	R/39
JA12.11	Kamchatka andesites and basalts 56N, 159E	Tp	(33)	*a, d*	84N, 309E	N	R/40
JA12.12	Kamchatka andesites and basalts 56N, 159E	Tp	(63)	*a, d*	73N, 359E	R	R/41
JA12.13	Kamchatka andesites and basalts 56N, 159E	Tp	(32)	*a, d*	84N, 17E	N	R/42
JA13.1	Japan and East Asia, igneous rocks	Tp–Qr	36 (39)	*r*	80N, 174E	M	4/2
JA13.2	Andesites, Japan 36.4N, 138.7E	Qp (1.0–1.4)	11 (57)	*a, r*	46N, 203E	M	10/15
JA13.3	Usami Volcano, Japan 35N, 139E	Qp	11 (72)	*a, r*	71N, 47E	M	10/12
JA13.4	Yamaguchi basalt, Japan 34.5N, 131.5E	Qp	85 (300)	*a, r*	87N, 132E	M	3/8
JA13.5	North Izu znd Hakone rocks, Japan 35N, 139E	Qp–r	42 (300)	*a, g, r*	78N, 46E	M	1/7–15
JA13.6	Kamchatka andesites and basalts 53N, 158E	Qr	(320)	*a, d*	84N, 159E	N	R/2

†(1) Kawai *et al.* (1969).　　†(2) Kumagai *et al.* (1950).

Note. Numbers in brackets after some rock units are the mean declination and inclination (D_m, I_m).

AFRICA

AF1.1	Modipe gabbro, Botswana 24.7S, 26.2E	Pc (2630)	10 (36)	*a, c, t, r*	33N, 211E	M	8/157
AF1.2	Great Dyke, Rhodesia 18.5S, 30.3E	Pc (2520)	9 (90)	*a*	21N, 62E	—	7/62
AF1.3	Great Dyke, Rhodesia 18.5S, 30.3E	Pc (2520)	6 (10)	*a*	11N, 69E	—	7/63
AF1.4	Gaberones granite 25S, 25.6E	Pc (2340)	7 (19)	*a, c, r*	34N, 103E	M	9/158
AF1.5	Ventersdorp lavas 28.7S, 24.8E	Pc (2200–2300)	2 (13)	*a, t*	55N, 175E	—	9/157
AF1.6	Vredefort Ring dykes 27S, 27E	Pc (1970)	4 (12)	*a*	22N, 27E	—	12/163
AF1.7	Bushveld gabbro 25.5S, 28E	Pc (1950)	5 (99)	*f*	23N, 36E	—	1/142
AF1.8	Mashonaland dolerites 18.5S, 31.5E	Pc (1880)	14 (121)	*a, r*	7N, 340E	M	8/151
AF1.9	Premier Mine Kimberlite 25.7S, 28.5E	Pc (1750)	(16)	*a, t*	51N, 38E	—	10/197
AF1.10	Van Dyke Mine dolerite dyke 26.2S, 28.2E	Pc (1650)	(8)	*a*	12N, 14E	—	8/156
AF1.11	Waterberg redbeds 22.9 – 25.8S, 28.4 – 29.9E	Pc (1950–1750)	12 (60)	*a, t, f, r*	†(1)	M	9/148–52
AF1.12	Waterberg diabases 23–27S, 26–30E	Pc (1750)	13 (90)	*a, c, r*	65N, 50E	M	8/155
AF1.13	Umkondo lavas and dolerites 20S, 33E	Pc (1750)	18 (98)	*a*	64N, 26E	—	8/154
AF1.14	Pilansberg dykes 26S, 28E	Pc (1300)	5 (169)	*a*	8N, 43E	—	1/141
AF2.1	Ntonya Ring structure 15.5S, 35.3E	(Pc–) €l (600)	7 (27)	*a*	28N, 345E	R	9/137
AF2.2	Sijarira Group 17.5S, 28.5E	(Pc–) €?	9 (40)	*t, g*	2N, 352E	R	12/149

302

Pole no.	Rock unit	Age (My)	Si (Sa)	Stab.	Pole Posn	P	GJ
AF3.1	Hook intrusives 15S, 27E	(€u–) Ol (500)	3 (10)	*a*	14N, 336E	R	9/132
AF3.2	Lower T.M.S. shales 34S, 18E	O	(8)	*a*	50N, 349E	N	4/32
AF6.1	Dwyka Varves 8.8 – 17.5S, 28.2 E – 32.9E	Cl	5 (29)	*t, f, r*	26N, 206E	M	9/117
AF6.2	K3 Galula redbeds 8.8S, 32.9E	Cu (–Pl)	5 (34)	*a, t*	46N, 220E	R	8/92
AF7.1	K3 Songwe–Ketewaka redbeds 10S, 34.5E	Pl	4 (27)	*a, t*	27N, 269E	R	8/91
AF8.1	Upper Beaufort sediments 30S, 28E	Trl	4 (19)	*a*	67N, 267E	N	8/73
AF8.2	Shawa ijolite 19.2S, 31.7E	Tru (197)	5 (13)	*a*	64N, 266E	N	8/72
AF8.3	Red sandstone formation 16.2S. 28.8E	Tru	6 (32)	*a, t*	68N, 230E	N	8/67
AF8.4	Marangudzi complex 22.1S, 30.7E	Tru (–Jl) (190)	8 (68)	*a, r*	70N, 285E	M	10/77
AF9.1	Karroo dolerites	Jl–m (154–190)	10 (67)	*a, c, r*	66N, 255E	M	8/59
AF9.2	Stormberg lavas 29.5S, 28.5E	Jl	4 (74)	*a, c, r*	71N, 269E	M	6/40–43
AF9.3	Karroo lavas, Central Africa 16–20S, 26–34E	Jl	9 (—)	*a, r*	57N, 264E	M	12/93
AF9.4	Mateke Hills complexes 21.8S, 31.2E	Jm (168)	6 (36)	*a, r*	58N, 260E	R	8/63
AF10.1	Mlange Massif 16S, 35.6E	Kl (128–116)	(8)	*a*	60N, 262E	N	9/40
AF10.2	Lupata volcanics 16.7S, 34.2E	Kl (110–106)	7 (61)	*a*	62N, 259E	N	7/21
AF11.1	Ethiopian traps 9N, 39E	Te–o †(2) (45±15)	20 (52)	*a, r*	81N, 168E	M	12/46
AF11.2	Ethiopian traps 10.1S, 38.3E	Te–o †(2) (45±15)	11 (21)	*t*	87N, 253E	N	8/36
AF12.1	Turkana lavas 4N, 35E	To–m (32–12)	62 (109)	*a, t, r*	85N, 163E	M	12/43
AF12.2	Rift Valley lavas 0–4S, 34–38E	Tm (14–7)	43 (119)	*a, t, r*	85N, 140E	M	12/31
AF12.3	Massif de Cavallo, Algeria 32N, 5E	Tm	13 (51)	*a*	88N, 144E	N	11/19
AF12.4	Rift Valley lavas 0–4S, 34–38E	Tp (7 – 2.2)	60 (142)	*a, t, r*	88N, 153E	M	12/27
AF13.1	Rift Valley lavas 0–4S, 34–38E	Qp–r (2.2 – 0)	60 (156)	*a, t, r*	88N, 144E	M	12/20

†(1) Polar-wandering recorded through the sequence of sediments.
†(2) More recent dates suggest these traps could be as young as Miocene.

	ARABIA						
AR2.1	Redbeds, Jordan 29.7N, 35.3E	€ (–O)	(20)	*a, t, **h**, r*	37N, 323E	M	12/147
AR7.1	Red sandstones, Turkey 41.6N, 32.5E	P	3 (16)	*a, f*	18N, 282E	R	8/84
AR9.1	Bayburt Volcanics and sediments, Turkey 40.4N, 39.9E	J	(8)	*a*	39N, 265E	R	10/71
AR9.2	Basalts and tuffs, Lebanon‡ 34N, 36E	Ju	(20)	*a*	1N, 120E	R	9/44
AR10.1	Volcanics and sediments, Lebanon 34N, 36E	Kl	5 (15)	*a*	38N, 282E	N	9/41
AR10.2	Volcanics, Israel 31.0N, 35E	Kl	(22)	*a*	53N, 265E	N	12/71
AR10.3	Niksar basalts, Turkey 40.7N, 37E	K	(9)	*a, f*	51N, 296E	R	10/62
AR10.4	Gumushane Volcanics and sediments, Turkey 40.5N, 39.3E	Km–u	6 (31)	*a, f*	61N, 276E	R	10/56
AR10.5	Volcanics, Israel 32.5N, 35E	Ku	(15)	*a*	42N, 264E	N	12/67

Appendix

Pole no.	Rock unit	Age (My)	Si (Sa)	Stab.	Pole Posn	P	GJ
AR11.1	Gumushane Volcanics and sediments, Turkey 40.8N, 38.9E	Ku–Te	6 (23)	*a, f*	69N, 261E	N	10/49
AR11.2	Sumgait River Suite, East Azerbaidzhan 40N, 49E	Tpa	(15)	*r*	67N, 234E	M	R/71
AR11.3	Tunceli Group, Turkey 39N, 39.5E	Te	4 (16)	*a, f, r*	65N, 294E	M	10/41
AR12.1	Jebel Khariz Volcanics, Aden 13N, 45E	Tm (10)	14 (47)	*a, r*	77N, 309E	M	9/14
AR12.2	Little Aden Volcanics, Aden 13N, 45E	Tp (5–6)	5 (16)	*a*	78N, 147E	R	9/15
AR12.3	Aden Volcanics 13N, 45E	Tp (5)	11 (—)	*a, r*	85N, 298E	M	9/17
AR12.4	Dolerites, South Georgia 41N, 44E	Tp	(53)	a, d	81N, 239E	R	R/50
AR12.5	Balakhany Suite, East Azerbaidzhan 40N, 49E	Tp	(16)	a, d	80N, 212E	R	R/37–8
AR12.6	Surakhany Suite, East Azerbaidzhan 40N, 49E	Tp	(10)	d	78N, 229E	N	R/31
AR12.7	Surakhany Suite, East Azerbaidzhan 40N, 49E	Tp	(53)	r, d	74N, 191E	M	R/32
AR12.8	Surakhany Suite, East Azerbaidzhan 40N, 49E	Tp	(42)	d	73N, 248E	R	R/33
AR12.9	Surakhany Suite, East Azerbaidzhan 40N, 49E	Tp	(10)	d	77N, 177E	R	R/35
AR12.10	Surakhany Suite, East Azerbaidzhan 40N, 49E	Tp	(10)	a, d	77N, 182E	N	R/36
AR13.1	Volcanics, Turkey 38.5N, 29–48E	Qp–r (1.5 – 0.4)	22 (145)	*a, t, r*	87N, 162E	M	10/11
AR13.2	Igneous rocks, South Georgia 41N, 44E	Q	(24)	a, d	86N, 174E	N	R/11

‡ Site located in the region of the Dead Sea fault system.

SOUTH AMERICA

SA1.1	Roraima dolerites (I) 6N, 61W	Pc (2070)	7 (34)	*a*	63N, 231E	—	10/160
SA1.2	Roraima dolerites (II) 6N, 61W	Pc (2070)	7 (35)	*a*	45N, 167E	—	10/161
SA2.1	Salta and Jujuy redbeds, Argentina 23S, 66W	€– (O)	(18)	*t*	12N, 329E	R	12/146
SA3.1	Jujuy redbeds, Argentina 24S, 65W	O	(26)	*t*	11N, 333E	R	12/141
SA3.2	Sediments, Bolivia 18S, 65W	O	(18)	*t, f, r*	4N, 302E	M	12/140
SA3.3	Urucum formation, Bolivia 19S, 58W	O (–S)	(10)	*t*	17N, 347E	R	†
SA5.1	Sediments, Bolivia 17S, 67W	Dl (O?)	(26)	*t*	7N, 307E	R	12/136
SA5.2	Picos and Passagem Series, Brazil 7S, 41.5W	Dl	(12)	*t, r*	30N, 133E	M	12/135
SA6.1	Taiguati formation, Bolivia 17.5S, 65W	C	2 (8)	d, t, h, r	45N, 160E	M	12/127
SA6.2	Tupambi formation, Bolivia 17.5S, 65W	C	(16)	f	85N, 315E	N	12/129
SA6.3	Piaui formation, Brazil 5S, 43W	Cu	(15)	t	50N, 165E	R	12/124
SA6.4	Pipiral formation, Colombia, 7.5N, 73W	Cu (–P)	(11)	f, r	52N, 182E	M	12/123
SA7.1	Paganzo II (Huaco), Argentina 30S, 68W	P–C	(9)	*t*	68N, 174E	R	12/117
SA7.2	Paganzo II (Los Colorados), Argentine 29.5S, 67W	P–C	(26)	*t*	60N, 178E	R	12/116
SA7.3	La Colina basalt, Argentina 30S, 67W	Pl	2(8)	*a, r*	59N, 183E	M	12/114

Pole no.	Rock unit	Age (My)	Si (Sa)	Stab.	Pole Posn	P	GJ
SA7.4	Paganzo II (Los Colorados) Argentina 29.5S, 67W	P (u)	(23)	*t*	74N, 128E	R	12/109
SA7.5	Sierra de la Ventana, Argentina 38S, 62W	Pu	(9)	*t, f*	78N, 39E	R	11/61
SA8.1	Paganzo III (La Rioja) Argentina 29S, 68W	(P–) Tr	(22)	*t*	77N, 355E	N	11/60
SA8.2	Mendoza lavas, Nihuil, Argentina 33S, 68W	Trl	(41)	*a, r*	81N, 102E	M	11/56
SA8.3	Giron formation, Colombia 8N, 73W	Tr	(11)	*t*	77N, 74E	N	12/102
SA8.4	Mendoza lavas, Cuesta los Teneros 33S, 68W	Tr	(91)	*a, r*	80N, 48E	M	11/49
SA8.5	Cacheuta Group basalts, Argentina 32.5S, 69.1W	Tru	(30)	*a, r*	74N, 86E	M	11/46
SA9.1	Chon Aike lavas, Argentina 48S, 66W	Jm (162)	(66)	*a, r*	84N, 256E	M	11/41
SA10.1	Serra Geral formation 30S, 55W	Kl (120–135)	30 (80)	*a, r*	78N, 236E	M	6/35
SA10.2	Sediments and volcanics, Peru 11S, 76W	K	(12)	*r*	63N, 210E	M	12/60
SA11.1	Boqueron sediments, Peru 9S, 76W	T	(9)	*t, f*	75N, 106E	R	12/51
SA11.2	Minor dyke suite, Guyana 6N, 61W	T	4 (20)	*a*	73N, 11E	N	10/48
SA12.1	Cainozoic basalts, Argentina 34–40S, 69–71W	Tm–Qp (27–1)	18 (158)	*a, r*	87N, 132E	M	11/15

† Creer (1967*b*).

AUSTRALIA

Pole no.	Rock unit	Age (My)	Si (Sa)	Stab.	Pole Posn	P	GJ
AU1.1	Widgiemooltha dyke suite 32S, 122E	Pc (2420)	11 (40)	*a, c, f, r*	8N, 337E	M	10/200
AU1.2	Hart dolerite 17.5S, 127E	Pc (1800)	12 (36)	*a, c, r*	29N, 46E	M	12/162
AU1.3	Edith River Volcanics 13S, 132E	Pc (1760)	10 (10)	*f*	6N, 346E	—	1/145
AU2.1	Antrim Plateau Volcanics 13.6 – 18.1S, 128.2 – 131.8E	℮	14 (52)	*a, t, r*	9N, 160E	M	12/148
AU2.2	Hudson formation 17S, 128E	℮m	11 (17)	*a, t*	18N, 19E	R	†(1)
AU3.1	Jinduckin formation 14.4S, 131.7E	(℮u–) Ol	8 (21)	*a, t, h, r*	13N, 205E	M	12/143
AU4.1	Duoro and Laidlaw Volcanics 35S, 149E	Sm–u	13 (42)	*a, r*	54N, 91E	M	†(1)
AU4.2	Mugga Mugga Porphyry 35.1S, 149.4E	Su (400)	(17)	*a, t*	80N, 160E	N	8/127
AU5.1	Bowning Group Volcanics 35S, 149E	Dl	7 (25)	*a*	64N, 225E	N	†(1)
AU5.2	Ainslie Volcanics 35S, 149E	Dl	7 (24)	*a, t*	71N, 173E	N	†(1)
AU5.3	Lochiel formation 37S, 150E	Du	12 (25)	*a, t, r*	58N, 140E	M	†(1)
AU6.1	Gilmore and Isismurra Volcanics 32S, 151E	Cl	11 (23)	*a, r*	73N, 34E	M	†(1)
AU6.2	Paterson toscanite 32.5S, 151.6E	Cu (298)	4 (12)	*a*	73N, 327E	N	8/108
AU6.3	Seaham formation 32.6S, 151.7E	Cu	5 (30)	*t, f*	53N, 329E	R	8/103
AU6.4	Rocky Creek conglomerate 30S, 150.3E	Cu	8 (32)	*g, t, f*	51N, 318E	R	8/104
AU6.5	Currabubulla formation 31.1S, 150.8E	Cu	5 (21)	*f, t*	42N, 315E	R	8/105
AU7.1	Lower Marine basalt/Moonbi lamprophyre 31S, 151E	P (l)	2 (9)	*a*	40N, 305E	R	{ 7/40, 8/79 }

305

Pole no.	Rock unit	Age (My)	Si (Sa)	Stab.	Pole Posn	P	GJ
AU7.2	Upper Marine latites 34.6s, 150.8E	Pm–u (248)	12 (43)	*a*	44N, 312E	R	7/39
AU7.3	Milton Monzonite 35.3s, 150.5E	Pu (240)	4 (9)	*a*	32N, 350E	R	8/80
AU8.1	Narrabeen shales 33.9s, 150.9E	Trl	4 (32)	*t*	49N, 340E	N	7/34
AU8.2	Brisbane tuff 27.8s, 153E	Trm	6 (12)	*a*	57N, 323E	N	7/33
AU9.1	Gibralter syenite 34.5s, 150.4E	Jl (178)	2 (10)	*a*	41N, 326E	N	5/20
AU9.2	Prospect dolerite 33.8s, 150.8E	Jm (168)	3 (10)	*a*	51N, 331E	N	5/21
AU9.3	Tasmanian dolerite 42s, 147E	Jm (167)	51 (132)	*a, g*	51N, 340E	N	7/26
AU9.4	Gingenbullen dolerite 34.4s, 150.3E	J	(8)	*a*	53N, 324E	R	5/19
AU9.5	Noosa Heads Complex 26.4s, 153.1E	Ju (140)	4 (10)	*a, r*	36N, 312E	M	7/28
AU10.1	Cygnet complex 43.2s, 147.1E	Kl (104)	15 (45)	*a, c*	50N, 338E	N	6/31
AU10.2	Mt Dromedary complex 36s, 150E	Ku (93)	22 (55)	*a, t, c, r*	56N, 318E	M	7/23
AU11.1	Older Volcanics, Victoria 38s, 145.5E	Tpa–e	20 (50)	*a, r*	63N, 320E	M	7/14
AU11.2	Barrington Volcano 32s, 151.4E	Te (52)	33 (—)	*a, r*	70N, 306E	M	11/27
AU11.3	Liverpool Volcano 31.7s, 150.2E	To (34)	36 (—)	*a, r*	71N, 276E	M	11/22
AU12.1	Main Range and Tweed Volcanos 27s, 152E	Tm (21–24)	54 (108)	*a, r*	75N, 292E	M	†(2)
AU12.2	SE Queensland igneous rocks 27s, 152E	Tm (21–24)	12 (33)	*a, r*	78N, 266E	M	8/29
AU12.3	Nandewar Volcano 30.3s, 150.3E	Tm (17)	34 (—)	*a, r*	78N, 264E	M	11/20
AU13.1	Newer Volcanics, Victoria 38s, 143.5E	Tp–Qp (0–4)	46 (133)	*a, r*	87N, 266E	M	†(3)

†(1) Luck (1971).   †(2) Wellman (1971).   †(3) Aziz-ur-Rahman (1971).

		NEW ZEALAND					
NZ12.1	Stoddart formation, South Island 43.7s, 172.7E	Tp (5.85)	44 (60)	*a*	61N, 10E	R	12/28
NZ12.2	Akaroa Volcano, South Island 43.7s, 172.7E	Tp (8.4 – 9.1)	70 (140)	*a, r*	81N, 158E	M	12/29
NZ13.1	Volcanics, North Island 35.5 – 39.3s, 173.6 – 174.1E	Qp–r (0.68)	22 (176)	*a*	82N, 346E	N	11/9

		INDIA					
IN1.1	Gwalior traps 26N, 78E	Pc (1815)	(7)	*a*	19N, 176E	—	7/60
IN1.2	Kaimur Sandstones 24.6N, 86.1E	Pc (1140)	20 (53)	*a, r*	82N, 286E	M	8/150
IN1.3	Malani rhyolites 26N, 73E	Pc (745)	9 (60)	*a, t*	78N, 45E	—	7/61
IN2.1	Purple Sandstone, Salt Range 32.7N, 73E	€l	10 (10)	*t, r*	28N, 212E	M	11/85
IN6.1	Talchir Series 21.4N, 79E	Cu	2 (23)	*a, f*	32N, 134E	R	10/144
IN7.1	Kamthi Beds (Wardha Valley) 20.1N, 79E	Pu	5 (57)	*a, t*	21N, 130E	R	11/64
IN8.1	Mangli Beds 20.5N, 79E	Trl	2 (23)	*a, t*	7N, 124E	R	11/57
IN8.2	Kamthi (Panchet?) Beds (Godavary Valley) 19N, 79.6E	Trl†	3 (52)	*a*	19N, 308E	N	9/66
IN8.3	Parsora Sandstones 23.4N, 81E	Tru	3 (49)	*a, t*	30N, 305E	N	11/45
IN8.4	Pachmarhi Beds 22.4N, 78.4E	Tru (–Jl)	3 (31)	*a, t*	10N, 310E	R	11/43

Pole no.	Rock unit	Age (My)	Si (Sa)	Stab.	Pole Posn	P	GJ
IN10.1	Rajmahal traps 24.5N, 87.5E	Kl (105)	15 (92)	*a, t*	12N, 294E	N	8/60
IN10.2	Rajmahal traps 24.7N, 87.6E	Kl (105)	8 (16)	*a, t*	3N, 298E	N	12/80
IN10.3	Sylhet traps 25N, 91E	Kl (105)?	(25)	*a, t*	16N, 300E	N	7/31
IN10.4	Rajmahendri traps 17N, 81.8E	K	3 (35)	*t*	20N, 306E	N	12/73
IN10.5	Gondwana dykes 23.8N, 85E	K	12 (98)	*a, t*	34N, 299E	N	12/72
IN10.6	Tirupati sandstones 16.8N, 81.2E	Ku†	4 (65)	*a, r*	30N, 287E	M	9/39
IN10.7	Satyavedu sandstones 13.5N, 80E	Ku†	(13)	*a, t, h*	26N, 293E	N	12/59
IN11.1	Lower Deccan traps 18N, 75E	Ku–Tpa (65)	63 (—)	*a, t*	30N, 281E	R	8/43
IN11.2	Upper Deccan traps 18N, 75E	Ku–Tpa (65)	29 (—)	*a, t*	37N, 282E	N	8/42
IN11.3	Deccan traps, Amarkantah 22.7N, 81.3E	Ku–Tpa (65)	14 (75)	*a, t, r*	33N, 294E	M	12/52

† Interpretations of ages follow McDougall & McElhinny (1970).

MADAGASCAR

MD10.1	Lavas and dykes	Ku†(1)	10 (—)	*t*	68N, 192E	N	1/150
MD10.2	Lavas 23.5S, 44.3E	Ku†(1)	2 (12)	*a*	46N, 182E	N	8/47
MD10.3	Massif d'Androy 24.3N, 46E	Ku†(2)	7 (36)	*a*	65N, 252E	N	11/33
MD10.4	Mangoky-Onilahy Volcanics 23S, 44E	Ku†(2)	9 (47)	*a*	77N, 250E	N	11/34

†(1) Turonian.   †(2) Santonian to Campanian.

EAST ANTARCTICA

AN(E)2.1	Charnockites (Mirnyy Station) 67S, 92E	€u (502)	5 (17)	*a*	2N, 208E	—	†
AN(E)3.1	Sør Rondane Intrusives 72S, 24E	Ol (480)	3 (16)	*a*	28N, 190E	R	10/140
AN(E)9.1	Ferrar dolerites (Ferrar glacier) 78S, 161E	Jm (147–163)	5 (57)	*a*	58N, 38E	N	2/27
AN(E)9.2	Ferrar dolerites (Wright and Victoria Valleys) 77.4S, 161.6E	Jm (147–163)	46 (83)	*a, d, c*	45N, 39E	N	6/36
AN(E)9.3	Ferrar dolerites (Beardmore glacier) 84S, 165E	Jm (147–163)	9 (13)	*a, d*	59N, 41E	N	10/70
AN(E)9.4	Dolerite intrusions 79–82S, 25W	J?	7 (8)	*r*	54N, 44E	M	2/26
AN(E)9.5	Dufek Massif 82.5S, 52W	(P) J?	43 (57)	*a, r*	48N, 16E	M	9/63
AN(E)12.1	Cape Hallet lavas 72S, 171E	T (m–p)?	(23)	*a, r*	81N, 94E	M	2/11

† McQueen *et al.* (1972).

WEST ANTARCTICA

AN(W)9.1	Plutons, Jones Mts and Thurston Is.	Tr–J	3 (9)	*a*	82N, 290E	N	†
AN(W)10.1	Dykes, Johes Mts and Thurston Is.	J–K	8 (24)	*a, r*	86N, 344E	M	†
AN(W)10.2	Plutons, Marie Byrd Land	Kl (118–98)	6 (15)	*a*	47N, 300E	N	†
AN(W)10.3	Andean Intrusive Suite	Ku (96)	12 (77)	*a, r*	86N, 178E	M	7/19
AN(W)11.1	Dykes, Marie Byrd Land	T	4 (12)	*a*	62N, 244E	N	†
AN(W)12.1	Volcanics	Tm–Q	29 (76)	*a, r*	82N, 125E	M	†

† Scharnberger (personal communication).

Pole no.	Rock unit	Age (My)	Si (Sa)	Stab.	Pole Posn	P	GJ
		OCEANIC ISLANDS					

(a) North Atlantic

Pole no.	Rock unit	Age (My)	Si (Sa)	Stab.	Pole Posn	P	GJ
OC11.1	Faeroes Volcanics, 62N, 7E	Te (53–59)	253 (1495)	*a, r*	77N, 161E	M	12/50
OC11.2	Iceland Volcanics 66N, 23.5W	Te–m	60 (175)	*a, t, r*	82N, 155E	M	10/39
OC12.1	Iceland Volcanics 65.2N, 20W	Tm	102 (102)	*r*	89N, 5E	M	1/25
OC12.2	Santa Antao, Cape Verde Is. 17.1N, 25.1W	Tm	40 (120)	*a, r*	84N, 168E	M	10/26
OC12.3	Sao Vicente, Cape Verde Is. 16.8N, 25W	Tm	46 (143)	*a, r*	83N, 87E	M	10/27
OC12.4	Sao Nicolau, Cape Verde Is. 16.6N, 24.3W	Tm	12 (36)	*a, r*	87N, 125E	M	10/28
OC12.5	Sal, Cape Verde Is. 16.7N, 22.9W	Tm	5 (16)	*a, r*	67N, 114E	M	10/29
OC12.6	Sao Tiago, Cape Verde Is. 15.1N, 23.6W	Tm	30 (93)	*a, r*	82N, 179E	M	10/31
OC12.7	Maio, Cape Verde Is. 15.2N, 23.2W	Tm	4 (18)	*a, r*	79N, 197E	M	10/30
OC12.8	Hierro Island, Canaries 28N, 18W	Tm–Qr	33 (66)	*a*	84N, 77E	N	8/14
OC12.9	Gomera Island, Canaries 28N, 17.5W	Tm–Qr	18 (36)	*a, r*	78N, 125E	N	8/15
OC12.10	Teneriffe Is., Canaries 28N, 17W	Tm–Qr	46 (92)	*a, r*	83N, 135E	M	8/16
OC12.11	Grand Canaria, Canaries 28N, 16W	Tm–Qr	39 (78)	*a, r*	79N, 109E	M	8/17
OC12.12	Lanzarotte, Canaries 29N, 13.5W	Tm–Qr	38 (76)	*a, r*	76N, 120E	M	8/18
OC12.13	Madeira, Volcanics 32.5N, 17W	≤ Tm	29 (58)	*a, r*	85N, 126E	M	8/19
OC12.14	Lavas, Baked laterites, Iceland 65N, 22W	Tm–p (2–18)	31 (47)	*a, t, c, r*	74N, 000E	M	9/19
OC13.1	Fogo, Cape Verde Is. 15N, 24.4W	Qp–r	(8)	*a*	75N, 162E	N	10/14
OC13.2	Flores Volcanics, Azores 39.4N, 31.2W	Qp–r (0.2–2.2)	24 (—)	*a, r*	88N, 64E	M	10/4
OC13.3	Postglacial lavas, Iceland 64N, 19W	Qr	29 (—)	*a*	88N, 135E	N	†(1)
OC13.4	Lavas, Iceland 65N, 22W	Tp–Qp	59 (59)	*a, r*	83N, 82E	M	1/19–20
OC13.5	Jan Mayen lavas and dykes 71.1N, 8.2W	Qr	10 (10)	*a*	83N, 326E	N	8/3

(b) South Atlantic

Pole no.	Rock unit	Age (My)	Si (Sa)	Stab.	Pole Posn	P	GJ
OC12.15	King George Is., South Shetlands 62S, 58W	To–m	20 (174)	*a, r*	83N, 117E	M	†(2)
OC12.16	Island of 25 May, South Shetlands 62.2S, 58.9W	Tm	(22)	*a, t, r*	86N, 54E	M	12/35
OC12.17	James Ross Is. South Shetlands 64S, 58W	Tm	3 (14)	*a, r*	83N, 294E	M	12/36
OC13.6	Deception Is., South Shetlands 63S, 60.7W	Qp	15 (35)	*a, t*	84N, 350E	N	12/12
OC13.7	South Sandwich Is., 58S, 27W	Qr	3 (14)	*a*	79N, 32E	N	12/3
OC13.8	Tristan da Cunha 37S, 13W	Qr	3 (8)	*a, t*	80N, 342E	N	8/1
OC13.9	Tristan da Cunha and Inaccessible Is., 37S, 13W	Qr	10 (30)	*a, r*	87N, 24E	N	8/2

(c) Indian Ocean

Pole no.	Rock unit	Age (My)	Si (Sa)	Stab.	Pole Posn	P	GJ
OC12.18	Mauritius Volcanics 20.3S, 57.5E	Tp (5.2 – 7.9)	26 (—)	*a, t, r*	86N, 257E	M	11/13
OC12.19	Heard Island 53S, 73.5E	T–Qr	9 (22)	*a, r*	79N, 45E	M	8/10
OC13.10	Mauritius Volcanics 20.3S, 57.5E	Tp–Qp (2 – 3.4)	10 (—)	*a, t, r*	78N, 217E	M	11/11
OC13.11	Mauritius Volcanics 20.3S, 57.5E	Qp–r	13 (—)	*a, t*	83N, 242E	N	11/10
OC13.12	Reunion Island 21S, 55.5E	Qp–r (0.6 – 0.1)	49 (59)	*a, t*	83N, 184E	N	10/6
OC13.13	Reunion Island 21S, 55.5E	Qp (1 – 1.3)	10 (13)	*a, t, r*	80N, 165E	M	10/7
OC13.14	Reunion Island 21S, 55.5E	Qp (1.9 – 2.0)	38 (58)	*a, t, r*	89N, 172E	M	10/8

Pole no.	Rock unit	Age (My)	Si (Sa)	Stab.	Pole Posn	P	GJ
(d) North Pacific							
OC11.3	Adak Island, Finger Bay Volcanics 52N, 177W	Te	5 (77)	*a, r*	61N, 74E	M	12/48
OC12.20	Shemya Is. andesites and basalts 53N, 174E	Tm (12–15)	7 (56)	*a, r*	73N, 189E	M	12/34
OC12.21	Chiapanshan basalts, Taiwan 24.8N, 121.3E	Tm	10 (67)	*a, r*	82N, 294E	M	9/21
OC12.22	Shiukuran River andesites, Taiwan 23.5N, 121.4E	Tm	4 (15)	*a*	80N, 293E	R	9/22
OC12.23	Loho tuffs, Taiwan 23.3N, 121.3E	Tm	3 (12)	*a*	40N, 192E	N	9/23
OC12.24	Ishigaki Is., Ryukyu Islands 24.4N, 124.2E	Tm	3 (16)	*a*	53N, 40E	N	10/33
OC12.25	Kume and Okinawa andesites, Ryukyu Is. 26.2N, 126.8E	Tm	7 (55)	*a, r*	85N, 252E	M	10/34
OC12.26	Adak Island, andesite dome 52N, 177W	Tp (5)	4 (37)	*a, r*	74N, 114E	M	12/25
OC13.15	Tatun Volcanic Group, Taiwan 25.3N, 121.5E	Tp–Qp	36 (161)	*a*	84N, 275E	N	9/6
OC13.16	Penglin Island basalts, Taiwan 23.6N, 119.5E	Tp–Qp	14 (165)	*a, r*	85N, 271E	M	9/7
OC13.17	Shiukuran River andesites, Taiwan 23.5N, 121.4E	Tp–Qp	14 (65)	*a*	86N, 283E	N	9/8
OC13.18	Keelung Volcano, Taiwan 25.3N, 121.8E	Qp	32 (79)	*a, r*	73N, 238E	M	9/5
OC13.19	Hawaiian lavas 19.6N, 155.6W	Tp–Qp	143 (—)	*a*	84N, 303E	N	5/9
OC13.20	Hawaiian lavas 19–22N, 155–160W	Tp–Qr	38 (106)	*a, r*	84N, 58E	M	8/5
OC13.21	Hawaiian lavas 19.5N, 155.5W	Qp–r (< 0.8)	112 (632)	*a, t*	84N, 313E	N	8/4
OC13.22	Kau Volcanic Series, Hawaii 19.5N, 155.6W	Qr	54 (419)	*a*	82N, 61E	N	11/1
OC13.23	Adagdak, Andrews lavas, Adak Is. 52N, 177W	Qp (< 0.5)	(61)	*a*	85N, 115E	N	12/15
OC13.24	Round Head basalts, Kananga Is. 52N, 177W	Qp	14 (—)	*a*	85N, 189E	N	12/14
OC13.25	Kanaton basalts, Kananga Is. 52N, 177W	Qp	(52)	*a*	84N, 216E	N	12/13
OC13.26	Driftwood Bay lavas, Unalaska Is. 54N, 167W	Q	20 (—)	*a*	89N, 351E	N	12/6
OC13.27	Crater Creek basalts, Umnak Is. 53N, 168W	Q	15 (—)	*a*	86N, 32E	N	12/5
OC13.28	New Jersey Creek lavas, Umnak Is. 53N, 168W	Q	19 (—)	*a*	80N, 66E	N	12/4
(e) South Pacific							
OC13.29	Rarotonga, Cook Is. 21.2S, 159.7W	Qp–r (2.3)	9 (—)	*a, t*	49N, 327E	R	10/18
OC13.20	Rarotonga, Cook Is. 21.2S, 159.7W	Qp–r (2.8)	3 (—)	*a, t*	82N, 20E	R	10/19
OC13.31	Volcanics, Samoa 13.5 – 14.5S, 170.5 – 173.0W	Tp–Qp	16 (—)	*a, r*	84N, 313E	M	8/12
OC13.32	Volcanics, Tonga 21S, 175W	Qp	5 (21)	*a, r*	86N, 86E	M	8/11
OC13.33	New Hebrides lavas, 16S, 168E	Qp	9 (31)	*a, t*	88N, 268E	N	9/3

†(1) Irving (1964). †(2) Scharnberger (personal communication).

References and author index

Italic numbers in brackets refer to the page numbers on which the reference is cited
Entries are arranged alphabetically per number of authors.

Adams, J. C. (*7*)

Ade-Hall, J. M. (1964a). A correlation between remanent magnetism and petrology and chemical properties of Tertiary basalt lavas from Mull, Scotland. *Geophys. J. Roy. Astron. Soc.* **8**, 403–23. (*112*)

Ade-Hall, J. M. (1964b). The magnetic properties of some submarine oceanic lavas. *Geophys. J. Roy. Astron. Soc.* **9**, 85–92. (*157, 158*)

Ade-Hall, J. M. & Watkins, N. D. (1970). Absence of correlations between opaque petrology and natural remanence polarity in Canary Island lavas. *Geophys. J. Roy. Astron. Soc.* **19**, 351–60. (*113*)

Ade-Hall, J. M. & Wilson, R. L. (1963). Petrology and natural remanence of the Mull lavas. *Nature*, **198**, 659–60. (*112*)

Ade-Hall, J. M. & Wilson, R. L. (1969). Opaque petrology and natural remanence polarity in Mull (Scotland) dykes. *Geophys. J. Roy. Astron. Soc.* **18**, 333–52. (*113*)

Ade-Hall, J. M. *see also* Dagley *et al.* (1967); Wilson & Ade-Hall (1970).

Afanasieva (*7*)

Aitken, M. J. (1970). Dating by archaeomagnetic and thermoluminescent methods. *Phil. Trans. Roy. Soc. London*, A**269**, 77–88. (*19, 20, 185, 186*)

Aitken, M. J. & Hawley, H. N. (1967). Archaeomagnetic measurements in Britain IV. *Archaeometry*, **10**, 129–35. (*19*)

Aitken, M. J. & Weaver, G. H. (1962). Magnetic dating: some archaeomagnetic measurements in Britain. *Archaeometry*, **5**, 4–22. (*19*)

Akimoto, S. *see* Nagata *et al.* (1952).

Allan, D. W. (1958). Reversals of the earth's magnetic field. *Nature*, **182**, 469–71. (*13*)

Allan, D. W. (1962). On the behaviour of systems of coupled dynamos. *Proc. Cambridge Phil. Soc.* **58**, 671–93. (*13*)

Allan, T. D. (1969). Review of marine geomagnetism. *Earth Sci. Rev.* **5**, 217–54. (*159, 162*)

Alldredge, L. D. & Hurwitz, L. (1964). Radial dipoles as sources of the earth's main magnetic field. *J. Geophys. Res.* **69**, 2631–40. (*14, 16, 141*)

Almond, M. *see* Clegg *et al.* (1954).

Anderson, C. N. *see* Vogt *et al.* (1970).

Anderson, D. L. & O'Connell, R. (1967). The viscosity of the earth. *Geophys. J. Roy. Astron. Soc.* **14**, 287–95. (*276*)

Anderson, E. M. (1951). *The dynamics of faulting.* Oliver and Boyd, Edinburgh, 2nd ed. (*151*)

Andreeva, O. L., Bucha, V. V. & Petrova, G. N. (1965). Laboratory evaluation of magnetic stability of rocks of the Czech massif. *Akad. Nauk SSSR Izv., Earth Phys. Ser.* 54–64. (*208, 289*)

Andrews, J. E. *see* Maxwell *et al.* (1970).

Andriamirado, R. & Roche, A. (1969). Étude paléomagnétique de formations volcaniques Crétacées de Madagscar. *C.R. Acad. Sci. Paris, Ser. D*, **269**, 16–19. (*234*)

Anuchin, A. V. *see* Pospelova *et al.* (1968).

Aparin, V. P. & Vlassov, A. Ya. (1965). Some questions in the history of the geomagnetic field at the end of the middle Paleozoic and in the upper Paleozoic. In *The present and past of the geomagnetic field*. Moscow, Nauka Press, 213–20. (Translated by E. R. Hope, Directorate of Scientific Information Services, DRB Canada T46IR, 1966.) (*214*)

Arkell, W. J. (1956). *The Jurassic geology of the world*. Oliver and Boyd, Edinburgh. (*267*)

As, J. A. (1960). Instruments and measuring methods in palaeomagnetic research. *Mededel. Verhandel, K.N.M.I.* **73**, 1–56. (*72*)

As, J. A. & Zijderveld, J. D. A. (1958). Magnetic cleaning of rocks in palaeomagnetic research. *Geophys. J. Roy. Astron. Soc.* **1**, 308–19. (*91, 93, 206*)

Athavale, R. N., Radhakrishnamurty, C. & Sahasrabudhe, P. W. (1963). Palaeomagnetism of some Indian rocks. *Geophys. J. Roy. Astron. Soc.* **7**, 304–13. (*232*)

Aumento, F. *see* Irving *et al.* (1970).

Aziz-ur-Rahman (1971). Palaeomagnetic secular variation for recent normal and reversed spochs, from the newer Volcanics of Victoria, Australia. *Geophys. J. Roy. Astron. Soc.* **24**, 255–69. (*306*)

Baadsgaard, H. *see* Folinsbee *et al.* (1968).

Babkine, J. *see* Bonhommet & Babkine (1967).

Bacon, Roger (*1*)

Balsley, J. R. & Buddington, A. F. (1958). Iron–Titanium oxide minerals, rocks and aeromagnetic anomalies of the Adirondack area, New York. *Econ. Geol.* **53**, 777–805. (*112*)

Bannerjee, S. K. (1971). New grain size limits for palaeomagnetic stability in haematite. *Nature Phys. Sci.* **232**, 15–16. (*60*)

Barazangi, M. & Dorman, J. (1969). World seismicity map compiled from ESSA Coast and Geodetic Survey epicentre data, 1961–7. *Bull. Seismol. Soc. Amer.* **59**, 369–80. (*155*)

Baron, J. G. *see* Heirtzler *et al.* (1966).

Basta, E. Z. (1960). Natural and synthetic titanomagnetites (the system Fe_3O_4-Fe_2TiO_4-$FeTiO_3$). *N. Jb. Miner.* **94**, 1017–48. (*41*)

Bauer, L. A. (1899). On the secular variation of a free magnetic needle. *Phys. Rev.* **3**, 34–48. (*19, 20*)

Beck, M. E. & Lindsley, N. C. (1969). Paleomagnetism of the Beaver Bay Complex, Minnesota. *J. Geophys. Res.* **74**, 2002–13. (*204*)

Beloussov, V. V. & Ruditch, E. M. (1961). Island arcs in the development of the earth's structure. *J. Geol.* **69**, 647–58. (*252*)

Belshé, J. C. *see* Uyeda *et al.* (1963).

Bergh, H. W. (1970). Palaeomagnetism of the Stillwater Complex, Montana. In Runcorn, S. K., ed., *Palaeogeophysics*. Academic Press, 143–58. (*30, 205*)

Beuf, S., Biju-Duval, B., Stevaux, J. & Kulbicki, G. (1966). Ampleur des glaciations Siluriennes au Sahara: Leurs influences et leurs conséquences sur la sédimentation. *Rev. Inst. Franc. Petrole Combust. Liquides*, **21**, 363–81. (*220*)

Bhalla, M. S. & Verma, R. K. (1969). Palaeomagnetism of Triassic Parsora Sandstones from India. *Phys. Earth Planet. Interiors*, **2**, 138–46. (*230*)

References and author index

Bhalla, M. S. *see also* Verma & Bhalla (1968).

Bhimasankaram, V. L. S. (1964). Partial self-reversal in pyrrhotite. *Nature*, **202**, 478–9. (*109*)

Biju-Duval, B. & Gariel, O. (1969). Nouvelles observations sur les phénomènes glaciares 'Eocambriens' de la bordure nord de la synéclise de Taoudeni, entre le Hanket le Tanzerouft, Sahara. *Palaeogeog. Palaeoclim. Palaeoecol.* **6**, 283–315. (*220*)

Biju-Duval, B. *see also* Beuf *et al.* (1966).

Bird, J. M. & Dewey, J. F. (1970). Lithospheric plate-continental margin tectonics and the evolution of the Appalachian orogen. *Geol. Soc. Amer. Bull.* **81**, 1031–59. (*184, 203, 271*)

Bird, J. M. *see also* Dewey & Bird (1970*a*, *b*).

Birkenmajer, K., Krs, M. & Nairn, A. E. M. (1968). A paleomagnetic study of Upper Carboniferous rocks from the Inner Sudetic Basin and the Bohemian Massif. *Geol. Soc. Amer. Bull.* **79**, 589–608. (*243*)

Black, D. I. (1967). Cosmic ray effects and faunal extinctions at geomagnetic field reversals. *Earth Planet. Sci. Letters*, **3**, 225–36. (*146, 147, 148*)

Black, R. F. (1964). Palaeomagnetic support of the theory of rotation of the western part of the island of Newfoundland. *Nature*, **202**, 945–8. (*203*)

Blackett, P. M. S. (1952). A negative experiment relating a magnetism and the earth's rotation. *Phil. Trans. Roy. Soc. London*, A**245**, 309–70. (*70, 72*)

Blackett, P. M. S. (1961). Comparison of ancient climates with the ancient latitudes deduced from rock magnetic measurements. *Proc. Roy. Soc. London*, A**263**, 1–30. (*193*)

Bochev, A. (1969). Two and three dipoles approximating the earth's magnetic field. *Pure & Appl. Geophys.* **74**, 29–34. (*16, 141*)

Bonhommet, N. & Babkine, J. (1967). Sur la présence d'aimantations inversées dans la Chaîne des Puys. *C.R. Acad. Sci. Paris*, **264**, 92–4. (*118*)

Borough, W. (*2, 5*)

Bowin, C. O. *see* Emery *et al.* (1970).

Boyce, R. E. *see* Maxwell *et al.* (1970).

Bracey, D. R. *see* Vogt *et al.* (1970).

Brailsford, F. (1966). *Physical principles of magnetism.* Van Nostrand, London. (*32*)

Brakl, J. *see* Rona *et al.* (1970).

Briden, J. C. (1965). Ancient secondary magnetizations in rocks. *J. Geophys. Res.* **70**, 5205–21. (*98, 99, 100, 227*)

Briden, J. C. (1966). Variation of intensity of the palaeomagnetic field through geological time. *Nature*, **212**, 246–7. (*29, 30*)

Briden, J. C. (1967). Recurrent continental drift of Gondwanaland. *Nature*, **215**, 1334–9. (*256, 259, 272, 279*)

Briden, J. C. (1968). Paleoclimatic evidence of a geocentric axial dipole field. In Phinney, R. A., ed., *History of the earth's crust.* Princeton Univ. Press, 178–94. (*198*)

Briden, J. C. (1972). A stability index of remanent magnetism. *J. Geophys. Res.* **77**, 1401–5. (*96, 98*)

Briden, J. C. & Irving, E. (1964). Palaeoclimatic spectra of sedimentary palaeo-climatic indicators. In Nairn, A. E. M., ed., *Problems in palaeoclimatology.* Interscience, N.Y., 199–250. (*194, 195, 197*)

Briden, J. C., Irons, J. & Johnson, P. A. (1971a). Palaeomagnetic studies of the Caerfai Series and Skomer Volcanic Group (Lower Palaeozoic, Wales). *Geophys. J. Roy. Astron. Soc.* **22**, 1–16. (*209, 289*)

Briden, J. C., Smith, A. G. & Sallomy, J. T. (1971b). The geomagnetic field in Permo-Triassic time. *Geophys. J. Roy. Astron. Soc.* **23**, 101–17. (*263*)

Briden, J. C. *see also* Irving & Briden (1962); McElhinny & Briden (1971); McElhinny *et al.* (1968).

Brock, A. (1968). The paleomagnetism of the Nuanetsi igneous province and its bearing upon the sequence of Karroo igneous activity in Southern Africa. *J. Geophys. Res.* **73**, 1389–97. (*132*)

Brock, A. (1971). An experimental study of palaeosecular variation. *Geophys. J. Roy. Astron Soc.* **24**, 303–17. (*28, 29, 139*)

Brock, A., Gibson, I. L. & Gacii, P. (1970). The palaeomagnetism of the Ethiopian flood basalt succession near Addis Ababa. *Geophys. J. Roy. Astron. Soc.* **19**, 485–97. (*222*)

Brock, A. *see also* McElhinny *et al.* (1968).

Broecker, W. S., Thurber, D. L., Goddard, J., Ku, T., Mathews, R. K. & Mesolelle, K. J. (1968). Milankovitch hypothesis supported by precise dating of coral reefs and deep-sea sediments. *Science*, **159**, 297–300. (*122*)

Brown, D. A., Campbell, K. S. W. & Crook, K. A. W. (1968). *The geological evolution of Australia and New Zealand*. Pergamon Press. (*125, 229, 258*)

Brown, D. A. *see also* Irving & Brown (1964, 1966).

Brunhes, B. (1906). Recherches sur la direction d'aimantation des roches volcaniques. *J. Phys.* **5**, 705–24. (*21, 108, 114*)

Brunton, C. H. C. *see* Cocks *et al.* (1970).

Brynjolfsson, A. (1957). Studies of remanent magnetism and viscous magnetism in the basalts of Iceland. *Phil. Mag. Supp. Adv. Phys.* **6**, 247–54. (*93*)

Bucha, V. V. (1961). Some results of paleomagnetic investigations of primary igneous rocks in Czechoslovakia. *Akad. Nauk SSSR Izv., Geophys. Ser.* 54–9. (*208, 289*)

Bucha, V. V. (1965). Results of the paleomagnetic research on rocks of Precambrian and Lower Palaeozoic age in Czechoslovakia. *J. Geomag. Geoelect.* **17**, 435–44. (*208, 289*)

Bucha, V. V. *see also* Andreeva *et al.* (1965).

Buddington, A. F. *see* Balsley & Buddington (1958).

Bull, C., Irving, E. & Willis, I. (1962). Further palaeomagnetic results from South Victoria Land, Antarctica. *Geophys. J. Roy. Astron. Soc.* **6**, 320–36. (*234*)

Bullard, E. C. (1949). The magnetic field within the earth. *Proc. Roy. Soc. London*, A**197**, 433–53. (*10*)

Bullard, E. C. (1955). The stability of a homopolar dynamo. *Proc. Cambridge Phil. Soc.* **51**, 744–60. (*13*)

Bullard, E. C. (1968). Reversals of the earth's magnetic field. *Phil. Trans. Roy. Soc. London*, A**263**, 481–524. (*11, 113, 148*)

Bullard, E. C. & Gellman, H. (1954). Homogeneous dynamos and terrestrial magnetism. *Phil. Trans. Roy. Soc. London*, A**247**, 213–78. (*12*)

Bullard, E. C., Everett, J. E. & Smith, A. G. (1965). A Symposium on Continental Drift. IV. The fit of the continents around the Atlantic. *Phil. Trans. Roy. Soc. London*, A**258**, 41–51. (*235, 246, 253, 255, 263, 266*)

Bullard, E. C., Maxwell, A. E. & Revelle, R. (1956). Heat flow through the deep sea floor. *Adv. Geophys.* **3**, 153–81. (*149*)

References and author index

Bullard, E. C., Freedman, C., Gellman, H. & Nixon, J. (1950). The westward drift of the earth's magnetic field. *Phil. Trans. Roy. Soc. London*, A243, 67–92. (*7, 8*)

Bunce, E. T. *see* Elvers *et al.* (1970).

Burckle, L. H. *see* Hays *et al.* (1969).

Burek, P. J. (1964). Korrelation revers magnatisierter Gesteinfolgen in Oberen Bundsandstein SW-Deutschlands. *Geol. Jahrb.* 84, 591–616. (*125, 129*)

Burek, P. J. (1970). Magnetic reversals: Their application to stratigraphic problems. *Amer. Assn. Petr. Geol. Bull.* 54, 1120–39. (*167*)

Burek, P. J. *see also* McElhinny & Burek (1971).

Cahen, L. & Snelling, N. J. (1966). *The geochronology of equatorial Africa*. North Holland, Amsterdam. (*258*)

Cain, J. C. (*7*)

Campbell, K. S. W. *see* Brown *et al.* (1968).

Carey, S. W. (1958). A tectonic approach to continental drift. In Carey, S. W., ed., *Continental Drift – A Symposium*. Univ. of Tasmania, Hobart, 177–355. (*246, 248, 249, 256, 277, 278, 279*)

Carmichael, C. M. (1959). Remanent magnetism of the Allard Lake ilmenites. *Nature*, 183, 1239–41. (*109*)

Carmichael, C. M. (1961). The magnetic properties of ilmenite–haematite crystals. *Proc. Roy. Soc. London*, A263, 508–30. (*43, 101, 109*)

Carmichael, C. M. (1967). An outline of the intensity of the paleomagnetic field of the earth. *Earth Planet. Sci. Letters*, 3, 351–4. (*29, 30, 31*)

Cattala, L. *see* Roche & Cattala (1959).

Chamalaun, F. H. (1964). Origin of the secondary magnetization of the Old Red Sandstones of the Anglo-Welsh cuvette. *J. Geophys. Res.* 69, 4327–37. (*98, 99, 100*)

Chamalaun, F. H. & Creer, K. M. (1964). Thermal demagnetization studies of the Old Red Sandstone of the Anglo-Welsh cuvette. *J. Geophys. Res.* 69, 1607–16. (*98, 270*)

Chamalaun, F. H. & Porath, H. (1968). Palaeomagnetism of Australian hematite ore bodies – I. The Middleback Ranges of South Australia. *Geophys. J. Roy. Astron. Soc.* 14, 451–62. (*230*)

Chamalaun, F. H. *see also* McDougall & Chamalaun (1966, 1969); Porath & Chamalaun (1968).

Chatterjee, J. S. (1956). The crust as the possible seat of the earth's magnetism. *J. Atmosph. Terr. Phys.* 8, 233–9. (*9*)

Chen Zhiqiang, Wang Cenghang & Deng Xinghui (1965). Some results of paleomagnetic research in China. In *The present and past of the geomagnetic field*. Moscow, Nauka Press, 309–11. (Translated by E. R. Hope, Directorate of Scientific Information Services, DRB Canada T462R, 1966.) (*216*)

Chevallier, R. (1925). L'aimantation des lavas de l'Etna et l'orientation du champs terrestre en Sicile du 12ᵉ au 17ᵉ siècle. *Ann. Phys. Ser. 10*, 4, 5–162. (*21*)

Clegg, J. A., Almond, M. & Stubbs. P. H. S. (1954). The remanent magnetism of some sedimentary rocks in Britain. *Phil. Mag.* 45, 583–98. (*206*)

Clegg, J. A., Deutsch, E. R. & Griffiths, D. H. (1956). Rock magnetism in India. *Phil. Mag.* 1, 419–31. (*230*)

Clegg, J. A., Radhakrishnamurty, C. & Sahasrabudhe, P. W. (1958). Remanent magnetism of the Rajmahal traps of north-eastern India. *Nature*, 181, 830–1. (*230*)

Clegg, J. A., Deutsch, E. R., Everitt, C. W. F. & Stubbs, P. H. S. (1957). Some recent palaeomagnetic measurements made at Imperial College, London. *Phil. Mag. Supp. Adv. Phys.* **6**, 219–31. (*206, 246*)

Clegg, J. A. *see also* Everitt & Clegg (1962).

Clifford, T. N. (1968). Radiometric dating and the pre-Silurian geology of Africa. In Hamilton, E. I. & Farquhar, R. M., eds., *Radiometric dating for geologists*. Interscience, N.Y., 299–416. (*258*)

Clifford, T. N. (1970). The structural framework of Africa. In Clifford, T. N. & Gass, I. G., eds., *African magmatism and tectonics*. Oliver and Boyd, Edinburgh, 1–26. (*258, 259*)

Cocks, L. R. M., Brunton, C. H. C., Rowell, A. J. & Rust, I. C. (1970). The first lower Palaeozoic fauna proved from South Africa. *Quart. J. Geol. Soc. London,* **125** *for 1969*, 583–603. (*221*)

Collinson, D. W. (1965a). The remanent magnetism and magnetic properties of red sediments. *Geophys. J. Roy. Astron. Soc.* **10**, 105–26. (*59, 61*)

Collinson, D. W. (1965b). Depositional remanent magnetization in sediments. *J. Geophys. Res.* **70**, 4663–8. (*62*)

Collinson, D. W. (1965c). Origin of remanent magnetization and initial susceptibility of certain red sandstones. *Geophys. J. Roy. Astron. Soc.* **9**, 203–17. (*101, 102, 103*)

Collinson, D. W. (1966). Magnetic properties of the Taiguati formation. *Geophys. J. Roy. Astron. Soc.* **11**, 337–47. (*101*)

Collinson, D. W. (1967). The design and construction of astatic magnetometers. In Collinson, D. W., Creer, K. M. & Runcorn, S. K., eds., *Methods in palaeoemagntism*. Elsevier, Amsterdam, 47–59. (*72*)

Collinson, D. W. (1970). An astatic magnetometer with rotating sample. *Geophys. J. Roy. Astron. Soc.* **19**, 547–9. (*73*)

Collinson, D. W. & Runcorn, S. K. (1960). Polar wandering and continental drift: Evidence of paleomagnetic observations in the United States. *Geol. Soc. Amer. Bull.* **71**, 915–58. (*201*)

Collinson, D. W., Creer, K. M. & Runcorn, S. K., eds. (1967). *Methods in palaeomagnetism*. Elsevier, Amsterdam. (*68*)

Collinson, D. W., Creer, K. M., Irving, E. & Runcorn, S. K. (1957). Palaeomagnetic investigations in Great Britain. I. The measurement of the permanent magnetization of rocks. *Phil. Trans. Roy. Soc. London,* A**250**, 73–82. (*72*)

Compston, W., McDougall, I. & Heier, K. S. (1968). Geochemical comparison of the Mesozoic basaltic rocks of Antarctica, South Africa, South America and Tasmania. *Geochim. Cosmochim. Acta,* **32**, 129–49. (*261*)

Cook, P. J. *see* Wells *et al.* (1970).

Cooper, C. *see* Vestine *et al.* (1947).

Cox, A. (1961). Anomalous remanent magnetization of basalt. *U.S. Geol. Surv. Bull.* **1083-E**, 131–60. (*70, 89*)

Cox, A. (1962). Analysis of present geomagnetic field for comparison with paleomagnetic results. *J. Geomag. Geoelect.* **13**, 101–12. (*29*)

Cox, A. (1968). Lengths of geomagnetic polarity intervals. *J. Geophys. Res.* **73**, 3257–60. (*14, 15, 20, 21, 141, 142*)

Cox, A. (1969). Geomagnetic reversals. *Science,* **163**, 237–45. (*115, 116, 117, 137, 142, 169*)

Cox, A. (1970). Latitude dependence of the angular dispersion of the geomagnetic field. *Geophys. J. Roy. Astron. Soc.* **20**, 253–69. (*29*)

References and author index

Cox, A. & Dalrymple, G. B. (1967). Statistical analysis of geomagnetic reversal data and the precision of Potassium–Argon dating. *J. Geophys. Res.* **72**, 2603–14. (*117, 136, 137*)

Cox, A. & Doell, R. R. (1960). Review of paleomagnetism. *Geol. Soc. Amer. Bull.* **71**, 647–768. (*22, 85, 88*)

Cox, A. & Doell, R. R. (1961). Palaeomagnetic evidence relevant to a change in the earth's radius. *Nature*, **190**, 36–7. (*278*)

Cox, A., Doell, R. R. & Dalrymple, G. B. (1963a). Geomagnetic polarity epochs and Pleistocene geochronometry. *Nature*, **198**, 1049–51. (*115, 158*)

Cox, A., Doell, R. R. & Dalrymple, G. B. (1963b). Geomagnetic polarity epochs: Sierra Nevada II. *Science*, **142**, 382–5. (*115*)

Cox, A., Doell, R. R. & Dalrymple, G. B. (1964). Reversals of the earth's magnetic field. *Science*, **144**, 1537–43. (*111, 115, 158*)

Cox, A., Doell, R. R. & Dalrymple, G. B. (1968). Radiometric time-scale for geomagnetic reversals. *Quart. J. Geol. Soc. London*, **124**, 53–66. (*115*)

Cox, A. *see also* Dalrymple *et al.* (1967); Doell & Cox (1965, 1967a, b).

Crawford, A. R. (1969). India, Ceylon and Pakistan: New age data and comparisons with Australia. *Nature*, **223**, 380–4. (*258*)

Crawford, A. R. (*257*)

Creer, K. M. (1958). Preliminary palaeomagnetic measurements from South America. *Ann. Geophys.* **14**, 373–90. (*225*)

Creer, K. M. (1959). A.C. demagnetization of unstable Triassic Keuper marls from S.W. England. *Geophys. J. Roy. Astron. Soc.* **2**, 261–75. (*91*)

Creer, K. M. (1962). Palaeomagnetism of the Serra Geral formation. *Geophys. J. Roy. Astron. Soc.* **7**, 1–22. (*225*)

Creer, K. M. (1964). A reconstruction of the Continents for the Upper Palaeozoic from palaeomagnetic data. *Nature*, **203**, 1115–20. (*256*)

Creer, K. M. (1965). A symposium on Continental drift. III. Palaeomagnetic data from the Gondwanic continents. *Phil. Trans. Roy. Soc. London*, A**256**, 569–73. (*225, 256*)

Creer, K. M. (1967a). Methods of measurement with the astatic magnetometer. In Collinson, D. W., Creer, K. M. & Runcorn, S. K., eds., *Methods in palaeomagnetism*. Elsevier, Amsterdam, 172–91. (*73*)

Creer, K. M. (1967b). Palaeomagnetism of some lower Palaeozoic South American rocks. *Symposium on Continental Drift* (UNESCO/IUGS), Montevideo, Uruguay, October 1967. [*Trans. Amer. Geophys. Un.* **53**, 172 (*1972*).] (*225, 305*)

Creer, K. M. (1967c). A synthesis of world-wide palaeomagnetic data. In Runcorn, S. K., ed., *Mantles of the earth and terrestrial planets*. Interscience, London, 351–82. (*256*)

Creer, K. M. (1968a). Arrangement of the continents during the Palaeozoic era. *Nature*, **219**, 41–4. (*256*)

Creer, K. M. (1968b). Palaeozoic palaeomagnetism. *Nature*, **219**, 246–50. (*270, 271*)

Creer, K. M. (1970a). A review of palaeomagnetism. *Earth Sci. Revs.* **6**, 369–466. (*202, 222, 227, 260, 272, 274*)

Creer, K. M. (1970b). Palaeomagnetic survey of South American rocks, Parts I–V. *Phil. Trans. Roy. Soc. London*, A**267**, 457–558. (*225*)

Creer, K. M. (1971). Mesozoic Palaeomagnetic Reversal Column. *Nature*, **233**, 545–6. (*132*)

Creer, K. M. & Embleton, B. J. J. (1967). Devonian palaeomagnetic pole for Europe and North America. *Nature*, **214**, 42–3. (*270*)

316

Creer, K. M. & Ispir, Y. (1970). An interpretation of the behaviour of the geomagnetic field during polarity transitions. *Phys. Earth Planet. Interiors*, **2**, 283–93. (*140, 141*)

Creer, K. M. & Sanver, M. (1967). The use of the sun compass. In Collinson, D. W., Creer, K. M. & Runcorn, S. K., eds., *Methods in palaeomagnetism*. Elsevier, Amsterdam, 11–15. (*69*)

Creer, K. M., Embleton, B. J. J. & Valencio, D. A. (1970). Permo-Triassic and Triassic palaeomagnetic data for South America. *Earth Planet. Sci. Letters*, **8**, 173–8. (*225*)

Creer, K. M., Irving, E. & Nairn, A. E. M. (1959). Palaeomagnetism of the Great Whin Sill. *Geophys. J. Roy. Astron. Soc.* **2**, 306–23. (*28*)

Creer, K. M., Irving, E. & Runcorn, S. K. (1954). The direction of the geomagnetic field in remote epochs in Great Britain. *J. Geomag. Geoelect.* **6**, 163–8. (*199, 206, 252*)

Creer, K. M., Irving, E. & Runcorn, S. K. (1957). Geophysical interpretation of palaeomagnetic directions from Great Britain. *Phil. Trans. Roy. Soc. London*, A**250**, 144–56. (*206*)

Creer, K. M., Irving, E., Nairn, A. E. M. & Runcorn, S. K. (1958). Palaeomagnetic results from different continents and their relation to the problem of continental drift. *Ann. Geophys.* **14**, 492–501. (*255*)

Creer, K. M. *see also* Chamalaun & Creer (1964); Collinson *et al.* (1957, 1967).

Crittenden, M. D. (1963). Effective viscosity of the Earth derived from isostatic loading of Pleistocene Lake Bonneville. *J. Geophys. Res.* **68**, 5517–30. (*276*)

Crook, K. A. W. *see* Brown *et al.* (1968).

Crowell, J. C. & Frakes, L. A. (1970). Phanerozoic glaciation and the causes of ice ages. *Amer. J. Sci.* **268**, 193–224. (*227, 230*)

Crowell, J. C. & Frakes, L. A. (1971). Late Palaeozoic glaciation of Australia. *J. Geol. Soc. Australia*, **17**, 115–55. (*230*)

Crowell, J. C. *see also* Frakes & Crowell (1968, 1969).

Cumming, G. L. *see* Folinsbee *et al.* (1968).

Dagley, P., Wilson, R. L., Ade-Hall, J. M., Walker, G. P. L., Haggerty, S. E., Sigurgeirsson, T., Watkins, N. D., Smith, P. J., Edwards, J. & Grasty, R. L. (1967). Geomagnetic polarity zones for Icelandic lavas. *Nature*, **216**, 25–9. (*118*)

Dalrymple, G. B., Cox, A., Doell, R. R. & Gromme, C. S. (1967). Pliocene geomagnetic polarity epochs. *Earth Planet. Sci. Letters*, **2**, 163–73. (*118*)

Dalrymple, G. B. *see also* Cox & Dalrymple (1967); Cox *et al.* (1963a, b, 1964, 1968); Doell & Dalrymple (1966).

David, P. (1904). Sur la stabilité de la direction d'aimantation dans quelques roches volcaniques. *C.R. Acad. Sci. Paris*, **138**, 41–2. (*21, 108*)

Davies, D. *see* McKenzie *et al.* (1970).

de Boer, J. (1963). The geology of the Vicentinian Alps (N.E. Italy) (with special reference to their palaeomagnetic history). *Geol. Ultraiectina*, **11**. (*211, 248*)

de Boer, J. (1965). Paleomagnetic indications of megatectonic movements in the Tethys. *J. Geophys. Res.* **70**, 931–44. (*248*)

de Jong, K. A. *see* Zijderveld & de Jong (1969); Zijderveld *et al.* (1970a).

Delesse, A. (*21*)

Deng Xinghui *see* Chen Zhiqiang *et al.* (1965).

Deutsch, E. R. (1969). Paleomagnetism and North Atlantic paleogeography. In Kay, M., ed., *North Atlantic – Geology and continental drift. Amer. Assn. Petr. Geol. Mem.* **12**, 931–54. (*254*)

References and author index

Deutsch, E. R. & Rao, K. V. (1970). Paleomagnetism of Ordovician sedimentary rocks from Bell Island, Newfoundland. *Trans. Amer. Geophys. Union*, **51**, 272. (*286*)

Deutsch, E. R., Kristjansson, L. G. & May, B. T. (1971). Remanent magnetism of Lower Tertiary lavas on Baffin Island. *Can. J. Earth Sci.* **8**, 1542–52. (*286*)

Deutsch, E. R., Radhakrishnamurty, C. & Sahasrabudhe, P. W. (1958). The remanent magnetism of some lavas in the Deccan traps. *Phil. Mag.* **3**, 170–84. (*230*)

Deutsch, E. R., Radhakrishnamurty, C. & Sahasrabudhe, P. W. (1959). Palaeomagnetism of the Deccan traps. *Ann. Geophys.* **15**, 39–59. (*230*)

Deutsch, E. R. *see also* Clegg *et al.* (1956, 1957); Murthy & Deutsch (1972).

Dewey, J. F. (1969). Evolution of the Appalachian/Caledonian orogen. *Nature*, **222**, 124–9. (*208, 254, 271*)

Dewey, J. F. & Bird, J. M. (1970*a*). Mountain belts and the new global tectonics. *J. Geophys. Res.* **75**, 2625–47. (*183, 184*)

Dewey, J. F. & Bird, J. M. (1970*b*). Plate tectonics and geosynclines. *Tectonophysics*, **10**, 625–38. (*183, 184*)

Dewey, J. F. & Horsfield, B. (1970). Plate tectonics, orogeny and continental growth. *Nature*, **225**, 521–5. (*258, 280*)

Dewey, J. F. *see also* Bird & Dewey (1970).

Dickson, G. O. & Foster, J. H. (1966). The magnetic stratigraphy of a deep sea core from the North Pacific Ocean. *Earth Planet. Sci. Letters*, **1**, 458–62. (*118*)

Dickson, G. O., Pitman III, W. C. & Heirtzler, J. R. (1968). Magnetic anomalies in the South Atlantic and ocean floor spreading. *J. Geophys. Res.* **73**, 2087–2100. (*161, 165, 167*)

Dickson, G. O., Everitt, C. W. F., Parry, L. G. & Stacy, F. D. (1966). Origin of thermoremanent magnetization. *Earth Planet. Sci. Letters*, **1**, 222–4. (*52, 57, 58*)

Dickson, G. O. *see* Heirtzler *et al.* (1968).

Dietz, R. S. (1961). Continent and ocean basin evolution by spreading of the sea floor. *Nature*, **190**, 854–7. (*149*)

Dietz, R. S. (1962). Ocean basin evolution by sea-floor spreading. In MacDonald, G. A. & Kuno, H., eds., *The crust of the Pacific basin. Geophys. Mono. Amer. Geophys. Union.* **6**. (*149*)

Dietz, R. S. & Holden, J. C. (1970). Reconstruction of Pangea: Breakup and dispersion of continents, Permian to present. *J. Geophys. Res.* **75**, 4939–56. (*269*)

Dietz, R. S. & Holden, J. C. (1971). Pre-Mesozoic oceanic crust in the eastern Indian Ocean (Wharton Basin)? *Nature*, **229**, 309–12. (*269*)

Dietz, R. S. *see also* Freeland & Dietz (1971); Sproll & Dietz (1969).

Dingle, R. V. & Klinger, H. C. (1971). Significance of Upper Jurassic sediments in the Knysna Outlier (Cape Province) for the timing of the breakup of Gondwanaland. *Nature Phys. Sci.* **232**, 37–8. (*267*)

Dodson, M. H. *see* Van Breemen *et al.* (1966).

Doell, R. R. & Cox, A. (1965). Measurement of the remanent magnetization of igneous rocks. *U.S. Geol. Surv. Bull.* **1203-A**, 1–32. (*69*)

Doell, R. R. & Cox, A. (1967*a*). Measurement of the natural remanent magnetization at the outcrop. In Collinson, D. W., Creer, K. M. & Runcorn, S. K., eds., *Methods in palaeomagnetism.* Elsevier, Amsterdam, 159–62. (*75*)

Doell, R. R. & Cox, A. (1967*b*). Analysis of alternating field demagnetization equipment. In Collinson, D. W., Creer, K. M. & Runcorn, S. K., eds., *Methods in palaeomagnetism.* Elsevier, Amsterdam, 241–53. (*92*)

Doell, R. R. & Dalrymple, G. B. (1966). Geomagnetic polarity epochs: A new polarity event and the age of the Brunhes–Matuyama boundary. *Science*, **152**, 1060–1. (*115, 137, 159*)

Doell, R. R. *see also* Cox & Doell (1960, 1961); Cox *et al.* (1963*a, b*, 1964, 1968); Dalrymple *et al.* (1967).

Dorman, J. *see* Barazangi & Dorman (1969).

Drake, C. L. & Nafe, J. E. (1969). The transition from ocean to continent from seismic refraction data. *Geophys. Monogr. Amer. Geophys. Union*, **12**, 174–86. (*167*)

Drake, C. L., Ewing, J. I. & Stockard, H. (1968). The continental margin of the eastern United States. *Can. J. Earth Sci.* **5**, 993–1010. (*266*)

Drake, C. L., Ewing, M. & Sutton, J. (1959). Continental margins and geosynclines: the east coast of North America, north of Cape Hatteras. In Ahrens, L. H., Rankama, K. & Runcorn, S. K., eds., *Physics and Chemistry of the Earth*. Pergamon, London, **5**, 110–98. (*183*)

Drake, C. L. *see also* Nafe & Drake (1969).

du Bois, P. M. (1959). Palaeomagnetism and the rotation of Newfoundland. *Nature*, **184**, 63–4. (*202*)

du Bois, P. M. (1962). Palaeomagnetism and and correlation of Keweenawan rocks. *Geol. Surv. Canad. Bull.* **71**. (*204*)

Dunlop, D. J. (1965). Grain distribution in rocks containing single domain grains. *J. Geomag. Geoelect.* **17**, 459–71. (*90*)

Dunlop, D. J. (1970). Hematite: Intrinsic and defect ferromagnetism. *Science*, **169**, 858–60. (*43, 100*)

Dunlop, D. J. & West, G. F. (1969). An experimental evaluation of single domain theories. *Rev. Geophys.* **7**, 709–57. (*48, 49*)

du Toit, A. L. (1937). *Our wandering continents*. Oliver and Boyd, London. (*246, 256, 270*)

du Toit, A. L. (1954). *The Geology of South Africa*. Oliver and Boyd, Edinburgh, 3rd edn. (*220, 221*)

Dymond, J. & Windom, H. L. (1968). Cretaceous K-Ar ages from Pacific ocean seamounts. *Earth Planet. Sci. Letters*, **4**, 47–52. (*173*)

Dyson, F. W. (*7*)

Early, T. *see* Scharon *et al.* (1970).

Edwards, J. *see* Dagley *et al.* (1967).

Egyed, L. (1957). A new dynamic conception of the internal constitution of the earth. *Geol. Rundschau*, **46**, 101–21. (*277*)

Egyed, L. (1960). Some remarks on continental drift. *Geofis. Pura Appl.* **45**, 115–16. (*277*)

Egyed, L. (1961). Palaeomagnetism and the ancient radii of the earth. *Nature*, **190**, 1097–8. (*278*)

Elsasser, W. M. (1946). Induction effects in terrestrial magnetism: Part I. Theory. *Phys. Rev.* **69**, 106–16. (*10*)

Elsasser, W. M. (1955). Hydromagnetism I: A review. *Amer. J. Phys.* **23**, 590–609. (*11*)

Elsasser, W. M. (1963). Early history of the earth. In *Earth Science and Meteoritics*. North Holland, Amsterdam, 1–30. (*31*)

Elvers, D. J., Matthewson, C. C., Kohler, R. E. & Moses, R. L. (1967). Systematic ocean surveys by the USC and GSS Pioneer 1961–3: *Coast and Geodetic Survey operational data report* 1 (U.S. Coast and Geodetic Survey). (*181*)

References and author index

Embleton, B. J. J. (1970). Palaeomagnetic results for the Permian of South America and a comparison with the African and Australian data. *Geophys. J. Roy. Astron. Soc.* **21**, 105–18. (*222, 225, 227, 256*)

Embleton, B. J. J. see also Creer & Embleton (1967); Creer et al. (1970).

Emery, K. O., Uchupi, E., Phillips, J. D., Bowin, C. O., Bunce, E. T. & Knott, S. T. (1970). Continental rise off eastern North America. *Amer. Assn. Petr. Geol. Bull.* **54**, 44–108. (*167*)

Ericson, D. B., Ewing, M., Wollin, G. & Heezen, B. C. (1961). Atlantic deep-sea sediment cores. *Geol. Soc. Amer. Bull.* **72**, 193–286. (*121, 122*)

Erman, A. (*7*)

Evans, M. E. & McElhinny, M. W. (1969). An investigation of the origin of stable remanence in magnetite bearing igneous rocks. *J. Geomag. Geoelect.* **21**, 757–73. (*48, 51, 52, 54, 90, 91*)

Evans, M. E. & Wayman, M. L. (1970). An investigation of small magnetic particles by means of electron microscopy. *Earth Planet. Sci. Letters*, **9**, 365–70. (*48, 52, 55*)

Evans, M. E., McElhinny, M. W. & Gifford, A. C. (1968). Single domain magnetite and high coercivities in a gabbroic intrusion. *Earth Planet. Sci. Letters*, **4**, 142–6. (*51, 52*)

Evans, M. E. see also McElhinny & Evans (1968); Murthy et al. (1971).

Everett, J. E. see Bullard et al. (1965).

Everitt, C. W. F. (1962). Self-reversal in a shale containing pyrrhotite. *Phil. Mag.* **7**, 831–42. (*109*)

Everitt, C. W. F. & Clegg, J. A. (1962). A field test for palaeomagnetic stability. *Geophys. J. Roy. Astron. Soc.* **6**, 312–19. (*86*)

Everitt, C. W. F. see also Clegg et al. (1957); Dickson et al. (1966).

Ewing, J. I. see Drake et al. (1968).

Ewing, M. see Drake et al. (1959); Ericson et al. (1961).

Fahrig, W. F. & Jones, D. L. (1969). Paleomagnetic evidence for the extent of the MacKenzie igneous events. *Can. J. Earth Sci.* **6**, 679–88. (*204*)

Fahrig, W. F., Gaucher, E. H. & Larochelle, A. (1965). Palaeomagnetism of diabase dykes of the Canadian shield. *Can. J. Earth Sci.* **2**, 278–98. (*205*)

Falvey, D. A. see Packham & Falvey (1971).

Faure, G. see Fenton & Faure (1969).

Fenton, M. D. & Faure, G. (1969). The age of the igneous rocks of the Stillwater Complex of Montana. *Geol. Soc. Amer. Bull.* **80**, 1599–1604. (*30*)

Finch, H. F. (*7*)

Fisher, R. A. (1953). Dispersion on a sphere. *Proc. Roy. Soc. London*, A**217**, 295–305. (*28, 78, 79*)

Folgheraiter, G. (1899). Sur les variations séculaires d'inclinaison magnétique dans l'antiquité. *Arch. Sci. Phys. Nat.* **8**, 5–16. (*19, 21*)

Folinsbee, R. E., Baadsgaard, H., Cumming, G. L. & Green, D. C. (1968). A very ancient island arc. *Amer. Geophys. Union Monogr.* **12**, 441–8. (*280*)

Forman, D. J. see Wells et al. (1970).

Foster, J. H. (1966). A paleomagnetic spinner magnetometer using a fluxgate gradiometer. *Earth Planet. Sci. Letters*, **1**, 463–6. (*76*)

Foster, J. H. & Opdyke, N. D. (1970). Upper Miocene to Recent magnetic stratigraphy in deep-sea sediments. *J. Geophys. Res.* **75**, 4465–73. (*118, 123, 124*)

Foster, J. H. see also Dickson & Foster (1966); Ninkovich et al. (1966); Opdyke et al. (1966); Smith & Foster (1969).

Fox, P. J., Heezen, B. C. & Johnson, G. L. (1970). Jurassic sandstone from the tropical Atlantic. *Science*, **170**, 1402–4. (*266*)

Frakes, L. A. & Crowell, J. C. (1968). Late Paleozoic glacial geography of Antarctica. *Earth Planet. Sci. Letters*, **4**, 253–6. (*234*)

Frakes, L. A. & Crowell, J. C. (1969). Late Paleozoic glaciation: I, South America. *Geol. Soc. Amer. Bull.* **80**, 1007–42. (*227*)

Frakes, L. A. *see also* Crowell & Frakes (1970, 1971).

Francheteau, J. (1970). Paleomagnetism and plate tectonics. *Rept. of the Marine Physical Laboratory, Scripps Institution of Oceanography, Univ. of Calif., San Diego. – SIO Ref.* 70–30. (*235, 236, 265*)

Francheteau, J. & Sclater, J. G. (1970). Comments on paper by E. Irving and W. A. Robertson, 'Test for polar wandering and some possible implications'. *J. Geophys. Res.* **75**, 1023–6. (*272*)

Francheteau, J., Harrison, C. G. A., Sclater, J. G. & Richards, M. L. (1970). Magnetization of Pacific seamounts: A preliminary polar curve for the north-eastern Pacific. *J. Geophys. Res.* **75**, 2035–61. (*171, 172, 173*)

Freedman, C. *see* Bullard *et al.* (1950).

Freeland, G. L. & Dietz, R. S. (1971). Plate tectonic evolution of Caribbean–Gulf of Mexico region. *Nature*, **232**, 20–3. (*266*)

Fritsche, H. (*7*)

Frost, D. V. *see* Nairn *et al.* (1959).

Fuller, M. D. (1963). Magnetic anisotropy and paleomagnetism. *J. Geophys. Res.* **68**, 293–309. (*65*)

Fuller, M. D. *see also* Uyeda *et al.* (1963).

Funnell, B. M. & Smith, A. G. (1968). Opening of the Atlantic Ocean. *Nature*, **219**, 1328–33. (*253, 266*)

Funnell, B. M. *see also* Harrison & Funnell (1964).

Furner, H. (*7*)

Gacii, P. *see* Brock *et al.* (1970).

Gaibar-Puertas, C. (1953). Varacion secular del campo geomagnetico. *Observ. del Elso*, Memo. No. 11. (*4*)

Gardner, W. K. *see* Mathews & Gardner (1963).

Gariel, O. *see* Biju-Duval & Gariel (1969).

Gaskell, T. F. *see* Irving & Gaskell (1962).

Gaucher, E. H. *see* Fahrig *et al.* (1965)

Gauss, K. F. (*5, 6, 7, 9, 115*)

Gellibrand, H. (*5*)

Gellman, H. *see* Bullard & Gellman (1954); Bullard *et al.* (1950).

Gibb, R. A. & Walcott, R. I. (1971). A Precambrian suture in the Canadian Shield. *Earth Planet. Sci. Letters*, **10**, 417–22. (*280*)

Gibson, I. L. *see* Brock *et al.* (1970).

Gibson, R. D. & Roberts, P. H. (1969). The Bullard–Gellman dynamo. In Runcorn, S. K. ed., *The application of modern physics to the earth and planetary interiors*. Wiley, London, 577–602. (*12*)

Gifford, A. C. *see* Evans *et al.* (1968).

Gilbert, William (*2, 4, 15*)

Girdler, R. W. *see* Uyeda *et al.* (1963).

Glass, B. P. *see* Opdyke & Glass (1969); Opdyke *et al.* (1966).

Goddard, J. *see* Broecker *et al.* (1968).

References and author index

Gold, T. (1955). The instability of the earth's rotational axis. *Nature*, **175**, 528–9. (*275*)

Goldich, S. S. (1968). Geochronology of the Lake Superior region. *Can. J. Earth Sci.* **5**, 715–24. (*204*)

Goldreich, P. & Toomre, A. (1969). Some remarks on polar wandering. *J. Geophys. Res.* **74**, 2555–67. (*275, 276, 277*)

Goldstein, M. A., Strangway, D. W. & Larson, E. E. (1969). Paleomagnetism of a Miocene transition zone in southeastern Oregon. *Earth Planet. Sci. Letters*, **7**, 231–9. (*134, 138*)

Goldstein, M. A. *see also* Strangway *et al.* (1968a).

Goodell, H. G. *see* Watkins & Goodell (1967).

Gordon, R. B. (1967). Discussion of paper by D. L. Anderson & R. O'Connell, 'Viscosity of the Earth'. *Geophys. J. Roy. Astron. Soc.* **14**, 295. (*277*)

Gough, D. I. (1956). A study of the palaeomagnetism of the Pilansberg dykes. *Mon. Not. Roy. Astron. Soc. Geophys. Suppl.* **7**, 196–213. (*219*)

Gough, D. I. (1964). A spinner magnetometer. *J. Geophys. Res.* **69**, 2455–63. (*76*)

Gough, D. I. & Opdyke, N. D. (1963). The palaeomagnetism of the Lupata alkaline volcanics. *Geophys. J. Roy. Astron. Soc.* **7**, 457–68. (*70*)

Gough, D. I. & Van Niekerk, C. B. (1959). On the palaeomagnetism of the Bushveld gabbro. *Phil. Mag.* **4**, 126–36. (*219*)

Gough, D. I., Opdyke, N. D. & McElhinny, M. W. (1964). The significance of paleomagnetic results from Africa. *J. Geophys. Res.* **69**, 2509–19. (*220, 256*)

Gough, D. I., *see also* McElhinny & Gough (1963); Murthy *et al.* (1971).

Graham, J. W. (1949). The stability and significance of magnetism in sedimentary rocks. *J. Geophys. Res.* **54**, 131–67. (*87, 89, 201*)

Graham, J. W. (1954). Rock magnetism and the earth's magnetic field during Paleozoic time. *J. Geophys. Res.* **59**, 215–22. (*201*)

Graham, J. W. (1955). Evidence of polar shift since Triassic time. *J. Geophys. Res.* **60**, 329–47. (*125, 201*)

Graham, J. W. (1956). Paleomagnetism and magnetostriction. *J. Geophys. Res.* **61**, 735–9. (*64, 201*)

Graham, J. W. (1967). Preliminary evaluation of a new resonance magnetometer. In Collinson, D. W., Creer, K. M. & Runcorn, S. K., eds., *Methods in palaeomagnetism*. Elsevier, Amsterdam, 96–9. (*73*)

Graham, K. W. T. (1961). The remagnetization of a surface outcrop by lightning currents. *Geophys. J. Roy. Astron. Soc.* **6**, 85–102. (*70, 89*)

Graham, K. W. T. & Hales, A. L. (1957). Palaeomagnetic measurements on Karroo dolerites. *Phil. Mag. Suppl. Adv. Phys.* **6**, 149–161. (*219*)

Graham, K. W. T. & Keiller, J. A. (1960). A portable drill rig for producing short oriented cores. *Trans. Geol. Soc. S. Africa*, **63**, 71–3. (*70*)

Graham, K. W. T., Helsley, C. E. & Hales, A. L. (1964). Determination of the relative positions of continents from paleomagnetic data. *J. Geophys. Res.* **69**, 3895–3900. (*239*)

Graham, K. W. T. *see also* Van Zijl *et al.* (1962a, b).

Grasty, R. L. *see* Dagley *et al.* (1967).

Green, D. C. *see* Folinsbee *et al.* (1968).

Green, D. H. & Ringwood, A. E. (1967). Genesis of basaltic magmas. *Beitr. Mineral. Petrog.* **15**, 103–90. (*151*)

Green, D. H. *see* Ringwood & Green (1966).

Green, R. *see* Irving & Green (1957, 1958).

Gregor, C. B. & Zijderveld, J. D. A. (1964). The magnetism of some Permian red sandstones from north-western Turkey. *Tectonophysics*, **1**, 289–306. (*225, 243*)

Gregory, J. W. (1921). *The rift valleys and geology of East Africa*. Seeley and Co., London. (*240*)

Griffiths, D. H., King, R. F., Rees, A. I. & Wright, A. E. (1960). The remanent magnetization of some recent varved sediments. *Proc. Roy. Soc. London*, A**256**. 359–83. (*62, 63*)

Griffiths, D. H. *see also* Clegg *et al.* (1956).

Griffiths, J. R. (1971). Continental margin tectonics and the evolution of southeast Australia. *APEA Jour.* 75–9. (*261, 262*)

Grommé, C. S. & Hay, R. L. (1971). Geomagnetic polarity epochs: Age and duration of the Olduvai normal polarity event. *Earth Planet. Sci. Letters*, **10**, 179–85. (*116, 117*)

Grommé, C. S. *see also* Dalrymple *et al.* (1967).

Gunst, R. H. *see* McDonald & Gunst (1968).

Gurariy, G. Z. (1969). Some information on the geomagnetic field on the Siberian platform in the early Paleozoic period. *Akad. Nauk SSSR Izv. Earth Phys. Ser. No. 6*, 105–13. (*214*)

Gurariy, G. Z., Kropotkin, P. N., Pevzner, M. A., Ro Vu Son & Trubikhin, V. M. (1966). Laboratory evaluation of the usefulness of North Korean sedimentary rocks for paleomagnetic studies. *Akad. Nauk SSSR Izv. Earth Phys. Ser.* 128–36. (*216*)

Gutenberg, B. & Richter, C. F. (1954). *Seismicity of the Earth*. Princeton Univ. Press. (*155*)

Haggerty, S. E. *see* Dagley *et al.* (1967); Irving *et al.* (1970); Watkins & Haggerty (1968); Wilson & Haggerty (1966).

Haigh, G. (1958). The process of magnetization by chemical change. *Phil. Mag.* **3**, 267–86. (*59, 61*)

Hales, A. L. *see* Graham & Hales (1957); Graham *et al.* (1964); Van Zijl *et al.* (1962a, b).

Hallam, A. *see* Smith & Hallam (1970).

Hamilton, W. (1967). Tectonics of Antarctica. *Tectonophysics*, **4**, 555–68. (*258, 262*)

Hamilton, W. (1970). The Uralides and the motion of the Russian and Siberian platforms. *Geol. Soc. Amer. Bull.* **81**, 2553–76. (*184, 245, 249, 271*)

Harland, W. B., Smith, A. G. & Wilcock, B., eds. (1964). *The Phanerozoic time-scale. Quart. J. Geol. Soc. London*, **120s**. (*167, 244*)

Harrison, C. G. A. (1966). The paleomagnetism of deep-sea sediments. *J. Geophys. Res.* **71**, 3033–43. (*118*)

Harrison, C. G. A. (1968). Evolutionary processes and reversals of the earth's magnetic field. *Nature*, **217**, 46–7. (*146*)

Harrison, C. G. A. & Funnell, B. M. (1964). Relationship of palaeomagnetic reversals and micropalaeontology in two late Cainozoic cores from the Pacific Ocean. *Nature*, **204**, 566. (*118, 143*)

Harrison, C. G. A. & Somayajulu, B. L. K. (1966). Behaviour of the earth's magnetic field during a reversal. *Nature*, **212**, 1193–5. (*134*)

Harrison, C. G. A. *see* Francheteau *et al.* (1970).

Hart, G. F. (1960). Microfloral investigations of the lower coal measures (K2), Ketewaka–Mchuchuma coalfield, Tanganyika. *Tanganyika Geol. Surv. Bull.* **30**. (*222*)

References and author index

Hartman, Georg (2)

Haughton, S. H. (1963). *Stratigraphic history of Africa south of the Sahara.* Oliver and Boyd, London. (220)

Hawley, H. N. *see* Aitken & Hawley (1967).

Hay, R. L. *see* Grommé & Hay (1971).

Hayes, D. E. *see* Heirtzler & Hayes (1967).

Hays, J. D. (1965). Radiolaria and late Tertiary and Quaternary history of Antarctic seas. In *Biology of the Antarctic Seas, II,* 5, *Antarctic Res. Ser. Amer. Geophys. Union Publ. No. 1297.* (143)

Hays, J. D. & Opdyke, N. D. (1967). Antarctic radiolaria, magnetic reversals and climatic change. *Science,* 158, 1001–11. (143, 144)

Hays, J. D., Saito, T., Opdyke, N. D. & Burckle, L. H. (1969). Pliocene–Pleistocene sediments of the equatorial Pacific: Their paleomagnetic, biostratigraphic and climatic record. *Geol. Soc. Amer. Bull.* 80, 1481–1514. (118, 143, 144)

Hays, J. D. *see also* Opdyke *et al.* (1966).

Hazeu, G. *see* Zijderveld *et al.* (1970b).

Hedley, I. G. (1968). Chemical remanent magnetization of the $FeOOH.Fe_2O_3$ system. *Phys. Earth Planet. Interiors,* 1, 103–21. (59)

Heezen, B. C. (1960). The rift in the ocean floor. *Sci. Amer.* 203 (4), 98–110. (149, 277)

Heezen, B. C. *see also* Ericson *et al.* (1961); Fox *et al.* (1970); Ninkovich *et al.* (1966).

Heier, K. S. *see* Compston *et al.* (1968).

Heirtzler, J. R. & Hayes, D. E. (1967). Magnetic boundaries in the North Atlantic Ocean. *Science,* 157, 185–7. (165, 167)

Heirtzler, J. R., Le Pichon, X. & Baron, J. G. (1966). Magnetic anomalies over the Reykjanes Ridge. *Deep-Sea Res.* 13, 427–43. (161)

Heirtzler, J. R., Dickson, G. O., Herron, E. M., Pitman III, W. C. & Le Pichon, X. (1968). Marine magnetic anomalies, geomagnetic field reversals and motions of the ocean floor and continents. *J. Geophys. Res.* 73, 2119–36. (161, 163, 164, 165, 166, 167, 168, 170)

Heirtzler, J. R. *see also* Dickson *et al.* (1968); Le Pichon & Heirtzler (1968); Pitman & Heirtzler (1966); Pitman *et al.* (1968, 1971); Rona *et al.* (1970).

Hekinian, R. *see* Opdyke & Hekinian (1967).

Helbig, K. (1965). Optimum configuration for the measurement of the magnetic moment of samples of cubical shape with a fluxgate magnetometer. *J. Geomag. Geoelect.* 17, 373–80. (74)

Helsley, C. E. (1965). Paleomagnetic results from the Lower Permian Dunkard Series of West Virginia. *J. Geophys. Res.* 70, 413–24. (128)

Helsley, C. E. (1969). Magnetic reversal stratigraphy of the Lower Triassic Moenkopi formation of western Colorado. *Geol. Soc. Amer. Bull.* 80, 2431–50. (129)

Helsley, C. E. & Spall, H. (1972). Paleomagnetism of 1140 to 1150 million year diabase sills from Gila County, Arizona. *J. Geophys Res.* 77, 2115–28. (286)

Helsley, C. E. & Steiner, M. D. (1969). Evidence for long intervals of normal polarity during the Cretaceous period. *Earth & Planet. Sci. Letters,* 5, 325–32. (128, 132)

Helsley, C. E. *see also* Graham *et al.* (1964).

Hendricks, S. J. (7)

Hendrix, W. C. *see* Vestine *et al.* (1947).

Henry, K. W. *see* Opdyke & Henry (1969).

Herron, E. M. *see* Heirtzler *et al.* (1968); Pitman *et al.* (1968).

Hess, H. H. (1954). Serpentines, Orogeny and epeiorogeny. In Poldervaart, A., ed., *The crust of the earth. Geol. Soc. Amer. Spec. Paper 62. (149)*

Hess, H. H. (1960). Evolution of ocean basins: Report to Office of Naval Research on research supported by ONR Contract Nonr 1858 (10). (*149, 150*)

Hess, H. H. (1962). History of ocean basins. In Engel, A. E. J., James, H. L. & Leonard, B. F., eds., *Petrologic Studies: a volume to Honor A. F. Buddington. Geol. Soc. Amer.* 599–620. (*ix, 149, 150*)

Hess, H. H. *see* Vine & Hess (1970).

Hess, W. N. (1964). Lifetimes and time histories of trapped radiation belt particles. In *Space Research IV*, North-Holland, Amsterdam. (*146*)

Hide, R. & Roberts, P. H. (1961). The origin of the main geomagnetic field. *Phys. Chem. Earth*, **4**, 27–98. (*137*)

Higgs, R. H. *see* Vogt *et al.* (1971).

Hilgenberg, O. (1962). Paläopollagen der Erde. *Neues Jahrb. Geol. Paläontol.* **116**, 1–56. (*277, 278, 279*)

Hirooka, H. *see* Kawai *et al.* (1969).

Holden, J. C. *see* Dietz & Holden (1970, 1971).

Holmes, A. (1965). Principles of Physical Geology. Thomas Nelson, London. (*x*)

Honea, R. M. *see* Strangway *et al.* (1967, 1968*b*).

Hope, E. R. (1967). The Baikalian Rift System (Translators comments). *Directorate of Scientific Information Services*, DRB Canada T435R. (*214*)

Hope, E. R. *see also* Aparin & Vlassov (1965); Chen Zhigiang *et al.* (1965); Khramov & Sholpo (1967); Khramov *et al.* (1965); Lee *et al.* (1963); Rodionov (1966); Vlassov & Kovalenko (1963); Vlassov & Popova (1963).

Horsfield, B. *see* Dewey & Horsfield (1970).

Hospers, J. & Van Andel, S. I. (1968). Palaeomagnetic data from Europe and North America and their bearing on the origin of the North Atlantic Ocean. *Tectonophys.* **6**, 475–90. (*253*)

Hospers, J. & Van Andel, S. I. (1969). Palaeomagnetism and Tectonics; A review. *Earth Sci. Revs.* **5**, 5–44. (*203, 241, 243, 248*)

Hospers, J. *see also* Van Andel & Hospers (1968*a, b*).

Howell, L. G. & Martinez, J. D. (1957). Polar movement as indicated by rock magnetism. *Geophysics*, **22**, 384–97. (*201*)

Howell, L. G., Martinez, J. D. & Statham, E. H. (1958). Some observations on rock magnetism. *Geophysics*, **23**, 285–98. (*201*)

Hsu, K. J. *see* Maxwell *et al.* (1970).

Hurley, P. M. & Rand, J. R. (1969). Pre-drift continental nuclei. *Science*, **164**, 1229–42. (*258*)

Hurwitz, L. (*7*); *see also* Alldredge & Hurwitz (1964).

Hutchings, A. M. J. *see* Tarling *et al.* (1967).

Inglis, D. R. (1957). Shifting of the earth's axis of rotation. *Rev. Mod. Phys.* **29**, 9–19. (*275*)

Irons, J. *see* Briden *et al.* (1971*a*).

Irving, E. (1956*a*). Palaeomagnetic and Palaeoclimatological aspects of polar wandering. *Geofis. Pura Appl.* **33**, 23–41. (*193*)

Irving, E. (1956*b*). The magnetization of the Mesozoic dolerites of Tasmania. *Pap. Proc. Roy. Soc. Tasmania*, **90**, 157–68. (*227*)

Irving, E. (1957*a*). The origin of the palaeomagnetism of the Torridonian Sandstone Series of Northwest Scotland. *Phil. Trans. Roy. Soc. London*, A**250**, 100–10. (*63*)

Irving, E. (1957*b*). Directions of magnetization in the Carboniferous glacial varves of Australia. *Nature*, **180**, 280–1. (*227*)

Irving, E. (1957*c*). Rock Magnetism: a new approach to some palaeogeographic problems. *Phil. Mag. Suppl. Adv. Phys.* **6**, 194–218. (*255*)

Irving, E. (1958*a*). Rock Magnetism: a new approach to the problems of polar wandering and continental drift. In Carey, S. W., ed., *Continental Drift – A Symposium*. Univ. of Tasmania, Hobart, 24–61. (*239, 255*)

Irving, E. (1958*b*). Palaeogeographic reconstruction from palaeomagnetism. *Geophys. J. Roy. Astron. Soc.* **1**, 224–37. (*255*)

Irving, E. (1959). Palaeomagnetic pole positions: A survey and analysis. *Geophys. J. Roy. Astron. Soc.* **2**, 51–79. (*255*)

Irving, E. (1960–5). Palaeomagnetic directions and pole positions. Parts I–VII. *Geophys. J. Roy. Astron. Soc.* **3**, 96–111 (1960*a*); **3**, 444–9 (1960*b*); **5**, 72–9 (1961); **6**, 263–7 (1962*a*); **7**, 263–74 (1962*b*) (with P. M. Stott); **8**, 249–57 (1963); **9**, 185–94 (1965). (*105, 282*)

Irving, E. (1963). Paleomagnetism of the Narrabeen chocolate shale and of the Tasmanian dolerite. *J. Geophys. Res.* **68**, 2283–7. (*125*)

Irving, E. (1964). *Paleomagnetism and its application to geological and geophysical problems*. Wiley, N.Y. (*22, 27, 28, 50, 80, 81, 86, 105, 107, 109, 110, 129, 135, 197, 206, 252, 270, 272, 289, 309*)

Irving, E. (1966). Paleomagnetism of some Carboniferous rocks from New South Wales and its relation to geological events. *J. Geophys. Res.* **71**, 6025–51. (*125, 227, 230, 256, 257, 260*)

Irving, E. (1967). Evidence for palaeomagnetic inclination error in sediment. *Nature*, **213**, 483–4. (*63*)

Irving, E. (1970). The Mid-Atlantic Ridge at 45°N. XVI. Oxidation and magnetic properties of basalt; review and discussion. *Can. J. Earth Sci.* **7**, 1528–38. (*167*)

Irving, E. & Briden, J. C. (1962). Palaeolatitude of evaporite deposits. *Nature*, **196**, 425–8. (*194*)

Irving, E. & Brown, D. A. (1964). Abundance and diversity of the labyrinthodonts as a function of paleolatitude. *Amer. J. Sci.* **262**, 689–708. (*196*)

Irving, E. & Brown, D. A. (1966). Reply to Stehli's discussion of Labyrinthodont abundance and diversity. *Amer. J. Sci.* **264**, 488–96. (*196, 198*)

Irving, E. & Gaskell, T. F. (1962). The palaeogeographic latitude of oil fields. *Geophys. J. Roy. Astron. Soc.* **7**, 54–64. (*194*)

Irving, E. & Green, R. (1957). The palaeomagnetism of the Kainozoic basalts of Victoria. *Mon. Not. Roy. Astron. Soc. Geophys. Suppl.* **7**, 347–59. (*227*)

Irving, E. & Green, R. (1958). Polar movement relative to Australia. *Geophys. J. Roy. Astron. Soc.* **1**, 64–72. (*227*)

Irving, E. & Major, A. (1964). Post-depositional detrital remanent magnetization in a synthetic sediment. *Sedimentology*, **3**, 135–43. (*62, 63*)

Irving, E. & Opdyke, N. D. (1965). The palaeomagnetism of the Bloomsburg redbeds and its possible application to the tectonic history of the Appalachians. *Geophys. J. Roy. Astron. Soc.* **9**, 153–67. (*96, 98, 99*)

Irving, E. & Parry, L. G. (1963). The magnetism of some Permian rocks from New South Wales. *Geophys. J. Roy. Astron. Soc.* **7**, 395–411. (*111, 125, 227, 230*)

Irving, E. & Robertson, W. A. (1969). Test for polar wandering and some possible implications. *J. Geophys. Res.* **74**, 1026–36. (*260, 272*)

Irving, E. & Runcorn, S. K. (1957). Analysis of the palaeomagnetism of the Torridonian Sandstone Series of North-west Scotland. *Phil. Trans. Roy. Soc.* A**250**, 83–99. (*88, 209*)

Irving, E. & Tarling, D. H. (1961). The Paleomagnetism of the Aden Volcanics. *J. Geophys. Res.* **66**, 549–56. (*241*)

Irving, E. & Ward, M. A. (1964). A statistical model of the geomagnetic field. *Pure & Appl. Geophys.* **57**, 47–52. (*28, 139*)

Irving, E., Robertson, W. A. & Stott, P. M. (1963). The significance of the paleomagnetic results from Mesozoic rocks of eastern Australia. *J. Geophys. Res.* **68**, 2313–17. (*227, 256*)

Irving, E., Stott, P. M. & Ward, M. A. (1961*a*). Demagnetization of igneous rocks by alternating magnetic fields. *Phil. Mag.* **6**, 225–41. (*93*)

Irving, E., Park, J. K., Haggerty, S. E., Aumento, F. & Loncarevic, B. (1970). Magnetism and opaque mineralogy of basalts from the mid-Atlantic Ridge at 45 °N. *Nature*, **228**, 974–6. (*159*)

Irving, E., Robertson, W. A., Stott, P. M., Tarling, D. H. & Ward, M. A. (1961*b*). Treatment of partially stable sedimentary rocks showing planar distribution of directions of magnetization. *J. Geophys. Res.* **66**, 1927–33. (*96*)

Irving, E. *see also* Briden & Irving (1964); Bull *et al.* (1962); Collinson *et al.* (1957); Creer *et al.* (1954, 1957, 1958, 1959); Watson & Irving (1957).

Isacks, B., Oliver, J. & Sykes, L. R. (1968). Seismology and the new global tectonics. *J. Geophys. Res.* **73**, 5855–99. (*155, 156, 157*)

Ishikawa, Y. & Syono, Y. (1963). Order-disorder transformation and reverse thermoremanent magnetism in the $FeTiO_3$–Fe_2O_3 system. *Phys. Chem. Solids*, **24**, 517. (*109*)

Ispir, Y. *see* Creer & Ispir (1970).

Ito, H. *see* Kawai *et al.* (1961).

Johnson, E. A. (1938). The limiting sensitivity of an alternating current method of measuring small magnetic moments. *Rev. Sci. Instr.* **9**, 263–6. (*76*)

Johnson, E. A., Murphy, T. & Torrenson, O. W. (1948). Prehistory of the Earth's magnetic field. *Terr. Magn. Atmos. Elec.* **53**, 349–72. (*62, 63, 104*)

Johnson, E. A. *see also* McNish & Johnson (1938).

Johnson, G. L. *see* Fox *et al.* (1970); Vogt *et al.* (1971).

Jones, H. S. (*7*)

Jones, D. L. & McElhinny, M. W. (1967). Stratigraphic interpretation of paleomagnetic measurements on the Waterberg redbeds of South Africa. *J. Geophys. Res.* **72**, 4171–9. (*97, 223*)

Jones, D. L. *see also* Fahrig & Jones (1969); McElhinny *et al.* (1968).

Jones, J. G. *see* Veevers *et al.* (1971).

Kalashnikov, A. G. (1961). The history of the geomagnetic field. *Akad. Nauk SSSR Izv., Geophys. Ser.* 1243–79. (*212*)

Karig, D. E. (1971). Origin and development of marginal basins in the Western Pacific. *J. Geophys. Res.* **76**, 2542–61. (*252, 258*)

Kawai, N., Hirooka, K. & Nakajima, T. (1969). Palaeomagnetic and potassium-argon age informations supporting Cretaceous–Tertiary hypothetic bending of the main island of Japan. *Palaeogeog. Palaeoclim. Palaeoecol.* **6**, 277–82. (*249, 251, 302*)

Kawai, N., Ito, H. & Kume, S. (1961). Deformation of the Japanese islands as inferred from rock magnetism. *Geophys. J. Roy. Astron. Soc.* **6**, 124–9. (*249*)

Kawai, N. *see also* Kumagai *et al.* (1950).

References and author index

Kay, M. (1951). North American geosynclines. *Geol. Soc. Amer. Mem.* **48**. (*184*)

Kay, M. (1969). Continental Drift in the North Atlantic Ocean. In Kay, M., ed., *North Atlantic – Geology and Continental Drift, A Symposium. Amer. Assn. Petr. Geol. Mem.* **12**, 965–73. (*266, 267*)

Keiller, J. A. *see* Graham & Keiller (1960).

Kennedy, G. C. (1959). The origin of continents, mountain ranges, and ocean basins. *Am. Sci.* **47**, 491–504. (*149*)

Kern, J. W. (1961). Effects of moderate stresses on directions of thermoremanent magnetization. *J. Geophys. Res.* **66**, 3801–5. (*64*)

Khramov, A. N. (1967). The earth's magnetic field in the late Paleozoic. *Akad. Nauk SSSR Izv. Earth Phys. Ser.* 86–108. (*125, 213*)

Khramov, A. N. & Sholpo, L. Ye. (1967). Synoptic tables of U.S.S.R. paleomagnetic data. Appendix I of *Paleomagnetism.* Nedra Press, Leningrad, 213–33. (Translated by E. R. Hope, Directorate of Scientific Information Services, DRB Canada T510R, 1970.) (*129, 212, 282*)

Khramov, A. N., Rodionov, V. P. & Komissarova, R. A. (1965). New data on the Paleozoic history of the geomagnetic field in the U.S.S.R. In *The present and past of the geomagnetic field.* Nauka Press, Moscow, 206–13. (Translated by E. R. Hope, Directorate of Scientific Information Services, DRB Canada T460R, 1966.) (*128*)

King, R. F. (1955). The remanent magnetism of artifically deposited sediments. *Mon. Not. Roy. Astron. Soc. Geophys. Suppl.* **7**, 115–34. (*62, 63*)

King, R. F. & Rees, A. I. (1966). Detrital magnetism in sediments: An examination of some theoretical models. *J. Geophys. Res.* **71**, 561–71. (*62*)

King, R. F. *see also* Griffiths *et al.* (1960).

Kip, A. F. (1962). *Fundamentals of electricity and magnetism.* McGraw-Hill, N.Y. (*32*)

Kittel, C. (1949). Ferromagnetic domain theory. *Rev. Mod. Phys.* **21**, 541–83. (*47*)

Klinger, H. C. *see* Dingle & Klinger (1971).

Klootwijk, C. T. *see* Wensink & Klootwijk (1968).

Knott, S. T. *see* Emery *et al.* (1970).

Kobayashi, K. (1959). Chemical remanent magnetization of ferromagnetic minerals and its application to rock magnetism. *J. Geomag. Geoelect.* **10**, 99–117. (*59, 92*)

Kobayashi, K. (1968). Paleomagnetic determination of the intensity of the geomagnetic field in the Precambrian period. *Phys. Earth Planet. Interiors*, **1**, 387–95. (*31*)

Koenisberger, J. G. (1938). Natural residual magnetism of eruptive rocks, parts I and II. *Terr. Magn. Atmos. Elec.* **43**, 119–127; 299–30. (*33*)

Kohler, R. E. *see* Elvers *et al.* (1967).

Komissarova, R. A. *see* Khramov *et al.* (1965).

Kotasek, J. & Krs, M. (1965). Palaeomagnetic study of tectonic rotations in the Carpathian Mountains of Czechoslovakia. *Palaeogeog. Palaeoclim. Palaeoecol.* **1**, 39–49. (*248, 292*)

Kovalenko, G. V. *see* Vlassov & Kovalenko (1963).

Krs, M. *see* Birkenmajer *et al.* (1968); Kotasek & Krs (1965).

Ku, T. *see* Broecker *et al.* (1968).

Kulbicki, G. *see* Beuf *et al.* (1966).

Kumagai, N., Kawai, N. & Nagata, T. (1950). Recent progress in Japan. *J. Geomag. Geoelect.* **2**, 61–5. (*302*)

Kume, S. *see* Kawai *et al.* (1961).

328

Kristjansson, L. G. *see* Deutsch *et al.* (1971).

Kropotkin, P. N. *see* Gurariy *et al.* (1966).

Lange, I. *see* Vestine *et al.* (1947).

Langseth, M. *see* Le Pichon & Langseth (1969).

Larionova, G. Ya. *see* Pospelova *et al.* (1968).

Larmor, J. (1919). How could a rotating body such as the sun become a magnet? *Rep. Brit. Assn.* 159–160. (*10*)

Larochelle, A. (1966). Palaeomagnetism of the Abitibi dyke swarm. *Can. J. Earth Sci.* **3**, 671–83. (*205*)

Larochelle, A. (1967). The palaeomagnetism of the Sudbury diabase dyke swarm. *Can. J. Earth Sci.* **4**, 323–32. (*205*)

Larochelle, A. *see also* Fahrig *et al.* (1965).

Larporte, L. *see* Vestine *et al.* (1947).

Larson, E. E. & Strangway, D. W. (1966). Magnetic polarity and igneous petrology. *Nature*, **212**, 756–7. (*112*)

Larson, E. E. & Strangway, D. W. (1968). Discussion on paper 'Correlation of petrology and natural magnetic polarity in Columbia plateau basalts' by R. L. Wilson & N. D. Watkins. *Geophys. J. Roy. Astron. Soc.* **15**, 437–41. (*113*)

Larson, E. E. *see also* Goldstein *et al.* (1969); Strangway *et al.* (1967, 1968a, b).

Lawley, E. (1970). The intensity of the geomagnetic field in Iceland during Neogene polarity transitions and systematic deviations. *Earth Planet. Sci. Letters*, **10**, 145–9. (*138*)

Leaton, B. R. (*7*)

Leaton, B. R. & Malin, S. R. C. (1967). Recent changes in the magnetic dipole moment of the earth. *Nature*, **213**, 1110. (*7*)

Lee, C., Lee, H., Liu, H., Liu, C. & Yeh, S. (1963). Preliminary study of paleomagnetism of some Mesozoic and Cenozoic redbeds of South China. *Acta Geologica Sinica*, **43**, 241–6. (Translated by E. R. Hope, Directorate of Scientific Information Services, DRB Canada T7C, 1966.) (*216*)

Lee, H. *see* Lee *et al.* (1963).

Le Pichon, X. (1968). Sea-floor spreading and continental drift. *J. Geophys. Res.* **73**, 3661–97. (*156, 170, 173, 175, 176, 235, 272*)

Le Pichon, X. & Heirtzler, J. R. (1968). Magnetic anomalies in the Indian Ocean and sea-floor spreading. *J. Geophys. Res.* **73**, 2101–17. (*161, 165, 260*)

Le Pichon, X. & Langseth, M. G. (1969). Heat flow from the mid-ocean ridges and sea-floor spreading. *Tectonophysics*, **8**, 319–44. (*157*)

Le Pichon, X. *see also* Heirtzler *et al.* (1966, 1968).

Light, B. G. *see* Nairn *et al.* (1959).

Lilley, F. E. M. (1970a). On kinematic dynamos. *Proc. Roy. Soc. London*, A**316**, 153–67. (*12*)

Lilley, F. E. M. (1970b). Geomagnetic reversals and the position of the North Magnetic Pole. *Nature*, **227**, 1336–7. (*12, 13*)

Lindsley, N. C. *see* Beck & Lindsley (1969).

Liu, C. *see* Lee *et al.* (1963).

Liu, H. *see* Lee *et al.* (1963).

Loncarevic, B. *see* Irving *et al.* (1970).

Lotze, F. (1957). *Steinsalz und Kalisalze.* Borntrager, Berlin. (*195*)

Lovering, J. F. (1958). The nature of the Mohorovicic discontinuity. *Trans. Amer. Geophys. Union*, **39**, 947–55. (*149*)

References and author index

Lowes, F. J. (1955). Secular variation and the non-dipole field. *Ann. Geophys.* **11**, 91–4. (*14, 16, 141*)

Lowes, F. J. & Wilkinson, I. (1963). Geomagnetic dynamo: a laboratory model. *Nature*, **198**, 1158–60. (*13, 15*)

Lowes, F. J. & Wilkinson, I. (1968). Geomagnetic dynamo: an improved laboratory model. *Nature*, **219**, 717–18. (*13*)

Luck, G. R. (1971). The palaeomagnetism of the Australian Palaeozoic. Ph.D. thesis, Australian National Univ. (*38, 228, 257, 306*)

Luck, G. R. *see also* McElhinny & Luck (1970a, b).

MacDonald, G. J. F. (1963). The deep structure of oceans and continents. *Rev. Geophys.* **1**, 587–665. (*275, 276, 277*)

MacDonald, G. J. F. *see also* Munk & MacDonald (1960).

Major, A. *see* Irving & Major (1964).

Malin, S. R. C. *see* Leaton & Malin (1967).

Martin, H. (1961). The hypothesis of continental drift in the light of recent advances of geological knowledge in Brazil and in southwest Africa. *Trans. Geol. Soc. S. Africa. Annexure to* **47**, 1–47. (*266*)

Martinez, J. D. *see* Howell & Martinez (1957); Howell *et al.* (1958).

Mathews, J. H. & Gardner, W. K. (1963). Field reversals of 'paleomagnetic' type in coupled disk dynamos. *U.S. Naval Res. Lab. Rept.* no. 5886. (*13, 14*)

Mathews, R. K. *see* Broecker *et al.* (1968).

Matthews, D. H. & Williams, C. A. (1968). Linear magnetic anomalies in the Bay of Biscay: a qualitative interpretation. *Earth Planet. Sci. Letters*, **4**, 315–20. (*246*)

Matthews, D. H. *see also* Vine & Matthews (1963).

Matthewson, C. C. *see* Elvers *et al.* (1967).

Matuyama, M. (1929). On the direction of magnetization of basalt in Japan, Tyosen and Manchuria. *Proc. Imp. Acad. Japan*, **5**, 203–5. (*22, 114*)

Maxwell, A. E., Von Herzen, R. P., Hsu, K. J., Andrews, J. E., Saito, T., Percival, S. F., Mitlow, E. D. & Boyce, R. E. (1970). Deep-sea drilling in the South Atlantic. *Science*, **168**, 1047–59. (*266*)

Maxwell, A. E. *see also* Bullard *et al.* (1956).

May, B. T. *see* Deutsch *et al.* (1971).

McConnell, R. K. (1965). Isostatic adjustment in a layered Earth. *J. Geophys. Res.* **70**, 5171–88. (*276*)

McDonald, K. L. & Gunst, R. H. (1968). Recent trends in the earth's magnetic field. *J. Geophys. Res.* **73**, 2057–67. (*7*)

McDougall, I. (1963). Potassium–Argon age measurements on dolerites from Antarctica and South Africa. *J. Geophys. Res.* **68**, 1535–45. (*267*)

McDougall, I. & Chamalaun, F. H. (1966). Geomagnetic polarity scale of time. *Nature*, **212**, 1415–18. (*115*)

McDougall, I. & Chamalaun, F. H. (1969). Isotopic dating and geomagnetic polarity studies on volcanic rocks from Mauritius, Indian Ocean. *Geol. Soc. Amer. Bull.* **80**, 1419–42. (*118*).

McDougall, I. & McElhinny, M. W. (1970). The Rajmahal traps of India – K–Ar ages and palaeomagnetism. *Earth Planet. Sci. Letters*, **9**, 371–8. (*232, 307*)

McDougall, I. & Tarling, D. H. (1964). Dating geomagnetic polarity zones. *Nature*, **202**, 171–2. (*115*)

McDougall, I. *see also* Compston *et al.* (1968); Wellman *et al.* (1969).

McElhinny, M. W. (1964). Statistical significance of the fold test in palaeomagnetism. *Geophys. J. Roy. Astron. Soc.* **8**, 338–40. (*89*)

McElhinny, M. W. (1966). An improved method for demagnetizing rocks in alternating magnetic fields. *Geophys. J. Roy. Astron. Soc.* **10**, 369–74. (*92*)

McElhinny, M. W. (1967a). Statistics of a spherical distribution. In Collinson, D. W., Creer, K. M. & Runcorn, S. K., eds., *Methods in palaeomagnetism.* Elsevier, Amsterdam, 313–21. (*83*)

McElhinny, M. W. (1967b). The paleomagnetism of the southern continents; a survey and analysis. *Symposium on Continental Drift* (UNESCO/IUGS). Montevideo, Uruguay, October 1967. [*Trans. Amer. Geophys. Un.* **52**, 176 (1972).] (*256*)

McElhinny, M. W. (1968–72). Palaeomagnetic directions and pole positions. Parts VIII–XII. *Geophys. J. Roy. Astron. Soc.* **15**, 409–30 (1968a); **16**, 207–24 (1968b); **18**, 305–27 (1969a); **20**, 417–29 (1970a); **27**, 237–57 (1972). (*105, 129, 282*)

McElhinny, M. W. (1968c). Northward drift of India – Examination of recent palaeomagnetic results. *Nature*, **217**, 342–4. (*230, 269*)

McElhinny, M. W. (1969b). The palaeomagnetism of the Permian of southeast Australia and its significance regarding the problem of intercontinental correlation. *Spec. Publ. Geol. Soc. Australia*, **2**, 61–7. (*125, 127*)

McElhinny, M. W. (1970b). The palaeomagnetism of the Cambrian Purple Sandstone from the Salt Range, West Pakistan. *Earth Planet. Sci. Letters*, **8**, 149–56. (*137, 232, 256*)

McElhinny, M. W. (1970c). Formation of the Indian Ocean. *Nature*, **228**, 977–9. (*260, 267, 269*)

McElhinny, M. W. (1971). Geomagnetic reversals during the Phanerozoic. *Science*, **172**, 157–9. (*128, 129, 131, 132, 133*)

McElhinny, M. W. & Briden, J. C. (1971). Continental drift during the Palaeozoic. *Earth Planet. Sci. Letters*, **10**, 407–16. (*208, 225, 272, 275*)

McElhinny, M. W. & Burek, P. J. (1971). Mesozoic palaeomagnetic stratigraphy. *Nature*, **232**, 98–102. (*131, 132, 167, 266*)

McElhinny, M. W. & Evans, M. E. (1968). An investigation of the strength of the geomagnetic field in the early Precambrian. *Phys. Earth Planet. Interiors*, **1**, 485–97. (*30*)

McElhinny, M. W. & Gough, D. I. (1963). The palaeomagnetism of the Great Dyke of Southern Rhodesia. *Geophys. J. Roy. Astron. Soc.* **7**, 287–303. (*84, 93*)

McElhinny, M. W. & Luck, G. R. (1970a). The palaeomagnetism of the Antrim Plateau Volcanics of Northern Australia. *Geophys. J. Roy. Astron. Soc.* **20**, 191–205. (*228, 230, 256*)

McElhinny, M. W. & Luck, G. R. (1970b). Paleomagnetism and Gondwanaland. *Science*, **168**, 830–2. (*256*)

McElhinny, M. W. & Opdyke, N. D. (1964). Paleomagnetism of the Precambrian dolerites of eastern Southern Rhodesia; an example of geologic correlation by rock magnetism. *J. Geophys. Res.* **69**, 2465–75. (*94, 95, 222, 257*)

McElhinny, M. W. & Opdyke, N. D. (1968). Paleomagnetism of some Carboniferous glacial varves from Central Africa. *J. Geophys. Res.* **73**, 689–96. (*256*)

McElhinny, M. W. & Wellman, P. (1969). Polar wandering and sea-floor spreading in the southern Indian Ocean. *Earth Planet. Sci. Letters*, **6**, 198–204. (*274*)

References and author index

McElhinny, M. W., Briden, J. C., Jones, D. L. & Brock, A. B. (1968). Geological and geophysical implications of paleomagnetic results from Africa. *Rev. Geophys.* **6**, 201–38. (*220, 222, 224*)

McElhinny, M. W. *see also* Evans & McElhinny (1969); Evans *et al.* (1968); Gough *et al.* (1964); Jones & McElhinny (1967); McDougall & McElhinny (1970); Opdyke & McElhinny (1965); Wellman & McElhinny (1970); Wellman *et al.* (1969).

McKenzie, D. P. (1966). The viscosity of the lower mantle. *J. Geophys. Res.* **71**, 3995–4010. (*275, 276, 277*)

McKenzie, D. P. (1967). The viscosity of the mantle. *Geophys. J. Roy. Astron. Soc.* **14**, 297–305. (*277*)

McKenzie, D. P. (1969). Speculations on the consequences and causes of plate motions. *Geophys. J. Roy. Astron. Soc.* **18**, 1–32. (*182*)

McKenzie, D. P. (1972). Plate tectonics. In Robertson, E., ed., *The nature of the solid earth*. McGraw-Hill, N.Y., 323–60. (*235, 272, 273, 275*)

McKenzie, D. P. & Morgan, W. J. (1969). Evolution of triple junctions. *Nature*, **224**, 125–33. (*177, 178, 179, 235*)

McKenzie, D. P. & Parker, R. L. (1967). The North Pacific: an example of tectonics on a sphere. *Nature*, **216**, 1276–80. (*157, 173, 176, 178, 235*)

McKenzie, D. P. & Sclater, J. G. (1971). The evolution of the Indian Ocean since the Late Cretaceous. *Geophys. J. Roy. Astron. Soc.* **24**, 437–528. (*164, 238, 267, 268, 269*)

McKenzie, D. P., Davies, D. & Molnar, P. (1970). Plate tectonics of the Red Sea and East Africa. *Nature*, **226**, 1–6. (*240, 241*)

McKenzie, D. P. *see also* Williams & McKenzie (1971).

McMahon, B. E. & Strangway, D. W. (1967). Kiaman Magnetic Interval in the western United States. *Science*, **155**, 1012–13. (*128*)

McMahon, B. E. & Strangway, D. W. (1968*a*). Investigation of Kiaman Magnetic Division in Colorado redbeds. *Geophys. J. Roy. Astron. Soc.* **15**, 265–85. (*125*)

McMahon, B. E. & Strangway, D. W. (1968*b*). Stratigraphic implications of paleomagnetic data from upper Paleozoic–Lower Triassic redbeds of Colorado. *Geol. Soc. Amer. Bull.* **79**, 417–28. (*125*)

McMahon, B. E. *see also* Strangway *et al.* (1967, 1968*b*).

McMurry, E. (1970). Palaeomagnetic results from Scottish lavas of Lower Devonian age. In Runcorn, S. K., ed., *Palaeogeophysics*. Academic Press, London, 253–62. (*270*)

McNish, A. G. & Johnson, E. A. (1938). Magnetization of unmetamorphosed varves and marine sediments. *Terr. Magn. Atmos. Elec.* **43**, 401–7. (*118*)

McQueen, D. M., Scharnberger, C. K., Scharon, L. & Halpern, M. (1972). Cambro-Ordovician Paleomagnetic Pole Position and Rubidium–Strontium total rock isochron for charnockitic rocks from Mirnyy Station, East Antarctica. *Earth Planet. Sci. Letters* (in press). (*307*)

Melloni, M. (*21*)

Melotte, P. J. (*7*)

Menard, H. W. (1967). Sea-floor spreading, topography and the second layer. *Science*, **157**, 923–4. (*164*)

Menard, H. W. (1969). Growth of drifting volcanos. *J. Geophys. Res.* **74**, 4827–37. (*172*)

Mercanton, P. L. (1926). Inversion d'inclinaison magnétique terrestre aux âges géologiques. *Terr. Magn. Atmos. Elec.* **31**, 187–90. (*22, 114, 125*)

Mercator, G. (*2*)

Merrill, R. T. (1970). Low temperature treatments of magnetite and magnetite bearing rocks. *J. Geophys. Res.* **75**, 3343–9. (*103*)

Mesolelle, K. J. *see* Broecker *et al.* (1968).

Mitlow, E. D. *see* Maxwell *et al.* (1970).

Molnar, P. *see* McKenzie *et al.* (1970).

Momose, K. (1963). Studies on the variations of the earth's magnetic field during Pliocene time. *Bull. Earthqu. Res. Inst. Tokyo*, **41**, 487–534. (*138, 140*)

Moorbath, S. (1969). Evidence for the age of deposition of the Torridonian sediments of north-west Scotland. *Scott. J. Geol.* **5**, 154–70. (*210*)

Morgan, W. J. (1968). Rises, trenches, great faults, and crustal blocks. *J. Geophys. Res.* **73**, 1959–82. (*156, 173, 174, 175, 176*)

Morgan, W. J. *see also* McKenzie & Morgan (1969).

Morris, W. A. (1972). Palaeomagnetism of the Northern Borrowdale Volcanics (Ordovician), English Lake District. *Geophys. J. Roy. Astron. Soc.* (in press). (*289*)

Morrish, A. H. (1965). *The physical principles of magnetism*. Wiley, N.Y. (*32*)

Morrish, A. H. & Yu, S. P. (1955). Dependence of the coercive force on the grain size and the density of some iron oxide powders. *J. Appl. Phys.* **26**, 1049–55. (*48, 54*)

Moses, R. L. *see* Elvers *et al.* (1967).

Munk, W. H. & MacDonald, G. J. F. (1960). *The rotation of the earth*. Cambridge Univ. Press. (*272, 275, 276, 277*)

Murthy, G. S. & Deutsch, E. R. (1972). Paleomagnetism of igneous rocks from the coast of Labrador. *Can. J. Earth Sci.* **9**, 207–12. (*286*)

Murthy, G. S., Evans, M. E. & Gough, D. I. (1971). Evidence of single domain magnetite in the Michikamau anorthosite. *Can. J. Earth Sci.* **8**, 361–70. (*48, 55*)

Murphy, T. *see* Johnson *et al.* (1948).

Nafe, J. E. & Drake, C. L. (1969). Floor of the North Atlantic–summary of the geophysical data. In Kay, M., ed., *North Atlantic – Geology and Continental Drift, A Symposium*. Amer. Assn. Petr. Geol. 59–87. (*266*)

Nafe, J. E. *see also* Drake & Nafe (1969).

Nagata, T. (1961). *Rock magnetism*. Maruzen, Tokyo (2nd edn.). (*41, 43, 62*)

Nagata, T. (1965). Main characteristics of recent geomagnetic secular variation. *J. Geomag. Geoelect.* **17**, 263–76. (*7, 13*)

Nagata, T., Uyeda, S. & Akimoto, S. (1952). Self-reversal of thermoremanent magnetism of igneous rocks. *J. Geomag. Geoelect.* **4**, 22–38. (*108*)

Nagata, T. *see also* Kumagai *et al.* (1950); Ozima *et al.* (1964).

Nairn, A. E. M. (1956). Relevance of palaeomagnetic studies of Jurassic rocks to continental drift. *Nature*, **178**, 935–6. (*219*)

Nairn, A. E. M. (1957a). Observations paléomagnetiques en France; roches Permiennes. *Bull. Soc. Géol. France*, **7**, 721–7. (*206*)

Nairn, A. E. M. (1957b). Palaeomagnetic collections from Britain and South Africa illustrating two points in weathering. *Phil. Mag. Suppl. Adv. Phys.* **6**, 162–8. (*219, 220*)

Nairn, A. E. M. (1960). A palaeomagnetic survey of the Karroo System. *Overseas Geol. Min. Res. Gt. Brit.* **7**, 398–410. (*219*)

References and author index

Nairn, A. E. M., Frost, D. V. & Light, B. G. (1959). Palaeomagnetism of certain rocks from Newfoundland. *Nature*, **183**, 596–7. (*202*)

Nairn, A. E. M. *see also* Birkenmajer *et al.* (1968); Creer *et al.* (1958, 1959).

Nakajima, T. *see* Kawai *et al.* (1969).

Nardin, M. *see* Zijderveld *et al.* (1970*b*).

Neckham, Alexander. (*1*)

Needham, J. (1962). *Science and civilization in China*, vol. 4. *Physics and Physical Technology*, 1. *Physics*. Cambridge Univ. Press. (*1, 2*)

Needham, J. *see* Smith & Needham (1967).

Néel, L. (1949). Théorie du traînage magnétique des ferromagnétiques aux grains fins avec applications aux terres cuites. *Ann. Geophys.* **5**, 99–136. (*49, 50, 90*)

Néel, L. (1955). Some theoretical aspects of rock magnetism. *Phil. Mag. Suppl. Adv. Phys.* **4**, 191–243. (*48, 49, 50, 54, 56, 108*)

Neumeyer, G. (*7*)

Newell, N. D. (1963). Crisis in the history of life. *Scient. Amer.* **208**(2), 76–92. (*143, 144, 145*)

Newton, Isaac. (*2*)

Nicholls, G. D. (1955). The mineralogy of rock magnetism. *Phil. Mag. Suppl. Adv. Phys.* **4**, 113–90. (*40*).

Ninkovich, D., Opdyke, N. D., Heezen, B. C. & Foster, J. H. (1966). Paleomagnetic stratigraphy, rates of deposition and tephrachronology in North Pacific deep-sea sediments. *Earth Planet. Sci. Letters*, **1**, 476–92. (*118, 119, 120, 121, 135, 136, 137*)

Nixon, J. *see* Bullard *et al.* (1950).

Nobili, A. (*70*)

Norman, Robert. (*2*)

O'Connell, R. *see* Anderson & O'Connell (1967).

Oersted, H. C. (*32*)

Oliver, J. *see* Isacks *et al.* (1968).

Oliver, R. L. (1964). Some basement rock relations in Antarctica. In Adie, R. J. ed., *Antarctic geology*. North Holland, Amsterdam, 259–64. (*258*)

Opdyke, N. D. (1961). The paleomagnetism of the New Jersey Triassic: A field study of the inclination error in red sediments. *J. Geophys. Res.* **66**, 1941–49. (*63*)

Opdyke, N. D. (1962). Palaeoclimatology and continental drift. In Runcorn, S. K., ed., *Continental drift*. Academic Press, N.Y., 41–65. (*194*)

Opdyke, N. D. (1964). The paleomagnetism of the Permian redbeds of southwest Tanganyika. *J. Geophys. Res.* **69**, 2477–87. (*97, 222*)

Opdyke, N. D. (1969). The Jaramillo Event as detected in oceanic cores. In Runcorn, S. K., ed., *The application of modern physics to the Earth and planetary interiors*. Wiley, Interscience, 549–52. (*121, 122, 123*)

Opdyke, N. D. & Glass, B. P. (1969). The paleomagnetism of sediment cores from the Indian Ocean. *Deep-Sea Res.* **16**, 249–61. (*118*)

Opdyke, N. D. & Hekinian, R. (1967). Magnetic properties of some igneous rocks from the mid-Atlantic ridge. *J. Geophys. Res.* **72**, 2257–60. (*157, 158*)

Opdyke, N. D. & Henry, K. W. (1969). A test of the dipole hypothesis. *Earth Planet. Sci. Letters*, **6**, 139–51. (*186*)

Opdyke, N. D. & McElhinny, M. W. (1965). The reversal at the Triassic–Jurassic boundary and its bearing on the correlation of Karroo igneous activity in southern Africa. *Trans. Amer. Geophys. Union*, **46**, 65. (*132*)

Opdyke, N. D. & Runcorn, S. K. (1960). Wind direction in the western United States in the late Paleozoic. *Geol. Soc. Amer. Bull.* **71**, 959–72. (*198*)

Opdyke, N. D., Glass, B., Hays, J. D. & Foster, J. H. (1966). Paleomagnetic study of Antarctic deep-sea cores. *Science*, **154**, 349–57. (*118, 143, 144*)

Opdyke, N. D. *see also* Foster & Opdyke (1970); Gough & Opdyke (1963); Gough *et al.* (1964); Hays & Opdyke (1967); Hays *et al.* (1969); Irving & Opdyke (1965); McElhinny & Opdyke (1964, 1968); Ninkovich *et al.* (1966).

Osborn, J. A. (1945). Demagnetizing factors of the general ellipsoid. *Phys. Rev.* **67**, 351–7. (*45*)

Oversby, B. (1971). Palaeozoic plate tectonics in the southern Tasman geosyncline, Australia. *Nature Phys. Sci.* **234**, 45–7. (*229, 258, 259*)

Ozima, M. & Ozima, M. (1965). Origin of thermoremanent magnetization. *J. Geophys. Res.* **70**, 1363–9. (*52*)

Ozima, M., Ozima, M. & Nagata, T. (1964). Low temperature treatment as an effective means of 'magnetic cleaning' of natural remanent magnetization. *J. Geomag. Geoelect.* **16**, 37–40. (*102*)

Packham, G. H. & Falvey, D. A. (1971). An hypothesis for the formation of marginal seas in the western Pacific. *Tectonophysics*, **11**, 79–109. (*258*)

Palmer, H. C. (1970). Paleomagnetism and correlation of some Middle Keweenawan rocks, Lake Superior. *Can. J. Earth Sci.* **7**, 1410–36. (*204*)

Papapetrou, A. (*72*)

Park, J. K. (1970). Acid leaching of redbeds, and its application to the relative stability of the red and black magnetic components. *Can. J. Earth Sci.* **7**, 1086–92. (*101*)

Park, J. K. *see also* Irving *et al.* (1970).

Parker, R. L. *see* McKenzie & Parker (1967).

Parry, L. G. (1965). Magnetic properties of dispersed magnetite powder. *Phil. Mag.* **11**, 303–12. (*57*)

Parry, L. G. *see also* Dickson *et al.* (1966); Irving & Parry (1963).

Pecherskiy, D. M. (1970a). Paleomagnetic studies of Mesozoic deposits of the north-east of the U.S.S.R. *Akad. Nauk SSSR Izv., Earth Phys. Ser.* 69–83. (*217, 218*)

Pecherskiy, D. M. (1970b). *Scientific works of the North East Complex Institute* (*SKVNEE*). U.S.S.R. Acad. Sci., Magadan, **37**, 58. (*132*)

Percival, S. F. *see* Maxwell *et al.* (1970).

Peregrinus, Petrus. (*1*)

Petrova, G. N. (1961). Various laboratory methods of determining the geomagnetic stability of rocks. *Akad. Nauk SSSR Izv., Geophys. Ser.* 1585–98. (*104*)

Petrova, G. N. *see also* Andreeva *et al.* (1965).

Pevzner, M. A. *see* Gurariy *et al.* (1966).

Phillips, J. D. (1967). Magnetic anomalies over the mid-Atlantic ridge near 27 °N. *Science*, **157**, 920–3. (*164*)

Phillips, J. D. *see also* Emery *et al.* (1967).

Picard, M. D. (1964). Paleomagnetic correlation of units within Chugwater (Triassic) formation, west-central Wyoming. *Amer. Assn. Petr. Geol. Bull.* **48**, 269–91. (*129*)

Piper, J. D. A. (1972). The palaeomagnetism of the Ordovician volcanics of Radnorshire, Wales. *Geophys. J. Roy. Astron. Soc.* (in press). (*289*)

Pitman III, W. C. & Heirtzler, J. R. (1966). Magnetic anomalies over the Pacific–Antarctic Ridge. *Science*, **154**, 1164–71. (*161, 162, 164*)

References and author index

Pitman III, W. C., Herron, E. M. & Heirtzler, J. R. (1968). Magnetic anomalies in the Pacific and sea-floor spreading. *J. Geophys. Res.* **73**, 2069–85. (*161, 165*)

Pitman III, W. C., Talwani, M. & Heirtzler, J. R. (1971). Age of the North Atlantic from magnetic anomalies. *Earth Planet. Sci. Letters*, **11**, 195–200. (*253, 266*)

Pitman III, W. C. *see also* Dickson *et al.* (1968); Heirtzler *et al.* (1968).

Plumstead, E. P. (1969). Gondwana floras, geochronology and glaciation in South Africa. *Internat. Geol. Cong. XXII, New Dehli 1964, Part 9*, 303–19. (*220*)

Popova, A. V. *see* Vlassov & Popova (1963).

Porath, H. (1968). Magnetic studies on specimens of intergrown maghemite and hematite. *J. Geophys. Res.* **73**, 5959–65. (*59, 61*)

Porath, H. & Chamalaun, F. H. (1968). Palaeomagnetism of Australian hematitite ore bodies–II. Western Australia. *Geophys. J. Roy. Astron. Soc.* **15**, 253–64. (*230*)

Porath, H. *see also* Chamalaun & Porath (1968).

Pospelova, G. A., Larionova, G. Ya. & Anuchin, A. V. (1968). Paleomagnetic investigations of Jurassic and Lower Cretaceous sedimentary rocks of Siberia. *Internat. Geol. Rev.* **10**, 1108–18. (*300*).

Prevot, M. & Watkins, N. D. (1969). Essai de determination de l'intensité du champ magnetique terrestre au cours d'un renversement de polarité. *Ann. Geophys.* **25**, 351–69. (*138*)

Pullaiah, G. *see* Verma & Pullaiah (1967).

Quennell, A. M. (1958). The structural and geomorphic evolution of the Dead Sea Rift. *Quart. J. Geol. Soc. London*, **114**, 1–24. (*225, 242*)

Radhakrishnamurty, C. (1963). Remanent magnetism of igneous rocks in the Gondwana formations of India. D.Sc. Thesis, Andra University, India. (*230*)

Radhakrishnamurty, C. *see also* Athavale *et al.* (1963); Clegg *et al.* (1958); Deutsch *et al.* (1958, 1959).

Rand, J. R. *see* Hurley & Rand (1969).

Ranford, L. C. *see* Wells *et al.* (1970).

Rao, K. V. *see* Deutsch & Rao (1970).

Raven, Th. *see* Van Dongen *et al.* (1967).

Rees, A. I. (1961). The effect of water currents on the magnetic remanence and anisotropy of susceptibility of some sediments. *Geophys. J. Roy. Astron. Soc.* **5**, 235–51. (*63*)

Rees, A. I. *see* Griffiths *et al.* (1960); King & Rees (1966).

Revelle, R. *see* Bullard *et al.* (1965).

Reyment, R. A. (1965). *Aspects of the geology of Nigeria: The stratigraphy of the Cretaceous and Cenozoic deposits*. Ibadan Univ. Press, Ibadan, Nigeria. (*266*)

Richards, M. L. *see* Francheteau *et al.* (1970).

Richter, C. F. *see* Gutenberg & Richter (1954).

Rikitake, T. (1958). Oscillations of a system of disk dynamos. *Proc. Cambridge Phil. Soc.* **54**, 89–105. (*13, 142*)

Rimbert, F. (1956). Sur l'action de champs alternatifs sur des roches portant une aimantation remanente isotherme de viscosité. *C.R. Acad. Sci. Paris*, **242**, 2536–8. (*93*)

Ringwood, A. E. (1969). Composition and evolution of the upper mantle. In Hart, P. J., ed., *The earth's crust and upper mantle. Amer. Geophys. Union, Geophys. Mono.* **13**, 1–17. (*150*)

Ringwood, A. E. & Green, D. H. (1966). An experimental investigation of the gabbro-eclogite transformation and some geophysical implications. *Tectonophysics*, **3**, 383–427. (*150*)

Ringwood, A. E. *see also* Green & Ringwood (1967).

Roberts, P. H. *see* Gibson & Roberts (1969); Hide & Roberts (1961).

Robertson, W. A. (1963). Paleomagnetism of some Mesozoic intrusives and tuffs from eastern Australia. *J. Geophys. Res.* **68**, 2299–2312. (*109*)

Robertson, W. A. (1969). Magnetization directions in the Muskox intrusion and associated dykes and lavas. *Geol. Surv. Can. Bull.* **167**. (*204*)

Robertson, W. A. *see also* Irving & Robertson (1969); Irving *et al.* (1961*b*, 1963).

Robinson, P. (1971). Paper presented at Royal Society London Symposium: *Plate tectonics and the evolution of the earth's crust.* (*199*)

Roche, A. (1957). Sur l'aimantation des roches volcaniques de l'Esterel. *C.R. Acad. Sci. Paris,* **244**, 2952–4. (*206*)

Roche, A. & Cattala, L. (1959). Remanent magnetism of the Cretaceous basalts of Madagascar. *Nature,* **183**, 1049–50. (*234*)

Roche, A. *see also* Andriamirado & Roche (1969).

Rodionov, V. P. (1966). Dipole character of the geomagnetic field in the Late Cambrian and Ordovician in the south of the Siberian platform. *Geolog. Geofis.* **1**, 94–101. (Translated by E. R. Hope, Directorate of Scientific Information Services, DRB Canada T460R, 1966.) (*214*)

Rodionov. V. P. *see also* Khramov *et al.* (1965).

Rona, P. A., Brakl, J. & Heirtzler, J. R. (1970). Magnetic anomalies in the northeast Atlantic between the Canary and Cape Verde Islands. *J. Geophys. Res.* **75**, 7412–20. (*167*)

Roquet, J. (1954). Sur les remanences des oxydes de fer et leur interêt en géomagnetisme (first and second parts). *Ann. Geophys.* **10**, 226–47; 282–325. (*59*)

Ro Vu Son *see* Gurariy *et al.* (1966).

Rowell, A. J. *see* Cocks *et al.* (1970).

Roy, J. L. (1963). The measurement of the magnetic properties of rock specimens. *Publ. Dominion Obs. Ottawa,* **27**, 420–39. (*72*)

Roy, J. L. (1969). Paleomagnetism of the Cumberland Group and other Paleozoic formations. *Can. J. Earth Sci.* **6**, 663–9. (*202*)

Ruditch, E. M. *see* Beloussov & Ruditch (1961).

Runcorn, S. K. (1955). Palaeomagnetism of sediments from the Colorado plateau. *Nature,* **176**, 505–6 (*201*)

Runcorn, S. K. (1956*a*). Paleomagnetic survey in Arizona and Utah: Preliminary results. *Geol. Soc. Amer. Bull.* **67**, 301–16. (*201, 252*)

Runcorn, S. K. (1956*b*). Paleomagnetic comparisons between Europe and North America. *Proc. Geol. Assoc. Canada,* **8**, 77–85. (*252*)

Runcorn, S. K. (1961). Climatic change through geological time in the light of the palaeomagnetic evidence for polar wandering and continental drift. *Quart. J. Roy. Met. Soc.* **87**, 282–313. (*194*)

Runcorn, S. K. (1962). Palaeomagnetic evidence for continental drift and its geophysical cause. In Runcorn, S. K., ed., *Continental drift.* Academic Press, N.Y., 1–39. (*252*)

Runcorn, S. K. (1964). A growing core and a convecting mantle. In Craig. H., Miller, S. L. & Wasserburg, G. L., eds., *Isotopic and cosmic chemistry.* North Holland, Amsterdam, 321–40. (*31*)

References and author index

Runcorn, S. K. (1965). A Symposium on Continental Drift I. Palaeomagnetic comparisons between Europe and North America. *Phil. Trans. Roy. Soc.* A**258**, 1–11. (*252*)

Runcorn, S. K. *see also* Collinson & Runcorn (1960); Collinson *et al.* (1957, 1967); Creer *et al.* (1954, 1957, 1958); Irving & Runcorn (1957); Opdyke & Runcorn (1960).

Rust, I. C. *see* Cocks *et al.* (1970).

Sahasrabudhe, P. W. (1963). Palaeomagnetism and the geology of the Deccan traps. *Semin. Geophys. Invest. Peninsular shield.* Osmania Univ. Hyderabad, 226–43. (*230*)

Sahasrabudhe, P. W. *see also* Athavale *et al.* (1963); Clegg *et al.* (1958); Deutsch *et al.* (1958, 1959).

Saito, T. *see* Hays *et al.* (1969); Maxwell *et al.* (1970).

Sanver, M. *see* Creer & Sanver (1967); Tarling *et al.* (1967).

Scharnberger, C. *see* McQueen *et al.* (1972); Scharon *et al.* (1969, 1970).

Scharon, L., Shimoyama, A. & Scharnberger, C. (1969). Paleomagnetic investigations in the Ellsworth Land area, Antarctica. *Antartic J. of the U.S.* **4**, 94–5. (*234*)

Scharon, L., Scharnberger, C., Early, T. & Shimoyama, A. (1970). International co-operation for paleomagnetic insight into Antarctic tectonic history. *Antarctic J. of the U.S.* **5**, 219–24. (*234*)

Schmidt, A. (*7*)

Schneider, E. D. *see* Vogt *et al.* (1970).

Schonland, B. F. (1953). *Atmospheric Electricity.* Methuen, London. (*69*)

Schopf, J. M. (1969). Ellsworth Mountains: Position in West Antarctica due to sea-floor spreading. *Science*, **164**, 63–5. (*232, 262*)

Schwarz, E. J. & Symons, D. T. A. (1969). Geomagnetic intensity between 100 million and 2500 million years ago. *Phys. Earth Planet. Interiors*, **2**, 11–18. (*30, 31*)

Sclater, J. G. *see* Francheteau & Sclater (1970); Francheteau *et al.* (1970); McKenzie & Sclater (1971).

Shimoyama, A. *see* Scharon *et al.* (1969, 1970).

Sholpo, L. Ye. *see* Khramov & Sholpo (1967).

Sigurgeirsson, T. *see* Dagley *et al.* (1967).

Simpson, J. F. (1966). Evolutionary pulsations and geomagnetic polarity. *Geol. Soc. Amer. Bull.* **77**, 197–203. (*144*)

Smith, A. G. & Hallam, A. (1970). The fit of the southern continents. *Nature*, **225**, 139–44. (*256, 257, 261, 262, 263*)

Smith, A. G. *see also* Briden *et al.* (1971*b*); Bullard *et al.* (1965); Funnell & Smith (1968); Harland *et al.* (1964).

Smith, J. D. & Foster, J. H. (1969). Geomagnetic reversal in Brunhes Normal Polarity Epoch. *Science*, **163**, 565–7. (*121, 122*)

Smith, P. J. (1967). The intensity of the ancient geomagnetic field: A review and analysis. *Geophys. J. Roy. Astron. Soc.* **12**, 213–362. (*20, 29, 30, 138, 140, 141*)

Smith, P. J. (1968). Pre-Gilbertian conceptions of terrestrial magnetism. *Tectonophysics*, **6**, 499–510. (*1*)

Smith, P. J. (1970). Petrus Peregrinus Epistola – The beginning of experimental studies of magnetism in Europe. *Atlas (News Supp. to Earth Sci. Revs.)*, **6** (1), A11–A17. (*1, 2*)

Smith, P. J. & Needham, J. (1967). Magnetic declination in medieval China. *Nature*, **214**, 1213–14. (*2*)

338

Smith, P. J. *see also* Dagley *et al.* (1967).

Snelling, N. J. *see* Cahen & Snelling (1966).

Somayajulu, B. L. K. *see* Harrison & Somayajulu (1966).

Sougy, J. (1962). West African fold belt. *Geol. Soc. Amer. Bull.* **73**, 871–6. (*259*)

Spall, H. (1971*a*). Precambrian apparent polar wandering evidence for North America. *Earth Planet. Sci. Letters*, **10**, 273–80. (*204, 205*)

Spall, H. (1971*b*). Paleomagnetism and K–Ar age of mafic dikes from the Wind River Range, Wyoming. *Geol. Soc. Amer. Bull.* **82**, 2457–72. (*286*)

Spall, H. *see also* Helsley & Spall (1972).

Sproll, W. P. & Dietz, R. S. (1969). Morphological continental drift fit of Australia and Antarctica. *Nature*, **222**, 345–8. (*256*)

Stacey, F. D. (1960*a*). Stress-induced magnetic anisotropy of rocks. *Nature*, **188**, 134–5. (*64, 65*)

Stacey, F. D. (1960*b*). Magnetic anisotropy of igneous rocks. *J. Geophys. Res.* **65**, 2429–42. (*64, 65*)

Stacey, F. D. (1961). Theory of the magnetic properties of igneous rocks in alternating magnetic fields. *Phil. Mag.* **6**, 1241–60. (*90*)

Stacey, F. D. (1963). The physical theory of rock magnetism. *Phil. Mag. Supp. Adv. Phys.* **12**, 46–133. (*47, 48, 49, 52, 54, 57, 58, 60, 62, 104*)

Stacey, F. D. (1967). The Koenigsberger ratio and the nature of thermoremanence in igneous rocks. *Earth Planet. Sci. Letters*, **2**, 67–8. (*57*)

Stacey, F. D. (1969). *Physics of the Earth.* Wiley, N.Y. (*122*)

Stacey, F. D. *see also* Dickson *et al.* (1966); Stott & Stacey (1959, 1960).

Statham, E. H. *see* Howell *et al.* (1958).

Stehli, F. G. (1966). Discussion: Labyrinthodont abundance and diversity. *Amer. J. Sci.* **264**, 481–7. (*196*)

Stehli, F. G. (1968). A paleoclimatic test of the hypothesis of an axial dipolar magnetic field. In Phinney, R. A., ed., *The history of the earth's crust*. Princeton Univ. Press, 195–207. (*196*)

Stehli, F. G. (1970). A test of the earth's magnetic field during Permian time. *J. Geophys. Res.* **75**, 3325–42. (*196*)

Steiner, M. B. *see* Helsley & Steiner (1969).

Steveaux, J. *see* Beuf *et al.* (1966).

Stockard, H. *see* Drake *et al.* (1968).

Stocklin, J. (1968). Structural history and tectonics of Iran: A review. *Amer. Assn. Petr. Geol. Bull.* **52**, 1229–58. (*242*)

Stoner, E. C. (1945). Demagnetizing factors for ellipsoids. *Phil. Mag.* **36**, 803–21. (*45*)

Storevedt, K. M. (1967). A synthesis of Palaeozoic palaeomagnetic data for Europe. *Earth Planet. Sci. Letters*, **3**, 444–8. (*206*)

Stott, P. M. & Stacey, F. D. (1959). Magnetostriction and palaeomagnetism of igneous rocks. *Nature*, **183**, 384–5. (*64*)

Stott, P. M. & Stacey, F. D. (1960). Magnetostriction and paleomagnetism of igneous rocks. *J. Geophys. Res.* **65**, 2419–24. (*64*)

Stott, P. M. *see also* Irving & Stott (1962); Irving *et al.* (1961*a, b*, 1963).

Strangway, D. W., Larson, E. E. & Goldstein, M. (1968*a*). A possible cause of high magnetic stability in volcanic rocks. *J. Geophys. Res.* **73**, 3787–95. (*52*)

Strangway, D. W., Honea, R. M., McMahon, B. E. & Larson, E. E. (1968*b*). The magnetic properties of naturally occurring geothite. *Geophys. J. Roy. Astron. Soc.* **15**, 345–59. (*59, 60*)

References and author index

Strangway, D. W., MacMahon, B. E., Honea, R. M. & Larson, E. E. (1967). Superparamagnetism in hematite. *Earth Planet. Sci. Letters*, **2**, 37–71. (*60*)

Strangway, D. W. *see also* Goldstein *et al.* (1969); Larson & Strangway (1966, 1968); McMahon & Strangway (1967, 1968*a*, *b*).

Stubbs, P. H. S. (1958). Continental drift and polar wandering, a palaeomagnetic study of British and European Trias and of the British Old Red Sandstones. Ph.D. thesis, London Univ. (*270*)

Stubbs, P. H. S. *see also* Clegg *et al.* (1954, 1957).

Sumner, J. S. (1954). Consequences of a polymorphic transition at the Mohorovicic discontinuity. *Trans. Amer. Geophys. Union*, **35**, 385. (*149*)

Sutton, J. *see* Drake *et al.* (1959).

Swift, W. H. (1961). An outline of the geology of Southern Rhodesia. *Southern Rhod. Geol. Surv. Bull.* **50**. (*280*)

Sykes, L. R. (1967). Mechanism of earthquakes and the nature of faulting on the mid-oceanic ridges. *J. Geophys. Res.* **72**, 2131–53. (*152*, *153*)

Sykes, L. R. *see also* Isacks *et al.* (1968).

Symons, D. T. A. *see* Schwarz & Symons (1969); Tarling & Symons (1967).

Syono, Y. *see* Ishikawa & Syono (1963).

Tachimaka, H. *see* Yukutake & Tachimaka (1968).

Talent, J. A. *see* Veevers *et al.* (1971).

Talwani, M. *see* Pitman *et al.* (1971).

Tarling, D. H. & Symons, D. T. A. (1967). A stability index of remanence in palaeomagnetism. *Geophys. J. Roy. Astron. Soc.* **12**, 443–8. (*95*)

Tarling, D. H., Sanver, M. & Hutchings, A. M. J. (1967). Further palaeomagnetic results from the Federation of South Arabia. *Earth Planet. Sci. Letters*, **2**, 148–54. (*241*)

Tarling, D. H. *see also* Irving & Tarling (1961); Irving *et al.* (1961*b*); McDougall & Tarling (1964).

Termier, H. & Termier, G. (1952). *Histoire géologique de la biosphere*. Masson, Paris. (*194*)

Thellier, E. (1951). Propriétés magnétiques des terres cuites et des roches. *J. de Phys. et Radium*, **12**, 205–18. (*56*)

Thellier, E. (1966). Le champ magnétique terrestre fossile. *Nucleus*, **7**, 1–35. (*19*)

Thellier, E. & Thellier, O. (1959). Sur l'intensité du champ magnétique terrestre dans le passé historique et géologique. *Ann. Geophys.* **15**, 285–376. (*19*)

Thurber, D. L. *see* Broecker *et al.* (1968).

Toomre, A. *see* Goldreich & Toomre (1969).

Trubikhin, V. M. *see* Gurariy *et al.* (1966).

Turnbull, G. (1959). Some palaeomagnetic measurements in Antarctica. *Arctic*, **12**, 151–7. (*234*)

Uchupi, E. *see* Emery *et al.* (1970).

Uffen, R. J. (1963). Influence of the earth's core on the origin and evolution of life. *Nature*, **198**, 143–4. (*142*)

Ulrych, T. (1972). Maximum entropy power spectrum of long period geomagnetic reversals. *Nature*, **235**, 218–19. (*132*)

Uyeda, S. (1958). Thermoremanent magnetism as a medium of palaeomagnetism with special reference to reverse thermoremanent magnetism. *Japan J. Geophys.* **2**, 1–123. (*43*, *109*)

Uyeda, S., Fuller, M. D., Belshé, J. C. & Girdler, R. W. (1963). Anistropy of magnetic susceptibility of rocks and minerals. *J. Geophys. Res.* **68**, 279–91. (*65*)

Uyeda, S. *see also* Nagata *et al.* (1962).

Vacquier, V. (1962). A machine method for computing the magnetization of a uniformly magnetized body from its shape and a magnetic survey. *Proc. Benedum Earth Magnetism Symp.*, Univ. of Pittsburgh, 123. (*170*)

Valencio, D. A. & Vilas, J. F. (1970). Palaeomagnetism of some Middle Jurassic lavas from south-east Argentine. *Nature*, **225**, 262–4. (*225, 266*)

Valencio, D. A. *see also* Creer *et al.* (1970).

Van Andel, S. I. & Hospers, J. (1968a). Palaeomagnetism and the hypothesis of an expanding earth: A new calculation method and its results. *Tectonophysics*, **5**, 273–85. (*278, 279*)

Van Andel, S. I. & Hospers, J. (1968b). A statistical analysis of ancient earth radii calculated from palaeomagnetic data. *Tectonophysics*, **6**, 491–7. (*278, 279*)

Van Andel, S. I. *see also* Hospers & Van Andel (1968, 1969).

Van Breemen, O., Dodson, M. H. & Vail, J. R. (1966). Isotopic age measurements on the Limpopo orogenic belt, southern Africa. *Earth Planet. Sci. Letters*, **1**, 401–6. (*223*).

Van der Voo, R. (1967). The rotation of Spain: Palaeomagnetic evidence from the Spanish Meseta. *Palaeogeog. Palaeoclim. Palaeoecol.* **3**, 393–416. (*211*)

Van der Voo, R. (1968). Paleomagnetism and the Alpine tectonics of Eurasia IV. Jurassic, Cretaceous and Eocene pole positions from North-eastern Turkey. *Tectonophysics*, **6**, 251–69. (*225, 242*)

Van der Voo, R. (1969). Paleomagnetic evidence for the rotation of the Iberian Peninsula. *Tectonophysics*, **7**, 5–56. (*211, 246*)

Van der Voo, R. *see also* Van Dongen *et al.* (1967); Zijderveld *et al.* (1970a, b).

Van Dongen, P. G. (1967). The rotation of Spain: Palaeomagnetic evidence from the eastern Pyrenees. *Palaeogeog. Palaeoclim. Palaeoecol.* **3**, 417–32. (*211*)

Van Dongen, P. G., Van der Voo, R. & Raven, Th. (1967). Paleomagnetism and the Alpine tectonics of Eurasia III. Paleomagnetic research in the Central Lebanon Mountains and the Tartous area of Syria. *Tectonophysics*, **4**, 35–53. (*225*)

Van Hilten, D. (1962a). A deviating Permian pole from rocks in northern Italy. *Geophys. J. Roy. Astron. Soc.* **6**, 377–90. (*211*)

Van Hilten, D. (1962b). Presentation of paleomagnetic data, polar wandering and continental drift. *Amer. J. Sci.* **260**, 401–26. (*248*)

Van Hilten, D. (1963). Palaeomagnetic indications of an increase in the earth's radius. *Nature*, **200**, 1277–9. (*277*)

Van Hilten, D. (1964). Evaulation of some geotectonic hypotheses by paleomagnetism. *Tectonophysics*, **1**, 3–71. (*248*)

Van Hilten, D. (1968). Global expansion and paleomagnetic data. *Tectonophysics*, **5**, 191–210. (*278, 279*)

Van Hilten, D. & Zijderveld, J. D. A. (1966). Palaeomagnetism and the Alpine tectonics of Eurasia II. The magnetism of the Permian porphyries near Lugano (northern Italy and Switzerland). *Tectonophysics*, **3**, 429–46. (*211*)

Van Houten, F. B. (1961). Climatic significance of redbeds. In Nairn, A. E. M., ed., *Descriptive palaeoclimatology*. Interscience, N.Y., 89–139. (*60, 270*)

Van Niekerk, C. B. *see* Gough & Van Niekerk (1959).

References and author index

Van Zijl, J. S. V., Graham, K. W. T. & Hales, A. L. (1962a). The palaeomagnetism of the Stormberg lavas of South Africa I. Evidence for a genuine reversal of the earth's magnetic field in Triassic–Jurassic times. *Geophys. J. Roy. Astron. Soc.* 7, 23–39. (*134, 135*)

Van Zijl, J. S. V., Graham, K. W. T. & Hales, A. L. (1962b). The palaeomagnetism of the Stormberg lavas II. The behaviour of the magnetic field during a reversal. *Geophys. J. Roy. Astron. Soc.* 7, 169–82. (*134, 135, 137, 140*)

Veevers, J. J., Jones, J. G. & Talent, J. A. (1971). Indo-Australia stratigraphy and the configuration and dispersal of Gondwanaland. *Nature*, 229, 383–8. (*269*)

Verhoogen, J. (1959). The origin of thermoremanent magnetization. *J. Geophys. Res.* 64, 2441–9. (*52*)

Verhoogen, J. *see also* Wells & Verhoogen (1967).

Verma, R. K. & Bhalla, M. S. (1968). Paleomagnetism of Kamthi sandstones of Upper Permian age from Godavary Valley, India. *J. Geophys. Res.* 73, 703–9. (*230*)

Verma, R. K. & Pullaiah, G. (1967). Paleomagnetism of Tirupati sandstones from Godavary Valley, India. *Earth Planet. Sci. Letters*, 2, 310–16. (*230*)

Verma, R. K. *see also* Bhalla & Verma (1969).

Vestine, E. H. (*7*)

Vestine, E. H., Laporte, L., Cooper, C., Lange, I. & Hendrix, W. C. (1947). Description of the earth's magnetic field and its secular change. *Carnegie Inst. Wash. Publ. No. 578.* (*4*)

Vilas, J. F. *see* Valencio & Vilas (1970).

Vine, F. J. (1966). Spreading of the ocean floor: new evidence. *Science*, 154, 1405–15. (*160, 161, 163, 164, 165, 267*)

Vine, F. J. (1968a). Magnetic anomalies associated with mid-ocean ridges. In Phinney, R. A., ed., *The history of the earth's crust.* Princeton Univ. Press, 73–89. (*158*)

Vine, F. J. (1968b). Paleomagnetic evidence for the northward movement of the North Pacific basin during the past 100 m.y. *Trans. Amer. Geophys. Union*, 49, 156. (*171, 173*)

Vine, F. J. (1970). The Geophysical Year. *Nature*, 227, 1013–17. (*155*)

Vine, F. J. & Hess, H. H. (1970). Sea-floor spreading. In *The Sea*, volume IV, part III, 587–622. (*149, 173, 181*)

Vine, F. J. & Matthews, D. H. (1963). Magnetic anomalies over oceanic ridges. *Nature*, 199, 947–9. (*158, 159*)

Vine, F. J. & Wilson, J. T. (1965). Magnetic anomalies over a young ocean ridge off Vancouver Island. *Science*, 150, 485–9. (*158, 159*)

Vlassov, A. Ya. & Kovalenko, G. V. (1963). Geomagnetic field inversion in the Lower Devonian. In Vlassov, A. Ya., ed., *Rock magnetism and paleomagnetism.* Siberian Acad. Sci., Krasnoyarsk, 429–45. (Translated by E. R. Hope, Directorate of Scientific Information Services, DRB Canada T430R, 1965.) (*134*)

Vlassov, A. Ya. & Popova, A. V. (1963). Position of the pole in the Permian, Triassic and Cretaceous periods according to the findings of paleomagnetic studies of sedimentary rocks in the Soviet Far Eastern Maritime Province. In Vlassov, A. Ya., ed., *Rock magnetism and paleomagnetism.* Siberian Acad. Sci., Krasnoyarsk, 333–9. (Translated by E. R. Hope, Directorate of Scientific Information Services, DRB Canada T428R, 1965.) (*217*)

Vlassov, A. Ya. *see also* Aparin & Vlassov (1965).

Vogt, P. R., Higgs, R. H. & Johnson, G. L. (1971). Hypothesis on the origin of the Mediterranean Basin: Magnetic data. *J. Geophys. Res.* **76**, 3207–28. (*248*)

Vogt, P. R., Anderson, C. N., Bracey, D. R. & Schneider, E. M. (1970). North Atlantic Magnetic Smooth Zones. *J. Geophys. Res.* **75**, 3955–68. (*167*)

Von Herzen, R. P. *see* Maxwell *et al.* (1970).

Von Humboldt, A. (*21*)

Waddington, C. J. (1967). Paleomagnetic field reversals and cosmic radiation. *Science*, **158**, 913–15. (*146*)

Wadia, D. N. (1953). *Geology of India*. MacMillan, London (3rd edn). (*269*)

Walcott, R. I. *see* Gibb & Walcott (1971).

Walker, G. P. L. *see* Dagley *et al.* (1967).

Wang Cenghang *see* Chen Zhiqiang *et al.* (1965).

Ward, M. A. (1963). On detecting changes in the earth's radius. *Geophys. J. Roy. Astron. Soc.* **8**, 217–25. (*278, 279*)

Ward, M. A. *see also* Irving & Ward (1964); Irving *et al.* (1961*a, b*).

Watkins, N. D. (1965). Frequency of extrusions of some Miocene lavas in Oregon during an apparent transition of the polarity of the geomagnetic field. *Nature*, **206**, 801–3. (*134*)

Watkins, N. D. (1968). Short period geomagnetic polarity events in deep-sea sedimentary cores. *Earth Planet. Sci. Letters*, **4**, 341–9. (*122*)

Watkins, N. D. (1969). Non-dipole behaviour during an Upper Miocene geomagnetic polarity transition in Oregon. *Geophys. J. Roy. Astron. Soc.* **17**, 121–49. (*134, 140*)

Watkins, N. D. & Goodell, H. G. (1967). Geomagnetic polarity change and faunal extinction in the southern ocean. *Science*, **156**, 1083–7. (*143*)

Watkins, N. D. & Haggerty, S. E. (1968). Oxidation and magnetic polarity in single Icelandic lavas and dykes. *Geophys. J. Roy. Astron. Soc.* **15**, 305–15. (*113, 114*)

Watkins, N. D. *see also* Ade-Hall & Watkins (1970); Dagley *et al.* (1967); Prevot & Watkins (1969); Wilson & Watkins (1967); Wilson *et al.* (1968).

Watson, G. S. (1956*a*). Analysis of dispersion on a sphere. *Mon. Not. Roy. Astron. Soc. Geophys. Supp.* **7**, 153–9. (*80*)

Watson, G. S. (1956*b*). A test for randomness of directions. *Mon. Not. Roy. Astron. Soc. Geophys. Supp.* **7**, 160–1. (*81*)

Watson, G. S. & Irving, E. (1957). Statistical methods in rock magnetism. *Mon. Not. Roy. Astron. Soc. Geophys. Supp.* **7**, 289–300. (*82*)

Wayman, M. L. *see* Evans & Wayman (1970).

Weaver, G. H. *see* Aitken & Weaver (1962).

Wegener, A. (1929). *Die Enstehung der Kontinente und Ozeane*. Vieweg & Sohn. Braunschweig (4th edn). (*270*)

Wellman, P. (1971). The age and palaeomagnetism of the Australian Cenozoic volcanic rocks. Ph.D thesis, Australian National Univ. (*306*)

Wellman, P. & McElhinny, M. W. (1970). K–Ar age of the Deccan traps, India. *Nature*, **227**, 595–6. (*232, 269*)

Wellman, P., McElhinny, M. W. & McDougall, I. (1969). On the polar wander path for Australia during the Cenozoic. *Geophys. J. Roy. Astron. Soc.* **18**, 371–95. (*18, 118, 135, 169, 230*)

Wellman, P. *see also* McElhinny & Wellman (1969).

Wells, A. T., Ranford, L. C., Cook, P. J. & Forman, D. J. (1970). The geology of the Amadeus Basin, N.T. *Bur. Miner. Resour. Aust. Bull.* **100**. (*228*)

References and author index

Wells, J. M. & Verhoogen, J. (1967). Late Paleozoic paleomagnetic poles and the opening of the Atlantic Ocean. *J. Geophys. Res.* **72**, 1777–81. (*253*)

Wensink, H. (1968). Paleomagnetism of some Gondwana redbeds from Central India. *Palaeogeog. Palaeoclim. Palaeoecol.* **5**, 323–43. (*230*)

Wensink, H. & Klootwijk, C. T. (1968). The paleomagnetism of the Talchir Series of the Lower Gondwana System, Central India. *Earth Planet. Sci. Letters*, **4**, 191–6. (*230, 232*)

West, G. F. *see* Dunlop & West (1969).

Wilcock, B. *see* Harland *et al.* (1964).

Wilkinson, I. *see* Lowes & Wilkinson (1963, 1968).

Williams, C. A. & McKenzie, D. (1971). The evolution of the north-east Atlantic. *Nature*, **232**, 168–73. (*248, 253, 266*)

Williams, C. A. *see also* Matthews & Williams (1968).

Willis, I. *see* Bull *et al.* (1962).

Wilson, J. T. (1960). Some consequences of expansion of the earth. *Nature*, **185**, 880–2. (*277*)

Wilson, J. T. (1965a). A new class of faults and their bearing on continental drift. *Nature*, **207**, 343–7. (*151, 152, 154, 175, 249*)

Wilson, J. T. (1965b). Transform faults, oceanic ridges and magnetic anomalies southwest of Vancouver Island. *Science*, **150**, 482–5. (*151, 152*)

Wilson, J. T. (1966). Did the Atlantic close and then reopen? *Nature*, **211**, 676–81. (*254, 271*)

Wilson, J. T. *see also* Vine & Wilson (1965).

Wilson, R. L. (1962a). The palaeomagnetic history of a doubly baked rock. *Geophys. J. Roy. Astron. Soc.* **6**, 397–9. (*111*)

Wilson, R. L. (1962b). The palaeomagnetism of baked contact rocks and reversals of the earth's magnetic field. *Geophys. J. Roy. Astron. Soc.* **7**, 194–202. (*110*)

Wilson, R. L. (1962c). An instrument for measuring vector magnetization at high temperatures. *Geophys. J. Roy. Astron. Soc.* **7**, 125–30. (*96*)

Wilson, R. L. (1964). Magnetic properties and normal and reversed natural magnetization in the Mull lavas. *Geophys. J. Roy. Astron. Soc.* **8**, 424–39. (*112*).

Wilson, R. L. (1966). Further correlations between the petrology and the natural magnetic polarity of basalts. *Geophys. J. Roy. Astron. Soc.* **10**, 413–20. (*112*)

Wilson, R. L. (1970a). Palaeomagnetic stratigraphy of Tertiary lavas from Northern Ireland. *Geophys. J. Roy. Astron. Soc.* **20**, 1–9. (*186*)

Wilson, R. L. (1970b). Permanent aspects of the earth's non-dipole magnetic field over Upper Tertiary times. *Geophys. J. Roy. Astron. Soc.* **19**, 417–37. (*189, 191*)

Wilson, R. L. (1971). Dipole offset – The time average palaeomagnetic field over the past 25 million years. *Geophys. J. Roy. Astron. Soc.* **22**, 491–504. (*189, 191, 192, 193*)

Wilson, R. L. & Ade-Hall, J. M. (1970). Palaeomagnetic indications of a permanent aspect of the non-dipole field. In Runcorn, S. K., ed., *Palaeogeophysics*. Academic Press, London, 307–12. (*189, 191, 192*)

Wilson, R. L. & Haggerty, S. E. (1966). Reversals of the earth's magnetic field. *Endeavour*, **25**, 104–9. (*39, 64*)

Wilson, R. L. & Watkins, N. D. (1967). Correlation of magnetic polarity and petrological properties in Columbia Plateau basalts. *Geophys. J. Roy. Astron. Soc.* **12**, 405–24. (*112, 113, 114*)

Wilson, R. L., Haggerty, S. E. & Watkins, N. D. (1968). Variation of palaeomagnetic stability and other parameters in a vertical traverse of a single Icelandic lava. *Geophys. J. Roy. Astron. Soc.* **16**, 79–96. (*95*)

Wilson, R. L. *see also* Ade-Hall & Wilson (1963, 1969); Dagley *et al.* (1967).

Windom, H. L. *see* Dymond & Windom (1968).

Wohlfarth, E. P. (1955). The remanent magnetization of haematite powders. *Phil. Mag.* **46**, 1155–64. (*60*)

Wollin, G. *see* Ericson *et al.* (1961).

Wright, A. E. *see* Griffiths *et al.* (1960).

Yeh, S. *see* Lee *et al.* (1963).

Yu, S. P. *see* Morrish & Yu (1955).

Yukutake, T. & Tachimaka, H. (1968). The non-dipole part of the earth's magnetic field. *Bull. Earthqu. Res. Inst. Tokyo,* **46**, 1027–74. (*7*)

Zijderveld, J. D. A. (1968). Natural remanent magnetizations of some intrusive rocks from the Sør Rondane Mountains, Queen Maud Land, Antarctica. *J. Geophys. Res.* **73**, 3773–85. (*256*)

Zijderveld, J. D. A. & de Jong, K. A. (1969). Paleomagnetism of some late Paleozoic and Triassic rocks from the eastern Lombardic Alps. *Geol. Mijnbouw,* **48**, 559–64. (*211*)

Zijderveld, J. D. A., de Jong, K. A. & Van der Voo, R. (1970*a*). Rotation of Sardinia: Palaeomagnetic evidence from Permian rocks. *Nature,* **226**, 933–4. (*211*)

Zijderveld, J. D. A., Hazeu, G. J. A., Nardin, M. & Van der Voo, R. (1970*b*). Shear in the Tethys and the Permian paleomagnetism in the southern Alps, including new results. *Tectonophysics,* **10**, 639–61. (*211, 248, 292*)

Zijderveld, J. D. A. *see also* As & Zijderveld (1958); Gregor & Zijderveld (1964); Van Hilten & Zijderveld (1966).

Index

Index

Index

Index

Index

354

Index